AF784456

Success and Pitfalls of Information Technology Management

Mehdi Khosrowpour
Pennsylvania State University

IDEA GROUP PUBLISHING
Hershey USA • London UK

Senior Editor: Mehdi Khosrowpour
Managing Editor: Jan Travers
Copy Editor: John Syphrit
Typesetter: Tamara Gillis
Printed at: BookCrafters

Published in the United States of America by
Idea Group Publishing
1331 E. Chocolate Avenue
Hershey PA 17033-1117
Tel: 717-533-8845
Fax: 717-533-8661
E-mail: jtravers@idea-group.com
Website: http://www.idea-group.com

and in the United Kingdom by
Idea Group Publishing
3 Henrietta Street
Covent Garden
London WC2E 8LU
Tel: 171-240 0856
Fax: 171-379 0609
Website: http://www.eurospan.co.uk

Library of Congress Cataloging-in-Publication Data

Khosrowpour, Mehdi, 1951-
Success and pitfalls of information technology management / Mehdi Khosrowpour.
p. cm.
Includes bibliographic references and index.
ISBN 1-878289-56X (soft cover)
1. Information technology--Management. I. Title
HD30.2.K47 1999 98-51081
658.4'038--dc21 CIP

British Cataloguing in Publication Data
A Cataloguing in Publication record for this book is available from the British Library.

Success and Pitfalls of Information Technology Management

Table of Contents

Foreword

I am pleased that Mehdi Khosrowpour and Idea Group Publishing have created the *Success and Pitfalls of Information Technology Management.* This collection of new cases is an important addition to the resources of the information technology field and will make valuable contributions to both teaching and research.

Case studies have long played an important role in business schools and IT programs, having been used in three primary ways:

1. To describe how interacting events, conditions, and decisions have caused a real-world organization to be in its current situation. The challenge for students is to analyze and understand how the situation came to be and to recommend appropriate courses of action for management to take. This is the classic use of a business case.

2. To describe current, leading-edge practices. Most of the innovative business practices come from practitioners who develop ways to effectively deal with a current situation, with all of its attendant challenges, problems, and opportunities. A case can be used to communicate the hard-earned knowledge of practitioners.

3. To help develop theory. Theory building is an important component of developing a cumulative body of knowledge in any field. Theories in IT can draw from reference disciplines (e.g., psychology, economics) but can also be developed from case studies that suggest relationships between the variables.

The first two uses of cases support teaching while the third is helpful in research.

Having good cases for classroom use is highly beneficial for students. The case gives students the chance to study and think about a phenomenon in its natural setting. A case typically does not have the artificial boundaries associated with other learning methods. For example, a case may involve a mix of strategy, marketing, IT and human-relations issues. With a case, student learning is more collaborative than the directive approach of lectures, when the instructor teaches and the student is expected to repeat what was taught. With collaborative learning through a case, the class sorts through the issues, comes to understandings, and develops shared insights and recommended courses of action.

Cases can be either the primary instructional method in some IT courses or be useful additions in others. Two courses in particular can be effectively taught using cases: (1) the core MBA course in IT and (2) an undergraduate or graduate course in information resource management (also called management of information systems). In these courses, IT is discussed in the context of business and organizational issues. Even in more "hands-on" courses such as systems analysis and design, database, telecommunications, and decision support systems, cases can be used to supplement and enhance other learning approaches. A good case helps put the technologies that students are learning in a real-world context. Finding good IT cases has been problematic in the past. Only a few schools and faculty members have been active in writing cases. When a good case is written, its existence is often passed on by word of mouth (more typically email) among those faculty who are part of an informal network of people that shares its knowledge about cases.

Unfortunately, even good IT cases have a short half life. Even if the issues in a case are more lasting than the technologies described, students often ask, "why are we studying this old case?" In a field like IT where change is constant and rapid, new cases are always needed.

I am excited about this new publication. It will support teaching and research, be beneficial to students, provide hard to find teaching materials, and encourage case research and publication.

Hugh J. Watson
C. Herman and Mary Virginia Terry
Professor of Business Administration
Terry College of Business
University of Georgia

Toy or Useful Technology? The Challenge of Diffusing Telemedicine in Three Boston Hospitals

Hüseyin Tanriverdi
Boston University

C. Suzanne Iacono
National Science Foundation[1]

EXECUTIVE SUMMARY

In response to increasing competition and cost pressures from managed-care practices, healthcare organizations are turning to information technology (IT) to increase efficiency of their operations and reach out to new patient markets. One promising IT application, telemedicine, enables remote delivery of medical services. Potentially, telemedicine could reduce costs and increase the quality and accessibility of medical services. However, the diffusion of telemedicine has remained low. We present case studies of telemedicine programs at three healthcare institutions in Boston, Massachusetts to better understand why telemedicine has not spread as quickly or as far as one would expect, given its promise. These case studies describe the environmental and organizational context of telemedicine applications, their champions, strategies and learning activities. Since the three cases represent varying levels of diffusion of telemedicine, they enable the reader to understand how and why some institutions, champions and approaches are more successful than others in diffusing a new technology like telemedicine.

BACKGROUND

Boston, Massachusetts is a city that hosts world-renowned medical schools such as Harvard, Boston University, and Tufts and 25 high-tech, specialty-care hospitals. While noted for their world-class medical expertise, Boston hospitals have also been notorious for oversupply of beds, high costs, and high-use rates. In the 1990s, cost pressures on Boston hospitals increased as many health plans instituted more stringent managed-care tactics and the selection of hospitals for patients. The hospitals responded by entering into merger and acquisition agreements to achieve economies of scale and increase their negotiating power (Katz, 1996). These agreements created integrated healthcare networks (IHCN) spanning the continuum of care from primary care practices to community and teaching hospitals. The challenge for IHCNs is to compete on cost and quality of care by consolidating resources and leveraging expertise within their networks.

One promising approach to these challenges is telemedicine. Broadly defined as the use of IT to deliver medical services at a distance (OTA, 1995), telemedicine is proposed as a solution to problems of accessibility, quality, and costs of medical care (Bashshur, Sanders, and Shannon, 1997). Telemedicine can increase access to care by eliminating distance barriers between patients and caregivers. It can improve quality by enabling medical experts to collaborate on complex clinical

Exhibit-1 Financial highlights of the case study sites, September 30, 1997, ($000)

	AlphaCare	BetaCare[1]	GammaCare
Revenues			
Net patient service revenue	$1,662,755	$607,200	$958,002
Research revenue	$364,672	$83,800	$85,020
Other	$181,578	—	$144,893
Total operating revenue	$2,209,005	$691,000	$1,187,915
Total operating expenses	$2,220,711	$705,303	$1,160,037
Net Income (Loss)	$(11,706)	$(14,303)	$27,878
Non operating revenue	$101,694	$42,381	0
Excess of revenue over expenses	$89,988	$28,078	$27,878

1 Figures are reported for LifeCare, parent company of BetaCare Sources: Annual Reports of AlphaCare, LifeCare, and GammaCare

problems when patients' disorders cannot be diagnosed or treated at referring sites. It can also reduce costs by enabling in-home monitoring of patients and by eliminating the need for on-call expertise, maintenance of expensive facilities, and transportation of physicians to patients or vice-versa.

Telemedicine systems are based on two types of technologies: (1) "Store and forward" telemedicine uses image capture, storage, and transmission technologies to enable asynchronous exchange of images (e.g., for radiology, dermatology, or pathology images); and (2) "real-time" telemedicine uses videoconferencing technologies to enable synchronous interactions between referring physicians, patients, and consulting physicians (e.g., in psychiatric or rare tumor consults).

Although telemedicine applications have proliferated in recent years (Grigsby & Allen, 1997), the volume of actual telemedicine consultations has remained low (Hassol, 1996). One of the challenges is to develop sustainable business models (i.e., finance systems which demonstrate the attainment and maintenance of profitability over time) for telemedicine applications. Capitated and fee-for-service payment systems are the traditional business models in healthcare. In the capitated payment system, physicians and hospitals agree to accept a set advance payment in exchange for providing healthcare services for a group of people, usually for a year. In the fee-for-service model, they receive a fee for each service they provide. Business models for telemedicine must balance the often conflicting demands of numerous stakeholders. Medical institutions, wishing to reach out to new patient markets, can invest in the development and support of telemedicine applications, but they cannot achieve profitability until physicians use them consistently and frequently. Physicians are primarily interested in models that guarantee their reimbursement for telemedicine consultations (Fendrick & Schwartz, 1994). Medical insurers, on the other hand, will only reimburse consultations once cost-effectiveness has been proven. And regulatory agencies will only certify a telemedicine application once it has been proven that it does not degrade the quality of medical diagnoses.

Aligning the incentives for all these stakeholders is a challenge and engenders a number of barriers to the diffusion of telemedicine. The most commonly cited barrier to regular use is restriction on physician reimbursement for telemedicine consultations. However, a nationwide survey of rural hospitals found no association between reimbursement and utilization of telemedicine (Hassol, 1996). Other commonly cited barriers include restriction of medical practice across state lines, risk of medical malpractice, and lack of high-bandwidth telecommunications infrastructure, especially in the rural areas targeted for telemedicine because of the shortages of caregivers (Bashshur, Sanders, and Shannon, 1997).

Some barriers are social and behaviorial. Telemedicine has been found to change some physician behaviors (Anderson, 1997). For example, psychiatrists may have to consult with their patients over a video link rather than face-to-face and cardiologists may have to base their diagnoses on digital rather than video images. Further, traditional practices and workflow may have to change (Anderson, 1997). For example, a physician's time has to be allocated efficiently between on-site

Exhibit-2. Utilization Statistics, 1997

	AlphaCare	BetaCare[1]	GammaCare
Employees	17,469	14,500	13,362
Physicians	1,663	2,152	NAV[2]
Beds	2,574	1,669	1,551
Admissions	95,176	62,258	62,291

[1] Figures are reported for LifeCare, parent company of BetaCare [2] NAV: Not available Sources: Annual Reports of AlphaCare, LifeCare, and GammaCare

patients and remote patients who request consultations via telemedicine. Real-time telemedicine consultations require the coordination of various schedules: those of the patient, the medical staff involved in the consultation, and the video conferencing staff and room. As the volume of telemedicine consultations increases, the coordination of schedules becomes a serious issue. In order to foster regular use, however, hospitals must figure out how to integrate telemedicine procedures within their extant workflows or develop new ones (Cooper & Zmud, 1990).

SETTING THE STAGE

Out of 25 hospitals in Boston, AlphaCare, BetaCare, and GammaCare[2] had the most active telemedicine programs in September 1996 when our study began. These three hospitals are all not-for-profit organizations and key players in the Boston market. Exhibit 1 provides their financial highlights whereas Exhibit 2 summarizes some of their utilization statistics.

AlphaCare was created in 1994 by the merger of Alpha-A and Alpha-B, academic medical centers affiliated with a world-renowned medical school in Boston. Alpha-A and Alpha-B have traditionally been fierce competitors for top accolades in academia and the tertiary care market. U.S. News and World Report consistently ranks them among the country's top ten hospitals. Since the merger, AlphaCare has evolved into an IHCN that includes primary care and specialty physicians, community hospitals, academic medical centers, specialty facilities, and community health centers. AlphaCare has a tradition of excellence in research and training. Its research revenues reached $340 Million in 1996. Its training programs attract top students from around the world.

BetaCare is a world-class academic medical center in Boston. It is home to two full-service hospitals serving adults and children respectively. It is the principal teaching hospital of a world-renowned medical school, and has a tradition of excellence in medical research. BetaCare experienced financial difficulties until early 1997 when it merged with LifeCare, a non-profit regional healthcare system. In addition to financial stability, merger with LifeCare has brought in the strengths of a broad geographic market, a comprehensive and complementary range of services, high quality teaching and medical resources, and strong managed care penetration.

GammaCare was created in 1996 by the merger of Gamma-A, an academic medical center, Gamma-B, a healthcare network including an academic medical canter, a teaching hospital, and three community hospitals, and Gamma-C, a community teaching hospital. GammaCare also includes physician groups and other caregivers. It was created on the belief that community-based hospitals and academic teaching centers can work together to provide high-quality, personalized health and medical services, while also maintaining excellence in medical education and research. GammaCare offers community-based primary care, specialty services, and health services ranging from wellness programs to hospice care.

AlphaCare, BetaCare, and GammaCare have explored the use of telemedicine to integrate their member institutions, to consolidate resources and leverage expertise within their networks, and to sell medical expertise to remote healthcare institutions. Exhibit 3 briefly describes telemedicine applications developed in the three sites.

Exhibit 3. Description of telemedicine applications at the case study sites

Applications	Technology	Description
Fetal Telemedicine	Asynchronous + synchronous	Ultrasound data of pregnant women is digitally captured off of ultrasound machines in private clinics, compressed, and sent to a tertiary care center for opinions.
Home health	Asynchronous	Vital signs (e.g., blood pressure, sugar level, weight, etc.) of chronic patients at home are regularly transmitted to a hospital's database to enable caregivers to monitor patients' health status.
ICU-ICU Consults	Synchronous	X-ray, CT, MRI, etc. data of patients at a community hospital are digitally captured and sent real-time to physicians at an academic medical center for second opinions.
International Telemedicine —AlphaCare	Asynchronous	X-ray, CT, MRI, etc. data of patients in international sites are digitally captured, compressed, and sent to AlphaCare over telephone lines together with patients' medical history files for second opinions.
International Telemedicine —BetaCare	Asynchronous + synchronous	Patients and physicians in international sites are interviewed by BetaCare physicians over videoconferencing links. X-ray, CT, MRI, etc. data can also be exchanged off-line prior to the consult or on-line during the consult.
Nursing home	Synchronous	Patients and nurses at nursing homes are interviewed by physicians at a tertiary care center over videoconferencing links.
Pre-admission Testing	Synchronous	Pre-surgery interviews of surgery patients in the community are conducted by nurses at the pre-admission clinic over the videoconferencing link.
Rare tumor Consults	Synchronous	A surgeon at a tertiary care center views X-rays of rare tumor patients in a community hospital and interviews them over videoconferencing link to render an opinion.
Telecardiology	Asynchronous	Full motion video of cardiac studies in a community hospital are captured, stored, and transmitted to a tertiary care center for second opinions.
Tele-dermatology	Asynchronous	Digital photographs of skin lesions are taken in a primary care setting and sent to dermatologists at a tertiary care center for opinions.
Tele-ophthalmology	Asynchronous + synchronous	Eye images are captured via cameras in a primary care setting, compressed, and sent to ophthalmologists at a tertiary care center for opinions.
Telepathology	Asynchronous	Sections of pathology slides are digitally captured off of a micro scope, compressed, and sent to pathologists at a tertiary care center for opinions.
Telepsychiatry	Synchronous	Patients in a primary care setting are interviewed by psychiatrists at a tertiary care center over videoconferencing links for psychiatric assessment.
Teleradiology	Asynchronous	Radiographic studies (e.g., X-ray, CT, MRI) are digitally captured and sent over high bandwidth lines to remote radiologists for first or second opinions.

CASE DESCRIPTIONS

AlphaCare

AlphaCare inherited its telemedicine program from the early initiatives of one of its members, Alpha-A. During the 1960s, Alpha-A pioneered telemedicine. However, when government grants were discontinued, the program could not sustain itself and interest waned. Interest in telemedicine didn't develop again until the late 1980s when advances in telecommunications and digital image capture and storage technologies allowed efficient storage and transmission of radiological images. Dr. Turner, the chief radiologist at Alpha-A, started a teleradiology project, RADCARE, in an attempt to use these new technologies to reduce film storage costs and sell their radiology expertise to remote locations. According to Dr. Turner, one of the challenges was to manage the transition to a new way of practicing radiology. Using RADCARE meant "converting the work process of the first 100 years of radiology into a totally different work process." Instead of viewing films on a light-box, radiologists would have to view digitized films on a computer screen. Acceptance of RADCARE by radiologists was crucial since "the most expensive component of a telemedicine system is the physician."

Dr. Turner selected his project team members from young and enthusiastic radiologists who were advanced users of medical imaging technology. The team started off by demonstrating a commercial picture archiving and communications system (PACS) to other radiologists in the hospital. Radiologists found the PACS difficult to use, too slow in displaying images, and poor in image quality. They were highly concerned about the image quality because poor quality could lead to incorrect medical diagnoses. They were also concerned that slow response time would reduce their efficiency. Consequently, they preferred viewing films the old way — on a light-box. The project team learned from this experience that the system had to maintain spatial and contrast resolution of films, provide a very simple user interface, and display requested images within a few seconds. Dr. Turner also believed that his team should do rigorous scientific studies to prove to themselves and to the larger medical community that digital images do not compromise diagnostic accuracy. These studies were necessary to address scientific and ethical concerns about the quality of digital images.

In order to address the technical challenges and develop a system that would satisfy the radiologists, the project team collaborated with the IS department, a local phone company, the MIT Media Lab, consultants, and several vendors. They used cutting-edge technologies such as film digitizers, an optical jukebox, high-resolution workstations, image servers, an Ethernet LAN coupled with a fiber optic LAN, and a bridge to interface with remote sites over T-1 [1.54 Mbps] lines. It took them about three years to develop the system. According to Dr. Turner, most radiologists were happy with image quality, response times and the system's ease of use.

The next step was to convince physicians, through a rigorous study, that the system does not compromise diagnostic image quality. The team conducted a study that showed a 98% concordance rate between interpretations of plain films and the corresponding digitized films on the RADCARE system. Dr. Turner reported, "All of the radiologists saw that they could interpret images off the workstation accurately, and we were able to begin inviting more and more radiologists to take part in teleradiology."

At the time, many radiology departments were losing revenue due to Medicare payment cutbacks. Radiologists saw RADCARE as an opportunity to generate new sources of revenue. Dr. Turner stated, "teleradiology was seen as a way of importing work into the department." However, they did not know how to commercialize RADCARE. After discussions with physician organizations at Alpha-A, Dr. Turner decided to spin-off two subsidiaries. One of them would further develop and commercialize the RADCARE system. The other, WeCare, would make a service business out of RADCARE by selling radiology expertise of Alpha-A to remote locations.

One of the challenges for Dr. Turner and WeCare was to explore and develop markets for teleradiology. In domestic markets, there were many regulatory barriers such as the prohibition of telemedicine across state lines and limited reimbursement for consultations. However, international markets (e.g., the Middle East) were relatively free of regulation. Moreover, affluent patients in international markets were willing to make out-of-pocket payments for high quality telemedicine

Exhibit 4: International Telemedicine at AlphaCare

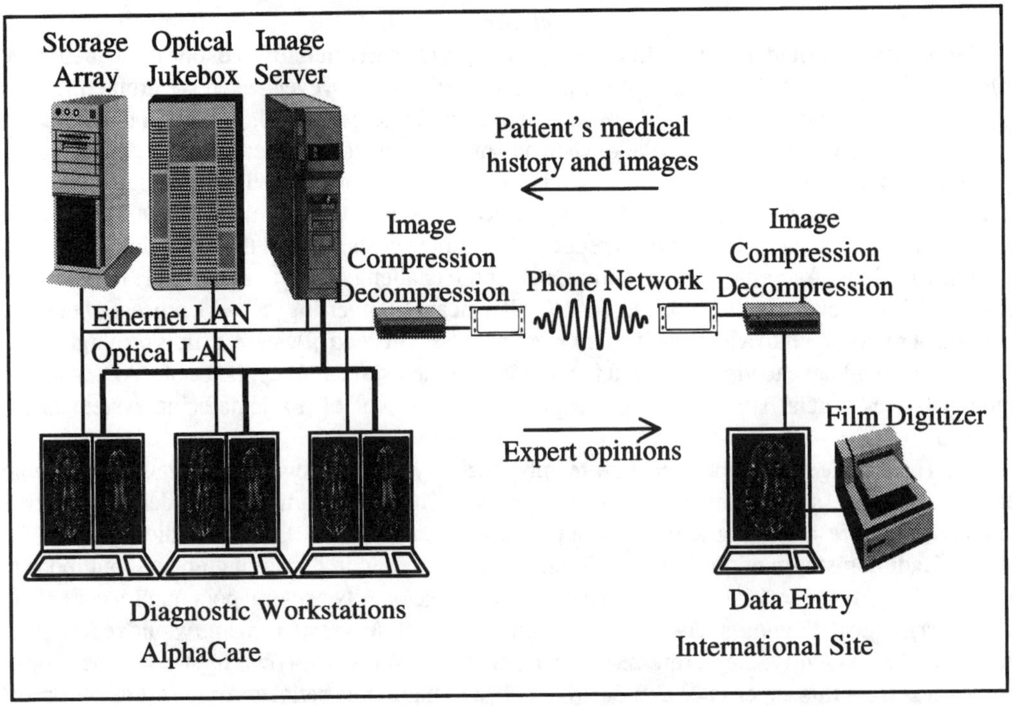

consultations. Dr. Turner and WeCare discovered an enormous potential for international telemedicine: international sites had a hunger for world-class medical expertise not only in radiology but also in other medical specialties. Dr. Turner was surprised to uncover such a demand for pathology expertise and convinced the pathology department to develop a system for doing international telepathology.

However, there was a serious technical barrier for doing telemedicine with international sites. Many international sites lacked high bandwidth T-1 (1.544 Mbits/sec) or T-3 (45 Mbits/sec) lines required by the RADCARE system. In order to exploit the market potential, Dr. Turner had to find a way to run RADCARE over low bandwidth plain ordinary telephone (POT) lines. Running RADCARE over the POT lines could be possible through compression of images. Thus, the radiology team undertook a research study to find appropriate compression techniques for compressing and decompressing radiographic images. By applying a wavelet-based compression technique, they were able to achieve an average compression ratio of 23:1. The results of the study were published in a radiology journal.

In the meantime, the pathologists had developed a telepathology system. Dr. Turner and his colleagues wanted to demonstrate that the systems could run over POT lines without compromising diagnostic quality of images. Alpha-A and WeCare demonstrated the two systems between, first, the U.S. and Saudi Arabia, and, then, the U.S. and the United Arab Emirates. The demonstrations showed that compression did not degrade diagnostic quality of radiographic images. The resolution of pathology images was also good. A study was published in a radiology journal as the first successful use of POT lines for international telemedicine. Exhibit 4 illustrates the configuration of the international telemedicine application.

All medical specialties which rely on radiographic images for diagnoses could now use RADCARE to sell their expertise to international sites. Within six months, WeCare linked eight international sites to Alpha-A via RADCARE. By March 1995, Alpha-A had conducted 750 consultations in radiology and 150 consultations in other medical specialties.

The next challenge was to mobilize more physicians to get involved in telemedicine applications and consultations. The publication of scientific studies about RADCARE and the revenues generated from international consultations helped to increase physician interest in telemedicine.

Having seen the success of teleradiology, the chiefs of the dermatology and psychiatry departments started research projects to assess whether and how telemedicine could be used in their own specialties. In the meantime, Dr. Turner was encouraging interested physicians to come together informally to talk about telemedicine. Over time, informal interactions among these physicians lead to the emergence of an ad-hoc telemedicine committee. This committee oversaw further development of telemedicine at the hospital.

In early 1995, Dr. King, a young dermatologist leading the teledermatology research, took the helm in the committee. The thriving international telemedicine business needed organizational support. The committee created three new positions: telemedicine coordinator, teleradiology coordinator, and teleradiology director. Operational procedures were developed for triaging incoming cases to physicians and ensuring that cases were turned around within set time limits. The committee invited other physicians to participate in order to meet the growing demand from international sites. Physician response was positive; they were able to earn consultation fees and gain access to intellectually challenging medical cases from around the world.

As the volume of consultations grew, it became clear that telemedicine would need better institutional support. Thus, the telemedicine committee prepared a strategic plan. In fall 1995, Alpha-A administration accepted the plan, approved establishment of a Telemedicine Center to streamline international telemedicine operations and to explore new telemedicine applications. Dr. King was appointed as the director of the center. After about a year, the CIO offered to make the Telemedicine Center a sub-unit of the IS department. He envisioned that telemedicine would be part of daily medical practice by the year 2000. According to Dr. King, this decision by the CIO moved telemedicine from being on the "lunatic fringe" to becoming an integral part of mainstream IS at Alpha-A.

In order to institutionalize telemedicine, the Telemedicine Center established departments to manage telemedicine operations, to conduct R&D on new telemedicine applications, to identify standards for telemedicine, and to explore remote education possibilities. It also established two telemedicine committees, the Alpha-A Telemedicine Committee and the AlphaCare Advisory Committee. Membership in the center and committees grew. A center manager commented on the Alpha-A Committee, "It is one of the very few committees that I have seen in my life that actually grew in time as opposed to turning inactive …." A research assistant reported, "At this center, every single department that we have is growing…."

As of May 1997, teleradiology was used routinely in domestic and international markets. All 65 radiologists were using the system to conduct about 3000 consultations annually. In other medical specialties, forty physicians out of a total of 400 were using international telemedicine to conduct about 400 consultations annually. Some specialties were less successful in developing telemedicine applications, however. In telepathology, pathologists were having problems with transmitting rich images within bandwidth constraints. In teledermatology and telepsychiatry, a number of studies had been published that clearly demonstrated that medical diagnoses were not degraded and that the technology worked in practice. Dr. King reported, however, that these applications were not deployed because they hadn't yet been shown to be profitable, "It [teledermatology] has to make sense either through a capitated model, or through a fee-for-service model. Currently, there are not either… So, it hasn't become part of the day-to-day for the [dermatology] department…." A telepsychiatry researcher reported, "In general, the problem for everyone today is proving cost-effectiveness…." The telecardiology application had been used only sporadically because cardiologists perceived that the amount of information they get from digital images is less than what they get from video tapes, which have traditionally been used to record echocardiography procedures. A cardiologist leading the teleradiology application commented, "So, they don't trust the technology… It is very hard to get physicians to change. That has probably been the biggest problem I have had… What it really takes, I think, is that you have to actually show them that this [telecardiology] does actually enhance evaluation of patients." Thus, he decided to conduct a research study to prove that telecardiology does not degrade the diagnostic content of echocardiography procedures.

By May 1997, AlphaCare was undertaking R&D on several new applications and was rolling out 25 videoconferencing units to its members. Dr. Turner commented on the evolution of telemedicine at AlphaCare, "AlphaCare is not an entrepreneurial innovative organization. It is very conservative. It is not going to get into a new concept or a new area like telemedicine as a leap of faith. The way AlphaCare is going to come into telemedicine is through the grass-roots interests of the younger, creative doctors and scientists in the system. We will be experimenting, testing, developing applications. As they mature, the business case can be made and we can get them implemented."

BetaCare

BetaCare became involved in telemedicine in 1990 when it received a grant from the local phone company to develop a telecardiology application for testing a new wide area fiber-optic network known as media broadband services (MBS). The CIO at the time and the chief of cardiology led the development effort. The idea was to exchange angiography images with remote locations using the MBS. The application would enable physicians to interact on-line through a voice channel with a capability to annotate on the exchanged images. The current CIO commented on the system problems, "It was big UNIX workstations with a T3 pipe [45 Mbps] to do cardiac angiography. It never worked! It was incredibly expensive! And it quickly missed the key point that nobody can afford a $75,000 system plus $2000-$3000 a month worth of network charges in the health economics environment today."

BetaCare learned from this experience that telemedicine solutions must be affordable if they are to be used widely in the hospital. The CIO went on, "In managed care, telemedicine has a very important role... [But] it is only there to solve specific problems, and it only makes sense if it solves them in a way that saves money and improves outcomes." The IS department partnered with an engineering firm to explore low-cost, standards-based telemedicine solutions. The technical telemedicine coordinator reported, "The idea was to make it [telemedicine] cheap enough so that experiments can be done without huge pockets of money." In six months, members from the IS department, the cardiology department and the engineering firm developed a desktop videoconferencing system that could digitize cardiac images off of angiography machines, compress them and send them to remote locations off-line as multimedia electronic messages. The new system was less expensive, standards-based, and required less bandwidth since it exchanged images off-line. However, it still ran on the expensive MBS infrastructure. The next challenge was to find cheaper ways of exchanging images.

The CIO explored the use of ISDN technology to meet their transmission needs at affordable costs. Tradeoffs between bandwidth and image quality were found through trial and error. The CIO asked several senior physicians to judge the quality of images for different bandwidths. When image quality was not satisfactory, more ISDN lines were used to increase bandwidth, and hence, the quality. For example, physicians were satisfied with the quality of details on an X-ray image achieved over a 128Kbps line. Psychiatrists, on the other hand, preferred to use a 384Kbps line when interviewing psychiatric patients in order to see them more clearly.

The CIO at BetaCare reported that unlike AlphaCare they did not normally undertake research studies to assess the quality of images or whether medical diagnoses were degraded or not, "We have never seen this [telemedicine] as a research project... [I]t is just part of our business." Their first priority was to make a business out of telemedicine. They did not see image quality as an important issue. The CIO reported, "I have never had one doctor ever complain about quality."

In early 1995, the CIO started demonstrating the system to the chiefs of clinics because "they were the guys who can make the correct decision [for their clinical telemedicine applications]." Many chiefs of clinics wanted to use the system to gain access to new patient markets. The CIO worked with them to conceptualize specific clinical telemedicine applications. He reported on the role of the IS department in this process, "We identify clinical needs and develop a business model." The IS department and engineering firm customized systems for telecardiology, international telemedicine, teleophthalmology, telepsychiatry, nursing home, home health, and fetal telemedicine applications. Each clinic was expected to pursue its own application. The CIO reported, "We have never tried to make telemedicine an institutional program.... There is no hospital-wide 'telemedicine budget' at BetaCare. Each department decides if telemedicine would work for them, and if so, if it can be worked into their budget."

Exhibit 5: Fetal Telemedicine at BetaCare

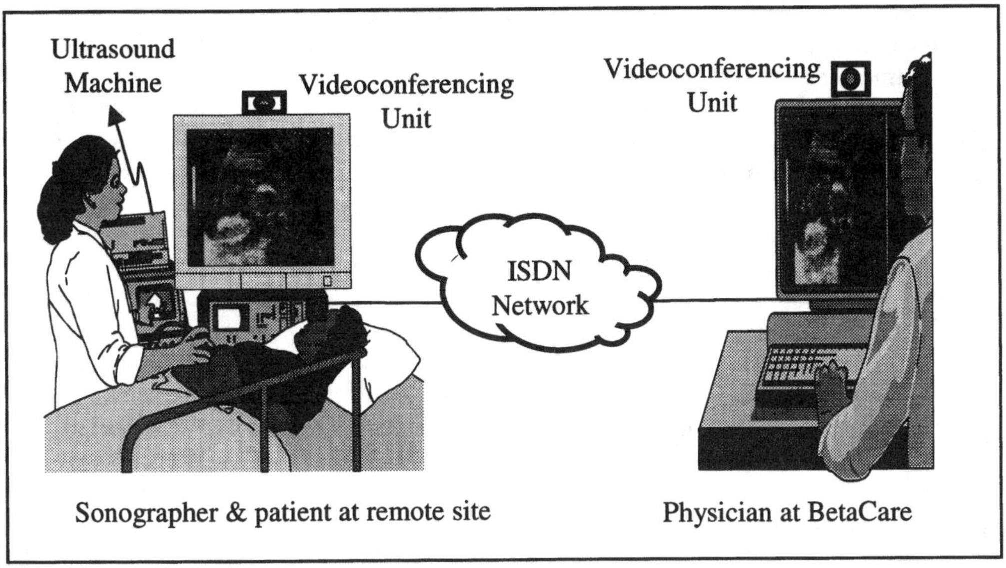

Initial trials showed that the business models envisaged for telecardiology, teleophthalmology, telepsychiatry, nursing home and home health applications could not generate reimbursement for consultations in domestic markets. Therefore, the chiefs of these clinics decided to discontinue the applications. However, all clinics continued to participate in international patient consultations. Physicians were motivated to participate in international telemedicine for two reasons. First, international telemedicine enabled them to see rare medical cases from around the world. The medical coordinator explained, "[An incentive for] physicians is to be stimulated by difficult [medical] problems. Being confronted with a broader range of things [difficult medical problems] is a stimulus." This is important for them to advance their expertise. Second, they were able to collect fees from international consultations. The CIO reported, "The fact that we are making so much money doing [international] telemedicine, has of course made it very popular." However, many physicians were not comfortable using the technology. The medical coordinator commented, "Physicians are some of the most technophobic people that you will ever meet." Technical and medical coordinators had to schedule the sessions, initiate the ISDN connections, and help the physicians to use the technology during consultations. As of March 1997, about 40 physicians (or 10% of the total at BetaCare) were using the application to conduct about 200 international consultations annually.

As Exhibit 5 illustrates, the fetal telemedicine application allows patients at remote sites to receive expert opinions from BetaCare obstetricians over videoconferencing links without having to travel to Boston. The CIO and some obstetricians became convinced of the profitability of this application during initial trials. However, before moving the application into clinical use, they had to show to the national society which accredits fetal telemedicine sites and the insurance companies which reimburse consultations that medical diagnoses were correct and that fetal telemedicine was cost-effective. Even though the CIO did not see the value of scientific studies, the physicians believed studies were necessary. One obstetrician reported, "The studies that we did to validate telemedicine have provisionally been accepted for two of our journals." They began to define how a typical fetal telemedicine session would be conducted. There were two possibilities: 1) on-line video connections; or 2) off-line image storage and transmission. Each had different implications for the daily routine of the clinic, the volume of cases that could be served, staffing requirements, and cost-effectiveness. One obstetrician commented, "We have to do traffic-flow type of studies before we can say whether it is cost-effective." As of March 1997, all eleven physicians in the clinic were using the application. They were conducting about 1000 consultations per year in their research studies. They expected the volume to grow dramatically when the application is moved from research into regular clinical use.

GammaCare

The telemedicine program at GammaCare is an extension of an early program developed by one of its founding members, Gamma-B Network. In early 1994, Gamma-B realized that its use of resources and methods of patient referral across its member institutions were inefficient. Due to a lack of appropriate expertise, the community hospitals were spending too much time and resources on difficult patient cases that could have been dealt with in a faster and more cost-effective way at their teaching hospitals. Under increasing pressure from managed care and capitated payment systems, Gamma-B began looking for ways to increase its operational efficiency.

In mid-1994, Henry, the head of media services, proposed to link member institutions via telemedicine (using videoconferencing technology) to enable "a seamless integration of care so that wherever the patients happen to be, they are part of the [Gamma-B] system." With the support of the CIO, a vice president and some physicians, he prepared a business plan that outlined a phased implementation starting with demonstrations of telemedicine between two hospitals and then expanding into all member institutions. The administration accepted the plan, provided funding for exploration of the technology, and appointed Henry as the telemedicine project manager.

In early 1995, Gamma-B acquired two group videoconferencing systems. Henry convinced a staff member of media services and a young cardiologist to serve as the technical and medical telemedicine coordinators. He also secured technical support for telemedicine from the IS department, but it was mainly limited to installation and maintenance of ISDN lines. Unlike AlphaCare and BetaCare, GammaCare did not get involved in the development or modification of telemedicine technology. Instead, it has tried to apply commercially available videoconferencing systems to their proposed applications. As champions of telemedicine, Henry and his technical and medical coordinators tried to raise awareness of the technology across the organization. They identified physician champions and site coordinators at member institutions, and provided them with administrative and technical support. They exposed physicians to the technology through demonstrations, and educational and administrative teleconferences.

Early adopters of the technology included a world-famous liver transplant surgeon at the academic medical center, two site coordinators at a community hospital, and a few young physicians. Two hospitals were linked in early 1995. Physicians at the community hospital now had access to educational tumor conferences at the medical center via teleconferencing. They conducted about 30 such tumor-board conferences until mid-1996. The surgeon also provided a few (3-4) consultations to patients at the community hospital by reviewing their X-rays and interviewing them over the video link. This application was called "rare tumor consults" since the surgeon specializes in rare tumors. The surgeon reported, "I look at this as really an incredible way to interview patients, look at their films, talk things over with them." Although he receives seven to eight rare tumor cases per week from around the world, he cannot use telemedicine since the referring physicians lack access to the technology.

In mid-1995, the community hospital lost its cardiology coverage. Third year fellows from the academic medical center commuted to the community hospital to cover emergency and intensive care units (ICU). These young, computer literate physicians used the technology to exchange X-rays, MRIs, echocardiography, etc. with the medical telemedicine coordinator at the ICU of the medical center. The president of the community hospital reported, "They were using the technology like the telephone... 'Here look at this, look at that, OK? I am going to send you the echo! Here is the EKG strip!'" They conducted about 30 to 40 ICU-to-ICU consults. The coordinator expressed satisfaction with the technology: "By and large, major decisions can be made using the current technology."

Another member teaching hospital soon acquired a videoconferencing unit and connected to the telemedicine network. Anesthesiologists, technicians, and nurses from the two teaching hospitals started doing teleconferences to conduct their regular educational and administrative meetings. They conducted about thirty such meetings.

The feasibility of conducting pre-surgery interviews with patients over videoconferencing was tested. This application would enable patients to go to the nearest member institution for their interviews, thus, enhancing the completion of interviews and reducing costly surgery cancellations.

Nurses interviewed two mock-patients at the community hospital. Although they were happy with the video interviews, they stated that incorporating telemedicine into their daily routines would be a challenge: "You couldn't just have so many [video] calls up in the middle [of serving patients here]." Cost-effectiveness of the application was not clear. The technical telemedicine coordinator reported, "At this point, we have not proved that [it] pays."

In summer 1996, Henry and coordinators evaluated their experiences, prepared a phase-II business plan, and put together a telemedicine executive committee. After two years with telemedicine, Henry reported that most physicians saw telemedicine as a "toy" rather than a clinically useful technology: "[T]he easy part is the technology... The much tougher part is really getting people to collaborate together on a regular basis, not just to play around with the toy but to do work together...." A site coordinator reported, "[Telemedicine] is still a technology that people are leery of in terms of how it can be used and how valuable it is."

Henry recognized challenges in making telemedicine work: "[I]t is really easy to do administrative applications.... Educational is also fairly easy.... But, clinical stuff is much more complex: You have to deal with reimbursement; you have to deal with clinical care; you have to deal with patients; you have to deal with physicians; you have to deal with infrastructure; you have to deal with..., you name it!" He described the next step: "We need to show [physicians] the real solutions as opposed to demonstrations...."

However, the telemedicine initiative started to drift in October 1996 when the academic medical center of Gamma-B merged with another academic medical center in Boston. During the merger turmoil, some physicians left the institution while others were busy with changes brought about by the merger. Teleconferences came to a complete halt. Henry spent this period expanding the telemedicine network to the remaining community hospitals. The initiative started up again in early 1997 with monthly meetings of the telemedicine committee.

CURRENT CHALLENGES/PROBLEMS

The extent to which telemedicine has diffused varies across the three sites. Diffusion was relatively high at AlphaCare. Teleradiology and international telemedicine applications were regularly used, had high volume consultations, and the number of physicians using them was increasing. At AlphaCare, the champions of telemedicine were physicians. Their grassroots efforts played an important role in making telemedicine part of mainstream IS and diffusing telemedicine applications throughout the hospital. They learned how to develop technically feasible telemedicine applications *de novo* in collaboration with the IS department and various external organizations. They proved the medical validity of their applications through scientific studies and developed business models that could generate reimbursement. Technically feasible, medically valid, and reimbursable applications engendered adoption and usage by more physicians and prompted the original champions to develop institutional support mechanisms for telemedicine. Future challenges for AlphaCare include showing cost-effectiveness for teledermatology and telepsychiatry; lowering physician resistance to telecardiology; further developing telemedicine business in international markets; and demonstrating that telemedicine could reduce costs and improve quality of care within its IHCN. Although they were happy with the achievements to date, Dr. Turner and Dr. King thought that the real potential of telemedicine would not be realized until all applications were integrated into one system. At the conclusion of the case study, they were wondering how this integration could be achieved across the organization.

At BetaCare, diffusion was moderate. International telemedicine was in regular use, but the volume of consultations was lower than that of AlphaCare. Fetal telemedicine was taking off, but there were organizational and economic barriers to its widespread deployment. One of the challenges was to decide whether real-time or store and forward type of consultations would be more efficient and cost-effective for BetaCare. Another challenge was to demonstrate the medical validity of the fetal telemedicine application to insurers and regulators. In other application areas, such as telecardiology and telepsychiatry, usage was sporadic because profitability was yet to be shown. BetaCare learned how to develop technically and economically feasible telemedicine applications by collaborating with an engineering firm. The CIO tried to diffuse the applications through a top-down approach of convincing various clinic chiefs of their usefulness. Under his guidance, BetaCare did not routinely undertake scientific studies. However, some physicians were beginning to understand the importance of scientific studies in convincing the external medical community (e.g., insurers and regulatory agencies) of the medical validity of their proposed telemedicine applications. Only the most economically promising applications were currently being considered for clinical use. At the end of the case study, the CIO was concerned about developing appropriate business models for the various applications and whether they could integrate telemedicine into the routines of the clinics.

At GammaCare, diffusion was low. At the end of two years, there was no active telemedicine application in use. Although the technology was being used for educational and administrative teleconferences, it had not yet diffused into medical practices. Physicians saw telemedicine as a nice "toy" rather than a clinically useful technology. At GammaCare, the telemedicine initiatives came from the media services department which focused primarily on promoting the technology rather than figuring out how it might be integrated into specific medical practices. Unlike AlphaCare or BetaCare, GammaCare did not engage in technology development or modification. Instead, it implemented commercially available technologies. The telemedicine program at GammaCare lacked innovative physicians who would develop new telemedicine applications or do telemedicine research. The involvement of the IS department was limited to technical support for the ISDN infrastructure. However, the telemedicine initiative was starting up again in early 1997. The primary challenges for GammaCare were to involve innovative physicians in its telemedicine program, develop clinical telemedicine applications, and demonstrate the value of these applications to its medical staff. At the end of the case, Henry was pondering how best to address these challenges based on the experiences they had gained in the last two years.

Telemedicine champions at AlphaCare, BetaCare, and GammaCare were aware that telemedicine is one of the keys to the survival and success of their organizations in the changing healthcare environment. However, exploitation of the opportunities offered by telemedicine depend heavily on its widespread adoption and use by physicians. Ad hoc demonstrations and experimentation with commercially available systems are not sufficient to convince physicians to routinely use telemedicine applications. The champions wondered how they could make telemedicine part of the daily life of more physicians in their hospitals.

NOTES

[1] The contents of this manuscript are the opinion of the authors, not of the National Science Foundation or the U.S. government.

[2] Pseudonyms have been used to disguise the names of all organizations and individuals.

REFERENCES

Anderson, J.G. "Clearing the way for physicians' use of clinical information systems," *Communications of the ACM,* (40:8), 1997, pp. 83-90.

Bashshur, R.L, Sanders, J.H, and Shannon, G.W (Eds.). *Telemedicine: Theory and Practice,* Charles C. Thomas Publisher, Ltd., Illinois, 1997.

Cooper, R.B., and Zmud, R.W. "Information technology implementation research: a technological diffusion approach," *Management Science,* (36:2), 1990, pp. 123-139.

Fendrick, A.M., and Schwartz, J.S. "Physicians' decisions regarding the acquisition of technology," in *Adopting New Medical Technology,* Institute of Medicine, National Academy Press, Washington, D.C., 1994, pp. 71-84.

Grigsby, B. and Allen, A. "4th Annual telemedicine program review," *Telemedicine Today,* August 1997, pp. 30-42.

Hassol, A. "Surprises from the rural telemedicine survey," *Telemedicine Today,* November/December 1996, pp. 5, 41.

Katz, A. "Boston, Massachusetts: site visit report," in *The Community Snapshot Project: Capturing Health System Change*, Ginsburg, P.B., and Fasciano, N.J., (eds.), The Robert Wood Johnson Foundation, Princeton, NJ, 1996, pp. 1-17.

OTA. *Bringing Health Care On Line: The Role of Information Technologies,* U.S. Congress, Office of Technology Assessment, OTA-ITC-624. US Government Printing Office, Washington, D.C, 1995.

BIOGRAPHICAL SKETCH

Huseyin Tanriverdi is a Doctoral Candidate and Teaching Fellow in the department of Management Information Systems, and a Research Associate at the Systems Research Center, in Boston University's School of Management. He received his M.Sc. degree in Information Systems from London School of Economics and Political Science, London, U.K; and B.Sc. and M.Sc. degrees in Electrical and Electronics Engineering from the Middle East Technical University, Ankara, Turkey. His research interests include diffusion of information technology innovations, knowledge work, and knowledge management.

C. Suzanne Iacono received her Ph.D. from the University of Arizona in Management Information Systems and her M.A. and B.A. from the University of California, Irvine in Social Ecology. She is currently the Program Director for Computation and Social Systems (Information and Intelligent Systems Division, Computer and Information Sciences and Engineering Directorate) of the National Science Foundation. Previously, she held a faculty position at Boston University and was a Visiting Scholar at the Sloan School, Massachusetts Institute of Technology. She is an Associate Editor for The Information Society and conducts research on social informatics and electronic communication.

The Benefits of Data Warehousing at Whirlpool

Barbara J. Haley
University of Virginia, USA

Hugh J. Watson
University of Georgia, USA

Dale L. Goodhue
University of Georgia, USA

EXECUTIVE SUMMARY

In today's competitive, high-velocity business environment, companies are focusing their attention on several key areas, including:

- incremental continuous quality improvement;
- more radical redesign of business processes;
- supply chain management;
- improved customer orientation; and
- globalization of business operations.

At Whirlpool, data warehousing is providing important support in all of these critical areas (see Table 1). To illustrate, Whirlpool's data warehouse enables quality engineers to easily track the performance of component parts. This allows the engineers to assess new components that are being field tested, to quickly detect problems with particular parts, and to identify the high and low quality suppliers. From a different perspective, suppliers can check on the performance of the parts they supply and, consequently, can manage proactively the quality provided to Whirlpool. Purchasing managers have parts information from around the world so that they can find the lowest-cost, highest quality part available on a global basis.

This case study briefly describes Whirlpool, the business need that suggested a data warehouse, the approval process, and the data warehouse that was built. It describes how the data warehouse is accessed, how users are trained and supported, and the major applications and benefits. The lessons learned also are described to benefit those companies that are implementing or thinking about implementing data warehousing.

Table 1: Key Strategic Areas for Whirlpool

incremental continuous quality improvement
more radical redesign of business processes
supply chain management
improved customer orientation
globalization of business operations

Like most companies, Whirlpool is continually changing. This case study describes Whirlpool and its data warehousing initiative through the end of 1997.

THE WHIRLPOOL CORPORATION

Whirlpool Corporation is the world's leading manufacturer and marketer of home appliances. The Whirlpool family consists of over 45,000 people who manufacture fine appliances in 12 countries and market them under 11 major brand names. The company is based in Benton Harbor, Michigan and reaches out to approximately 140 countries around the world. It is the only major home appliance company with a leadership position in North America, Europe, and Latin America, plus a growing presence in Asia.

Whirlpool began as a small family-owned business in 1911, and it now ranks 159 in the Fortune 500. The corporate vision for the company fosters growth and progress: *Whirlpool, in its chosen lines of business, will grow with new opportunities and be the leader in an ever-changing global market.* This vision is manifested in Whirlpool's Worldwide Excellence System (WES), its blueprint for approaching quality, customers, and continuous improvement. Initiated in 1991, WES incorporates the best of all Whirlpool quality programs, worldwide, with Malcolm Baldrige Award and International Standards Organization criteria to establish a common approach to quality, one that dedicates the company to the pursuit of excellence and total customer satisfaction.

Whirlpool is an information-intensive business. In North America, it has three or four thousand products that it sells at any point in time. Every one of the products has hundreds or thousands of components that are assembled every day in 12 major factories. The products are stored in 28 places. Over 16 million appliances are sold a year and they are tracked throughout their lifetime.

THE BUSINESS NEED FOR DATA WAREHOUSING

One of the keys to thriving in this information-intensive environment is the ability to effectively coordinate and control its myriad processes and activities. This can be challenging from an information systems perspective. Business units need a complete understanding of the processes for which they are responsible, and the diversity and heterogeneity among systems make it difficult for them to get the information they need and to manipulate it in a useful, timely manner.

In the early 90's, several business units identified a variety of specific information needs. For example, Quality wanted to create an application (later called Customer Quality Information System (CQIS)) that would proactively identify quality problems based on customer complaints. Data existed in several places, including Whirlpool's OneCall System that allowed any customer to take care of any necessary business, be it services, product information, or complaints with one phone call. CQIS was to provide an environment in which the data could be queried and analyzed.

These applications had obvious value to the business, and senior management understood that the information systems infrastructure was inadequate to effectively support the various initiatives. In fact, around that timeframe an expensive executive information system initiative had just been discontinued after several years of trying to combine data from multiple data sources and manipulating the data in a meaningful way for its users.

It was apparent that an infrastructure had to be put in place at Whirlpool to support the numerous decision support initiatives that its business units had identified and were expected to demand in the near future. At that time, the marketplace was promoting data warehousing as a viable alternative to organizations that wanted to create a decision support infrastructure. Data warehousing is the process of creating, maintaining, and using quality data for decision support purposes, and its technology had become cost effective and mature enough for organizations to implement. In the spring of 1993, the first efforts to use data warehousing at Whirlpool were approved, and CQIS was the first application to utilize the new infrastructure. It was expected that data warehousing would allow IS to provide business units with their applications quickly, with less cost, and with a greater likelihood of meeting their needs.

Many business initiatives, which rely on data warehousing, have emerged since 1993. Currently, the data warehouse contains 14 specific collections of data (i.e., subject areas) that describe important facets of Whirlpool's business, like competitors, business partners, and facilities (see Table 2). Table 3 presents a listing of the ten primary applications that use information from the subject areas along with the sponsors of the applications and the business needs that they address. With enhancements and alterations, the data warehouse has effectively evolved to support each of the applications.

THE DATA WAREHOUSING APPROVAL PROCESS

The business needs for Whirlpool's data warehouse and its applications were examined and assessed throughout their approval processes. Whirlpool gives approval for systems initiatives through a unique process called Value Oriented Systems Planning (VOSP). The idea of VOSP is to

Table 2: Subject Areas Represented in Whirlpool's Data Warehouse

BUSINESS PARTNER	Information about an organization or individual with whom Whirlpool has a business relationship (e.g., customers and vendors).
BUSINESS TRANSACTION	Information about the transaction of business between Whirlpool and a business partner (customer, supplier, etc.) such as sales, purchases or shipments; and the documentation of those transactions (e.g., orders, invoices, bills, manifests, etc.).
COMPETITOR	Individuals or organizations that presently offer, or may in the future offer, competitive products and services to Whirlpool's customers.
CONSUMER	A person who owns and/or uses a unit of any brand of product sold by Whirlpool.
EQUIPMENT	All the fixed assets other than land and buildings of a business enterprise. This includes machines used with tools in the factory.
FACILITY	Information about property and plants owned or used by Whirlpool Corporation. Includes location, size, cost, value, depreciation, usage, classification, etc.
FINANCIAL	Information about the assets and liabilities of the Corporation, the internal and external reporting and management thereof. Includes general ledger and sub-ledger information, income and expense, taxes on profits, information about currency, and all financial reports.
GEOGRAPHIC AREA	An area of jurisdiction; may be state, province, county, township, city, etc.
HUMAN RESOURCE	Information about rewards and services provided by Whirlpool for its employees.
MANAGEMENT SUPPORT	Information about the mechanisms used by management to perform day-to-day activities. Includes communication, computer systems and related components, messages, projects, records, reports and surveys.
MANUFACTURING	Information about the procurement of parts and materials, and assembly of parts and materials into other parts, assemblies or finished products.
ORGANIZATION	Information about the structure of the company. Includes business components, organizational units, branches, divisions, parts distribution centers, and departments.
SERVICE	Information about the provision of product diagnostic and repair services to the customer or consumer.
PRODUCT	A standard group of items that is, or has been in the past, manufactured or purchased by Whirlpool and competitors for distribution and sale.

Table 3: Data Warehousing Applications and the Business Needs That They Satisfy

Application	Sponsor	Business Need
Consumer Quality Information System (CQIS)	Product Quality	Early warning system for product quality issues
Customer and Model Profit Analysis (CAMPA)	Sales	Analyze profitability by specific dimensions
Global Procurement (GPDW)	Global Procurement	Identify lowest cost part available globally
Logistics Information for Everyone (LIFE)	Logistics	Provide availability analysis, inventory analysis, inventory turn analysis, and analyze overall cost of delivery
Online Waste Elimination Network (OWEN)	Manufacturing	Identify and eliminate inefficient processes through scrap analysis
Product Cost Analysis System (PCAS)	Manufacturing	Analyze costs associated with production by design changes and production mix
Super Efficient Refrigeration Program (SERP)	Sales	Track sales and customers of the high efficiency refrigerator in order to qualify for rebates from the SERP association
Marketing (DIALS)	Marketing	Analyze external market information on price points and demographics
SEARS POS	Sales	Analyze Sears Point of Sale Information
Service Management Analysis and Reporting Tool (SMART)	Consumer Service	Analyze cost and quality of service
Parts and Accessory Information System (PAID)	Consumer Service	Analyze profitability of part and accessory business by specific parts and customers

identify the value of an initiative to the company and the funds necessary to provide that value. The VOSP document has two parts. The first part is owned by the customer and identifies the functionality that is needed to meet some business need. The second part is completed by IS and describes the specific actions that IS will take to address the specified business needs and the funding required to accomplish this. The executive committee then ranks the VOSPs and decides whether or not to fund each one. A blanket IS VOSP is approved for significant hardware upgrades. The VOSP process continues after a system or application is in place; a post-implementation audit is conducted to ensure that the stated business needs are met.

Because data warehousing is an infrastructure investment, with benefits typically coming from subsequent applications, it was included under the blanket IS VOSP. Therefore, data warehousing per se has not received a post-implementation audit. However, individual VOSPs were created for each data warehousing business application. In keeping with the VOSP approach, each of these applications was reviewed for its value after its implementation. The successful audits of these dependent applications show data warehousing's importance. In addition, it is thought that the benefits from the data warehousing applications are much greater than the post-implementation audits demonstrate. (Interestingly, business units admit to downplaying benefits for fear of a negative impact on budget levels or future funding.)

CONSTRUCTION OF THE DATA WAREHOUSE

Before data warehousing, Whirlpool traditionally relied on IBM mainframe systems running DB2 for major systems initiatives. When data from these operational systems were needed, Whirlpool used IBM's QMF as the data access tool. However, Whirlpool's new business needs created data access requirements that were not achievable on operational data stores that included data in raw form, in inconsistent formats, and without the ability to support queries, reports, and historical requests. Whirlpool needed to handle large amounts of data, support batch loads, and perform hundreds of complex, often ad hoc, queries from large numbers of users.

NCR was selected to provide Whirlpool with a total data warehousing solution that included hardware and software. A careful review of multiple data warehouse solutions and technology alternatives led to the best solution for Whirlpool. This decision was based on several points:

- the ability to handle complex, ad hoc, multi-table queries against data stored in third normal form;
- ease of data warehousing administration;
- low cost of database management (e.g., database administration resources over a long period of time);
- superior mainframe integration for data loading, archiving, and user access;
- the ability to integrate a multidimensional dependent data mart (Pilot Analysis Server) into the same data warehouse hardware and software environment; and
- a proven data warehousing methodology and expertise in implementing successful data warehousing environments.

The decision has been reexamined a number of times since NCR was first introduced, and each time NCR was found to be the most cost-effective solution and one that best meets Whirlpool's business needs.

Whirlpool's production data warehouse platform in North America is the NCR WorldMark 5100MPP system configured with three 8-way SMP nodes running the Teradata Relational Data Base Management System (RDBMS) and one 4-way SMP node running Pilot Analysis Server. The 5100M is directly channel connected to Whirlpool's MVS mainframe and LAN connected to the corporate and remote LAN sites. This configuration provides over 200 gigabytes of usable storage to Whirlpool whose current levels of raw data exceed 90 gigabytes. For development, Whirlpool uses an NCR 4500 4-way SMP server also running the Teradata RDBMS, Pilot Analysis Server, and a development Web server that provides limited browser access to the data warehouse. The 4500 supports 40 gigabytes of RAID-protected disk storage.

Until 1994, data warehousing activity was limited to North America; however, global initiatives prompted the creation of a second regional data warehouse in Italy to support European information requirements. Ultimately, Whirlpool hopes to integrate regional data warehouses to meet global information needs. The North American data warehousing team members were instrumental in communicating the existing data warehousing methodology, practices, and technology choices to their European counterparts. However, to date the two warehouses remain segregated. One global application (i.e., Global Procurement) is running on the North American data warehouse, and is accessed by procurement users from all regions.

Figure 1 illustrates Whirlpool's data warehousing environment. Currently, the North American data warehouse is fed by at least six external data sources (e.g., data for appliance newspaper advertisements; POS information from appliance retailers) and over one hundred mainframe sources. Additionally, the warehouse provides data to dependent data marts that support marketing, logistics, manufacturing, and procurement applications. Overall, 1,700 users (from Logistics, Sales, Quality, Engineering, etc.) have access to warehouse applications, and 400 to 600 unique users access the

Figure 1: Whirlpool's Data Warehousing Environment

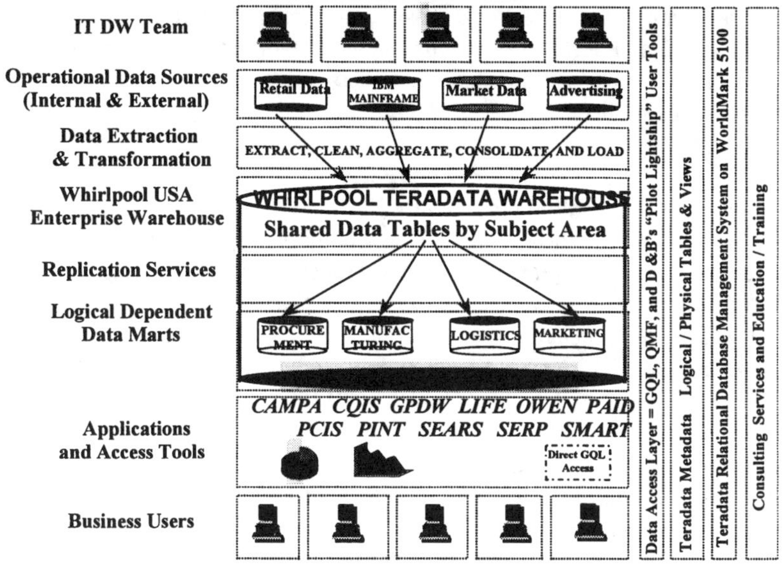

Table 4: Use of the Data Warehouse as of Final Quarter 1997

Total users	1,700
Active users per month	400-600
Queries per month	> 40,000
Views	500

applications each month (see Table 4). The volume of queries has risen to 40,000 in one month, and this figure continues to grow. Six full-time Whirlpool employees comprise the team that supports all of the North American data warehousing initiatives. This includes the team that designs the applications, develops the designs, loads the data, manages the Teradata RDBMS, provides security to the data warehouse, and assists the business users with their queries and future requirements.

USE OF THE DATA WAREHOUSE

Accessing Data

Three primary access tools are used to access Whirlpool's data warehouse. Originally, QMF emulation was used because QMF already existed at Whirlpool. However, QMF has a number of limitations, including its unwieldy user interface and its requirement for advanced knowledge of structured query language (SQL). In 1992, a Windows-based managed query environment called GQL from Andyne Computing was introduced to serve as a generic access tool for all applications that were supported by the data warehouse. The tool has become accepted as the ad hoc access tool of choice because it has an improved graphical user interface, and it supports flexible, ad hoc queries. However, users still need to know the data well before they can execute effective queries. Recently, Pilot Corporation's Decision Support Suite (desktop software) and Analysis Server (a multi-dimensional database) were implemented to complement GQL, and these will be used for predefined queries. Pilot Decision Support Suite does not require users to be quite as skilled in order to access data.

Whirlpool's Marketing Department has taken a different approach than other organizational units in terms of how warehouse data are accessed. First, most marketing users access data marts instead of the warehouse. They cite two reasons for this – better response time and less data complexity. But, data in the data mart are updated only monthly; consequently, marketing users still must turn to the data warehouse for the most current data. The big difference between Marketing and other areas lies in the user interface tool. For standard marketing applications, PowerBuilder has been employed to develop simple interfaces. This approach has evolved because of a highly skilled and respected application developer within Marketing who has learned to create intuitive marketing applications using PowerBuilder.

Educating Users

The data warehousing education provided to Whirlpool users varies greatly depending on the area within the company. In the Washers area, all new employees get an overview of CQIS as a part of orientation. However, one manager explained that in Quality, *"there are engineers that don't have the foggiest idea of what the data warehouse is about. So we haven't communicated that it exists and what its contents and potential benefits are."*

Overall, there are three approaches to warehousing education at Whirlpool:

- no education;
- education during employee orientation; and
- classes available for employees to sign-up on an as-needed basis.

A combination of the two latter approaches has the best results. Education allows employees to understand the ways in which data warehousing can support their jobs and to understand how to use the data warehouse tools effectively. Mary Schmidke, Decision Support Analyst, performs warehousing training: *"I find that if I get one or two people from a department that is not using the data warehouse to come to training – all of a sudden I have a massive entourage of the whole department coming and wanting to find out how to use the warehouse that same day."* Education improves users' understanding of how data warehousing can meet business needs, and educated users have great enthusiasm for using the data warehouse to do their jobs.

Supporting Users

A six-member group exists within IS to support warehousing and to help with future initiatives. But, in terms of supporting users, the most effective support has been found to come from the business units. Several business units have identified a functional person or persons to help in the use of the warehouse. The profile of such a person includes functional knowledge, technical know-how, and the ability to communicate with the users. This approach works well. The CAMPA application has a business analyst appointed to be its decision support analyst. Former data warehousing manager Laura Sager explains: *"She is in the business, she does all of the training for the CAMPA application, and she is the center of CAMPA's data warehousing support."*

Informal power users have sprung up for areas that do not have formalized data warehousing support in place. For example, in Washer Engineering, quite a few people try to access the warehouse. A power user in this group states, *"Once they make mistakes then I end up running all of the queries for them. They come to me and I write them. They know that I am the expert at it and I can do it better. Now I'm a product engineer, and I do very little product engineering because I'm doing this."* Whirlpool is working to minimize this situation. For example, Sager has made concerted efforts to identify business areas that heavily use the data warehouse and to encourage them to formally assign business users to data warehouse support roles.

THE BENEFITS FROM DATA WAREHOUSING APPLICATIONS

The benefits from data warehousing can be considered in a variety of ways. Figure 2 provides a generic framework that is used to organize the discussion of the data warehousing benefits at Whirlpool.

Time Savings

Now you can access a whole division in a matter of seconds. Now you can do just about anything you want to do to analyze the production for that plant in a matter of minutes; whereas,

Figure 2. The Benefits from Data Warehousing

- Time savings → for data suppliers; → for users
- More and better information
- Better decisions
- Improvement of business processes
- Support for the accomplishment of strategic business objectives

somebody would have keyed in for hours to provide that to you before we had the warehouse.
– Jason Denman, Financial Analyst

Time savings can occur for two groups: data suppliers and end users. While developing a data warehouse is time consuming for IS, once it is in place, there should be less time spent responding to ad hoc requests for data because users can help themselves. More importantly, data warehousing creates the decision support infrastructure upon which future applications are built. If data warehousing is conducted effectively, the start-up costs associated with new decision-support initiatives are dramatically reduced. On the users' side, business analysts spend less time accessing data, processing it, and putting it in a format appropriate for their needs.

Before the data warehouse was in place, IS and functional area personnel were often called upon to make data available. This required a combination of downloading files, re-keying data, and creating extract files. These time-consuming tasks are no longer required because of the existence of the data warehouse.

Jerry Briney is a former manager in Quality, and he uses CQIS to investigate quality problems. Before CQIS, Briney would read upwards of twenty thousand service call tickets a month looking for and investigating problems. Tickets had to be (1) sorted by product and brand, (2) sorted by defect, (3) manually counted, and (4) read thoroughly. This was a mind-numbing task, as Jerry explains, *"After a while you don't know what you are reading. You would read a ticket and not pay any attention to it."* With CQIS, Briney can specify a problem of interest and access all of the service tickets that report that problem. About 30 to 40 times as many service tickets can be checked using the data warehouse. Briney explains how this translates to the bottom line: "*We produce 17,000 washing machines each day. If we find a problem as small as a .1% service incident rate (SIR), we save 17 service calls per day (17,000 * .001). Each call is $75 which results in $1,275 saved in service calls per day and in a $38,250 savings per month."*

More and Better Information

The data warehouse gives us backorder information so that we can expedite what we need to expedite. There was no backorder report in the company before the data warehouse.
- Bill Friend, Logistics Manager

A data warehouse lets users get at data that were previously locked away in legacy systems or that did not exist. Through cleansing, aggregating, and possibly augmenting (e.g., external data) processes, warehouse data are used to create better information than was previously available. John Doyle is a manager in Procurement Quality. He helps insure that parts received from suppliers conform to standards. The data warehouse makes it feasible for him to access and use data in new ways. Previously, Doyle had difficulties getting information that required aggregation. As he describes it, *"I couldn't really use or summarize information without manually going through all the data. It was very difficult and cumbersome to do that. With the data warehouse, I can tell it what I want to see. I want to see an alphabetical listing of all the suppliers, and I want to know how many receipts that we have had from them, how many rejects we have, and what their PPM (parts per million) value is."* With this information, Doyle is able to track the performance of individual suppliers. Only suppliers who meet expected performance levels are given additional business. Prior to the warehouse, Doyle had to write complex queries to a mainframe, and even then could not access the information that he receives today.

Warehouse data make it possible to think about, ask questions, and explore issues in ways that were not previously possible. Sue Bailey describes her experience with the warehouse in the following way: *"In the past, you asked standard questions that were related to the reports that you got. Now we are able to think about our business differently. I can get information in response to very specific questions. If I see a problem, I can ask a series of questions to get a much better understanding of the dynamics of what is going on."* For example, using the data warehouse, Bailey and colleagues obtained actual sales and margin impacts by contract segmentation for a contract strategy study. The actual margin was obtainable rather than estimates based on samples. The savings in manual effort plus the actual data provided insights that would have been missed using small sample sizes. Additionally, they could change segmentation codes within the data warehouse and observe the impact on the prior year sales. Bailey explains that *"in essence, the data warehouse allowed us to test an approach prior to implementation."*

Better Decisions

> *Our analyst really doesn't spend less time analyzing. The data warehouse is providing for a much more thorough analysis and a much better understanding of what is happening in our business. Our top-level analysts are no longer just cranking data.*
> - Jason Denman, Financial Analyst

The fact that employees now access more and better information impacts the quality and process of decision making. People can ask questions that they could never ask before, rely on facts instead of intuition, and understand situations at a much lower level of detail. Sue Bailey explains that the data warehouse has made people at Whirlpool *"rethink the way we solve problems."*

Whirlpool is continually looking for ways to produce a higher quality, lower cost product. A major way to do this is through the component parts used in its appliances. When a potentially better component is identified, it is placed in a test run (possibly 100,000) of appliances, and then monitored using CQIS. Before the warehouse, it took up to a year to learn about its performance. Now it is possible to more quickly decide whether to put it into all of the appliances. As Jerry Briney describes it: *"You want either a quality improvement or a cost improvement. Either one, you want to get it as fast as possible, and it (CQIS) makes it much faster to do this. If you don't see any failures for six months in that 100,000, you say 'let's put it in four million per year.'"*

The CAMPA application supported hard dollar benefits through improved decision making. Bailey describes the process: "*Analysis of floor planning costs compared to sales of those floor planned models provided hard data to support change in flooring policy. This was one area that the field operations managers actually tracked hard costs reductions of over $1.0 M in the first year to the change in policy."*

Improvement in Business Processes

> *We have realized that now with these tools we can detect a hazard that was not recalled two years before, versus two years after, it was out in the field.*
> - Larry Latack, Product Safety Manager

Many of the biggest returns from the use of IT is when it is used in the redesign of business processes. This redesign can take place at the individual worker level. Before the warehouse, most of the quality data were based on 3 months of rolling data. Averages computed over the 3-month horizon tended to hide important developments. With the warehouse, monthly data became available for control chart purposes. According to John Doyle, *"when you get into actually charting the monthly data, you can see the swings in the data. When something 'pops' on the control chart, you can take immediate action and find out what is going on with that particular activity."* When Doyle sees that a part is out of control, he is able to drill down into the data warehouse to see whether the problem is occurring in a particular plant or is due to a particular supplier. The ability to quickly identify and correct the problem source results in tremendous savings. Before the warehouse, problems were slow to detect and difficult to correct.

Some business processes are being changed to include a broadening of the organizational boundary. As mentioned earlier, parts suppliers can access quality information by querying data from the warehouse and analyzing it using GQL. This allows suppliers to see the entire history of a part failure from the initial customer call to the final resolution by a service technician. In addition to studying a single part incident, suppliers can look at the performance for all parts of a single type in order to gain a higher-level perspective on parts performance.

Supplier access to the warehouse has also affected the cost recovery process for failed parts. Because supplier and Whirlpool personnel have access to the same parts failure data, it is easier to determine and agree upon the reason for a failure; decide who is responsible for it; and if it is the supplier, decide upon the suppliers' share of the warranty cost payment.

Support for Strategic Business Objectives

The most significant impact of IT is when it becomes an integral part of corporate strategy. At Whirlpool, senior management recognizes the need to align IT with business needs and corporate strategy (as identified earlier in Table 1). Like many companies, Whirlpool's systems were developed separately and in an unintegrated manner, which limits Whirlpool's ability to operate. Even though Whirlpool is in the early stages, senior management feels that the warehouse will ultimately allow the company to function more effectively as a "global, customer-oriented" company.

Whirlpool is beginning to work toward making IT and the data warehouse important parts of its corporate strategy. There are several reasons that make this a challenging undertaking: multiple production and business processes that frequently differ from facility to facility (i.e., plant or warehouse), different data definitions, and a variety of computing platforms and databases. For example, it is a challenge to develop performance metrics for various processes (since they differ) and then compute the associated performance measures (since the data are not easily accessed). The data warehouse is a partial solution to this problem since it intended to provide a single source of clean, consistent data to support decision making and planning and controlling activities.

INSIGHTS ABOUT DATA WAREHOUSING

Other companies' experiences are helpful to organizations that are involved in similar activities. Through the evolution of data warehousing at Whirlpool, ideas surfaced and practices were put in place that have made the data warehouse more effective. For companies that are planning to build a data warehouse, the insights gained at Whirlpool can help guide their efforts.

End Users Are Heterogeneous

End users differ in ways that should be considered when designing a data warehouse. Some users, for example, need narrow slices of detailed data that they manipulate further using a spreadsheet. For them, a dependent data mart and copy and paste capability may be the best solution. Other users may have broad data needs that are best served by a comprehensive "enterprise data warehouse" that provides a cross-organizational and cross-functional view of Whirlpool information. Users also differ in their computer skills, how often they need data in the warehouse, and their willingness and ability to be power users. At one extreme are users who simply want to click on a button to receive a report based on predefined queries. Power users, on the other hand, are more willing to deal with complexity in order to be able to access and analyze data in flexible ways.

Carefully Understand Users' Information Needs

Like any systems analysis and design effort, the starting point is to understand the existing system and the requirements of the new one. This is done through individual and group interviews with future users. There are several requirements for doing this well. Designers should understand that users and their needs are heterogeneous. Designers must also have the ability and willingness to communicate using the users' mental models and terminology. Users have little understanding and interest in the data models, relational tables, and primary and foreign keys that are so important to the world of database professionals. Users will not be able to articulate all of their data needs initially; many needs will surface only after they have used the data warehouse for awhile.

Provide Appropriate User Interfaces

Because users and their data needs differ, it should be expected that different interfaces (and the associated underlying software) will be needed. Managed query environment software, such as GQL, provides a friendlier interface than using straight SQL, but it is still perceived by many users to be difficult to use correctly. The result may be that either the data warehouse is not used or someone emerges (either formally or informally) who prepares the queries for their technically challenged colleagues. From an IS perspective, it is tempting to blame users for being unwilling and unable to master what is believed to be easy-to-use data access software, while from the users' perspective, they find the software difficult and time consuming to learn. For many users, the simple point-and-click access to information that is associated with executive information systems is a better solution when their information needs can be well defined.

Meta Data for Users Is Important

Being computer literate and able to use data access tools are necessary but not sufficient conditions for successful data warehousing. Users must also understand the data that they are accessing, including the data definitions, the source systems used, when the data were last updated, and peculiarities (e.g., the result of merging two data sources). Most data access tools provide a semantic layer that shields users from having to know cryptic database table and attribute names, but they do not eliminate completely the need to understand the database. Users without this understanding either refrain from using the warehouse, spend inordinate amounts of time developing and testing queries, or ask someone more skilled to write the queries. Well-developed meta data can address these inefficiencies by giving users a helpful tool that explains the relevant pieces of the data warehousing environment. This should be viewed as an important task, and personnel who provide the meta data must be end user oriented, knowledgeable about the subject area databases, and willing to actively involve users in the development process.

Business Units Need to Support Their Data Warehousing Applications

A skilled data warehousing team is critical to provide effective technical solutions that meet users needs; however, it is the business unit that understands the full potential of the data warehouse for its members. A data warehouse will best meet the business needs through the involvement of the business area during all phases of data warehousing, including the post-implementation phases. Too often the assumption is made that technology can meet needs without much effort. Instead the business units need to devote resources to continuous education and internal marketing so that warehousing can support the necessary business objectives. In fact, business users should be assigned permanent roles that support data warehousing efforts.

Benefits from Time Savings, More Information, and Better Decisions are Easy to Realize

Virtually everyone who has access to a data warehouse can give examples of these benefits being realized. These benefits are intertwined. Because data can be more quickly and easily accessed, it is possible to do a more thorough analysis, which results in higher quality decisions. Whirlpool's data warehouse has delivered significant benefits on this front.

Benefits from Changing Business Processes and Support for Strategic Objectives are More Difficult to Actualize

The benefits from the improvement of business processes and support for the accomplishment of strategic business objectives are more difficult to realize. At the individual level, a warehouse affects how workers perform their jobs. It's more difficult to affect changes, however, at the organizational level. Changes require a clear vision of what should be done, a champion, commitment, groups to work together, the handling of political resistance, as well as the data and technical infrastructure required to support the change. The potential benefits are great, however.

CONCLUSIONS

The evolution of data warehousing at Whirlpool reflects significant growth in data warehouse usage, support for an increasing number of business needs, a variety of benefits to the company, and valuable learnings. As mentioned earlier, nearly six hundred business users access data from the data warehouse each month, and this number should continue to grow. The data warehouse supports many business applications, and interest in these applications deepens with training and internal marketing efforts.

Most importantly, Whirlpool has experienced numerous benefits from data warehousing at the operational and strategic levels, in both quantifiable and intangible forms. An in-depth, exhaustive analysis of these benefits has not been conducted, per se, primarily because the analysis would be time-consuming and expensive. The identified benefits and high satisfaction with the data warehouse make this investment unnecessary. Instead, assessments of benefits are conducted during the post-audit process on an application-by-application basis. There are opportunities for Whirlpool to increase the return on investment from data warehousing by focusing on the high-level benefits derived from changing business processes and alignment with corporate strategy. In this way, Whirlpool can move forward with its business objectives supported by a sound and responsive information infrastructure.

ADDITIONAL SOURCES

Gray, Paul and Hugh J. Watson, *Decision Support in the Data Warehouse*, Prentice-Hall, Upper Saddle River, NJ, 1998.

Watson, Hugh. J. and Barbara J. Haley, "Managerial Considerations," *Communications of the ACM*, September 1998, pp. 32-37.

Devlin, Barry, *Data Warehouse: From Architecture to Implementation*, Addison-Wesley, Reading, Mass., 1997.

Kelly, Sean, *Data Warehousing in Action*, Wiley, NY, 1997.

BIOGRAPHICAL SKETCHES

Barbara J. Haley is an Assistant Professor of Commerce at the University of Virginia's McIntire School of Commerce. She received her Ph.D. in MIS from the University of Georgia, completing a large-scale study on data warehousing success as her Ph.D. dissertation. She is an Associate Editor for the Journal of Data Warehousing, has published in journals that include Information Systems Research, Communications of the ACM, Journal of Data Warehousing, and Information Systems Management, and has presented her work at national and international conferences. In 1998, Barbara was bestowed the title of Fellow by The Data Warehousing Institute for her research in data warehousing.

Hugh J. Watson is a Professor of MIS and a holder of a C. Herman and Mary Virginia Terry Chair of Business Administration at the Terry College of Business at the University of Georgia. He is the author of over 100 articles and 22 books, including Decision Support in the Data Warehouse, Prentice-Hall, 1998 (with Paul Gray). He currently specializes in data warehousing and is a Fellow of The Data Warehousing Institute and Senior Editor of the Journal of Data Warehousing.

Dale L. Goodhue is an Associate Professor of MIS at the University of Georgia's Terry College of Business. He received his Ph.D. in MIS from MIT, and has published in Management Science, MIS Quarterly, Decision Sciences, Sloan Management Review and other journals. His research interests include measuring the impact of information systems, the impact of task-technology fit on individual performance, and the management of data and other IS infrastructures/resources.

Client - Serve Yourself

Sorel Reisman
California State University, Fullerton, USA

Roger G. Dear
California State University, Fullerton, USA

Amir Dabirian
California State University, Fullerton, USA

EXECUTIVE SUMMARY

The Office of Information Technology at California State University, Fullerton is responsible for campus administrative mainframe and programming support, as well as for all mainframe applications such as the student records and financial accounting systems. Two campus organizations, University Extended Education and the Office of Analytical Studies rely heavily on the timeliness and accuracy of administration and student record data. These organizations have identified process improvements that are best met by client/server database applications. Due to limitations in the university's Office of Information Technology support capabilities (budgetary and legacy system related), both University Extended Education and Analytical Studies have opted for internal database development while still relying on the Office of Information Technology's mainframe data. This approach has resulted in increased local capabilities without the uncertainties related to working within an overloaded campus-wide Information Technology organization. Whether this approach is advantageous from an enterprise perspective remains to be seen.

BACKGROUND

California State University, Fullerton (CSUF) is one of 23 universities in the California State University System. Located approximately 30 miles southeast of Los Angeles (and about 8 miles north of Disneyland), CSUF provides instruction in a wide array of full- and part-time undergraduate and graduate degree programs to over 24,000 students. There are four Vice Presidents, a Chief Financial Officer, and a Chief Information Technology Officer who report directly to the President of the University (see Figure 1.)

The Office of Information Technology (OIT) has four main divisions: Administrative Computing (AC); Campus Help Desk; Data Network; and Telephone Services. Administrative Computing is responsible for campus administrative mainframe and programming support as well as for all mainframe applications such as the student records and financial accounting systems. AC currently has nine staff members who are responsible for all administrative computing tasks. The staff includes a database administrator, four application programmers, and four system programmers.

OIT Business Process Redesign Activities

Over the past few years OIT has supported the President in a series of major initiatives that include:

Figure 1: Top Level Organization Chart

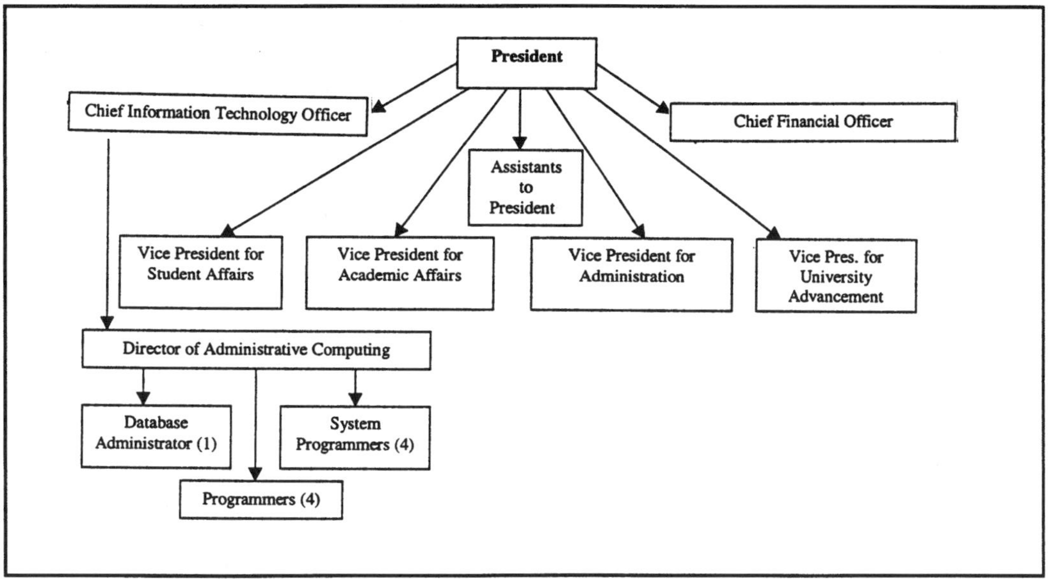

1) New desktop computers for all full-time faculty and staff
2) Email accounts for all students, faculty and staff
3) Campus Help Desk for students, faculty and staff support
4) Remote computer access for all students, faculty and staff
5) Expansion of main campus student computer laboratories

From an information systems theory standpoint, these initiatives fall under the category of business process redesign (BPR), since BPR consists of "radically transforming organizational processes through the optimal use of information technology to achieve major improvements in quality, performance, and productivity." (Raymond, et. al., 1998.) BPR is innovative. Participation tends to be top-down, with a broad, cross-functional scope. (Davenport, 1993.) These characteristics are consistent with the implementation activities at CSUF where the OIT successfully worked with the university community to implement these major campus-wide initiatives. In fact, the degree of innovation of some of these initiatives can be seen through the fact that within the CSU System, CSUF was the first campus i) to have a state-of-the-art computer and telephone network; ii) to place networked Windows NT Pentium II computers with Microsoft Office/Outlook on the desktop of all faculty, staff and administrators (in Spring 1998); iii) where every student is now given a permanent e-mail address while enrolled at CSUF; iv) where Web-based registration will take place (scheduled for a Spring 1999 implementation); and v) where most correspondence with students will eventually be accomplished via e-mail.

These initiatives were successful for a variety of reasons. First, the President spear-headed these changes, and top-level support is a necessary, but not sufficient condition for success. (Caron et. al., 1994). The President ensured that adequate financial/resource support was available. Second, OIT was operating in new territory and was not restricted by current/past practices. Processes and procedures were created, not revised. Existing infrastructures related to the mainframe environment were not impacted, other than to free up resources (such as when the VAX-based campus e-mail system was replaced.)

OIT Process Improvement Activities

Technological changes over the past decades have changed the fundamental way that computers are used. In the past, nearly all requests from administrative units were developed and performed on mainframe computers. In today's environment computation has been decentralized. Change requests tend to request AC to supply data rather than to perform computations. These requests are better classified as process improvement projects since they typically focus at the process level and do not include the entire value chain. (Recardo, et. al., 1997.) These are essentially incremental changes to existing processes where participation tends to be bottom-up, with a narrow functional scope. (Davenport, 1993.)

From an accounting standpoint, CSUF's OIT staff does not operate a charge-back system for campus consulting and support. Instead, the entire OIT operation is funded by the President. While each administrative unit has its own computer support team, the total number of employees campus-wide in these positions is fewer than 30. The OIT is relied upon to manage and oversee these process improvement projects. Procedures developed by AC require requests for new application development and report generation to be submitted to AC through a formal meeting of interested parties resulting in a specific development proposal. An approved proposal is scheduled for implementation in a priority order that does not depend on the requesting organization's ability to fund. These projects tend to be less successful than the BPR initiatives. One possible reason for this situation is that the existing AC infrastructure and small staff emphasize support for the mainframe applications with little time to support external requests. The absence of a charge-back system increases the likelihood that priorities shift as other projects take on greater urgency based on campus-wide objectives. Thus a project originally estimated to take three months may be pushed to 6-18 months, depending on the workload and priorities of the programmers.

In their defense, AC staff face several obstacles that hamper their productivity. There are several legacy systems that are difficult to maintain. These systems typically have cumbersome/limited/outdated user interfaces. Also, customized code must be tested (and often revised) whenever operating systems or applications are upgraded. Moreover, since most of the campus applications require COBOL programs to manipulate the master databases, the test/revision process is particularly inefficient because much of the code is not well written and/or is poorly documented. These applications are well entrenched and there are no plans for replacing them in the foreseeable future.

Organizationally, AC sees itself as the "keepers of the data," and, since AC is responsible for time-critical campus-wide systems there is a natural reluctance to investigate solutions that impact their applications. One fundamental issue behind the success/failure of process improvements is the centralization/decentralization of data and information processing. Both approaches have advantages and disadvantages. Centralization offers standardization and data integrity. Decentralization offers greater innovation and faster response. An enterprise-wide perspective is needed to resolve these issues. Presently, however, AC offers two options to users needing access to their data: database duplication or screen scraping. Database duplication is an off-line method of data processing which downloads the data periodically and uploads the updates, with potential synchronization problems (Raghavan, 1997). Screen scraping takes place at a terminal screen when user-entered data are "scraped" out of data entry fields before the transaction screen is transmitted to the main application. Screen scraping does not affect the processing of the main applications. Although these solutions might not be globally optimal, the users are left with no other options.

The problems confronting Administrative Computing and its end-users are not unique. In recent years one of the key issues in the development of information systems has been "building a responsive IT infrastructure" (Brancheau *et al.*, 1996). Unfortunately, unless sufficient resources are expended, legacy systems in large organizations often impede responsiveness. The perceived dissatisfaction with delays and the inflexibility of the traditional IS department has resulted in the emergence of end-user computing (Kawalek *et al.*, 1997). In the past, corporate cultures determined the priority that computer centers gave to departmental requests for information services (Basu *et al.*, 1995, Watson *et al.*, 1998). More recently, however, the corporate trend has been towards leveraging information technology to gain a competitive edge, and information service priorities, more and more are being directed towards business opportunities that contribute to the bottom line (Wheatley, 1998).

Setting the Stage

Historically, like most large institutions, CSUF performed all of its administrative computing using large mainframes. The mainframes varied from the IBM 3090 to the recent IBM 9672. The current campus mainframe utilizes an IBM legacy operating system MVS (recently re-released as OS390) and has the following configuration: IBM 9672-R32 (Approximately 36 MIPS - 3 Processors each @ 12 MIPS), 512MB Memory, 160 GB Disk Space (RAID with Hardware compression 4:1). A primary responsibility of Administrative Computing is the operation and maintenance of the student services applications which use a third party product called SIS+ (Information Associates, 1991), which was developed by SCT, a software company that specializes in enterprise solutions. SCT Software Groups include SCT Education Systems, SCT Utility Systems, SCT Manufacturing and Distribution Systems, and SCT Government Systems. At CSUF, the SCT applications are time-critical, and system administrators continually monitor the mainframe so that auxiliary applications do not impact its performance. Campus legacy systems are scheduled to remain intact for the foreseeable future, accompanied by all of their associated maintenance difficulties.

Two organizations requiring AC support are University Extended Education (UEE) and the Office of Analytical Studies. For both of these organizations AC may be considered the "server" while UEE and Analytical Studies may be considered to be "clients." This case study describes how both of these organizations cope in this "client-server" environment. The unifying theme between the two organizations is their desire for expanded computing capabilities and a reliance on AC to provide data. As will be seen, both UEE and Analytical Studies have selected the data duplication option over screen scraping.

UEE, which reports to the Vice President for Academic Affairs, offers credit courses to matriculated students during the intersession (January, and June through August), mainly for students who wish to accelerate their graduation. Non-credit courses are offered throughout the year to members of the community. All programs offered through UEE must be financially self-supporting. Thus the organization strives to operate as effectively and efficiently as possible, like a business in the private sector, or a non-profit organization in the public sector. UEE was established in the early 1980s, and during its first decade the department remained relatively small, focusing mainly on servicing intersession credit courses. A more aggressive approach was adopted in the early 1990s, and as of 1996/97 CSUF's University Extended Education program consisted of more than 125 full- and part-time employees (not including instructors) offering 950 non-credit courses, 45 certificates, and more than 55,000 extension enrollees.

UEE has its own Marketing department with a set of responsibilities that are unique for most publicly funded universities, but less so for private colleges and fund raising arms of public sector organizations. Marketing in UEE is responsible for promoting new (non-credit) courses and certificates that meet the needs of the local community. Their activities include market research, advertising, promotions, production, etc. To accomplish their goals, the Marketing department must operate in a traditional marketing fashion, one that requires a significant amount of data upon which to base their proposals for new instructional offerings. Programs that are undertaken by UEE, based on Marketing projections and proposals, are turned over to UEE Academic Programming staff to develop, offer, and maintain. UEE also has its own Student Services staff, Accounting department, and Information Systems (IS) staff to support all its academic functions.

Analytical Studies is an all-university central resource for institutional facts and figures, methodological advice, and related problem-solving. The office is responsible for organizing and presenting research results, models, alternative scenarios, and other information essential to campus policy formulation, academic planning, resource allocation, assessment and quality, and institutional effectiveness. Within Analytical Studies' purview are enrollment trends and forecasting, student and faculty demography, recruitment and retention, resource allocation and utilization, Academic Affairs' budget and faculty flow, curricular change and student learning, and any other topic relevant to documenting and improving institutional effectiveness. The Office of Analytical Studies consists of a Director and a small staff. A substantial amount of information generated by Analytical Studies is derived from data sources under the aegis of Administrative Computing. As with UEE, a need for added data requirements, etc. must be coordinated with AC.

CASE DESCRIPTIONS

Case 1 - University Extended Education

Approximately four years ago, with the advent of a new Dean, UEE decided to expand its instructional offerings and its associated marketing activities. A new director of marketing determined that there was a need to gather data about its potential "customers" in the local community. An apparent likely source of data was the student record database that had been accumulating for years via the student registration system, SIS+. SIS+ was originally designed as an online student registration, billing, and grade reporting system for full-time credit students registered in 'traditional' undergraduate or graduate programs. SIS+ had previously been modified by Administrative Computing to permit UEE to also register students into non-credit extension programs.

When the UEE Information Systems staff examined the feasibility of using the existing student database for marketing purposes, it soon became obvious that the nature of student records in the database was unsuitable. There were essentially two types of student records in SIS+. The first consisted of the academic record of an undergraduate or graduate student from initial enrollment in the university through the granting of a degree (or withdrawal). The second type consisted of information necessary to register and invoice non-credit extension students.

For the most part, extension students do not require detailed student records that track an academic career through multiple semesters/years. Extension courses are often not part of lengthier programs where more extensive tracking towards a degree is required. One exception is a non-credit certificate program that is comprised of a series of courses in which the monitoring of academic success is far less stringent than that of a degree program. In the case of these certificate programs, SIS+ had been modified slightly to track student progress towards successful completion of all courses in a certificate.

The single greatest drawback of the SIS+ student database, however, was that there was no facility for relating credit and non-credit students. This meant that a previously registered credit student's record could not be made directly available to UEE if that student returned to the university and registered for UEE-offered non-credit courses or certificates. In an effort to address this deficiency, discussions were undertaken between UEE's IS group and Administrative Computing to determine how the system could be altered to track a student's history from first registration, past withdrawal or granting of a degree, to also include future non-credit program registrations. UEE was informed that this was not technically possible, and even if it were, the task would take an inordinate amount of time to complete.

UEE's IS staff suggested other solutions, such as a variety of inelegant sorts and merges of student records, but Marketing had more grandiose plans to provide an easy-to-use, integrated system for targeting students/customers, and for determining demographic and academic trends. In particular, Marketing was interested in building student records that tracked the complete history of every student that ever took a course, credit or not, at CSUF. In addition, Marketing wanted to be able to add to that record, every instance of contact made by that student to UEE, in which the student/customer contacted the institution and requested information concerning a UEE instructional offering.

One of Marketing's primary sales strategies, typical of university extension programs, is "focused" mailings of promotional catalogs and brochures. In addition to purchasing mailing lists, Marketing wished to be able to access the university's own database to develop proprietary mailing lists for special CSUF events and offerings. This required the development of a facility that would capture "call-in and mail-in" requests by potential students for information regarding non-credit instructional offerings. AC and UEE determined that, without a massive overhaul, the SIS+ database would not be able to provide these additional data elements.

During the time that the analyses of SIS+ student records was taking place, other parts of UEE's operation were being examined to determine if other information systems needs could be addressed in order to improve UEE's overall operation and to provide improved service for students and instructors. The analysis, which was conducted over a three-month period, determined that UEE's operation could indeed benefit from a complete system overhaul. Accordingly, a structured analysis of the entire operation was undertaken wherein all the functions of the organization were examined.

Figure 2: Proposed System Implementation

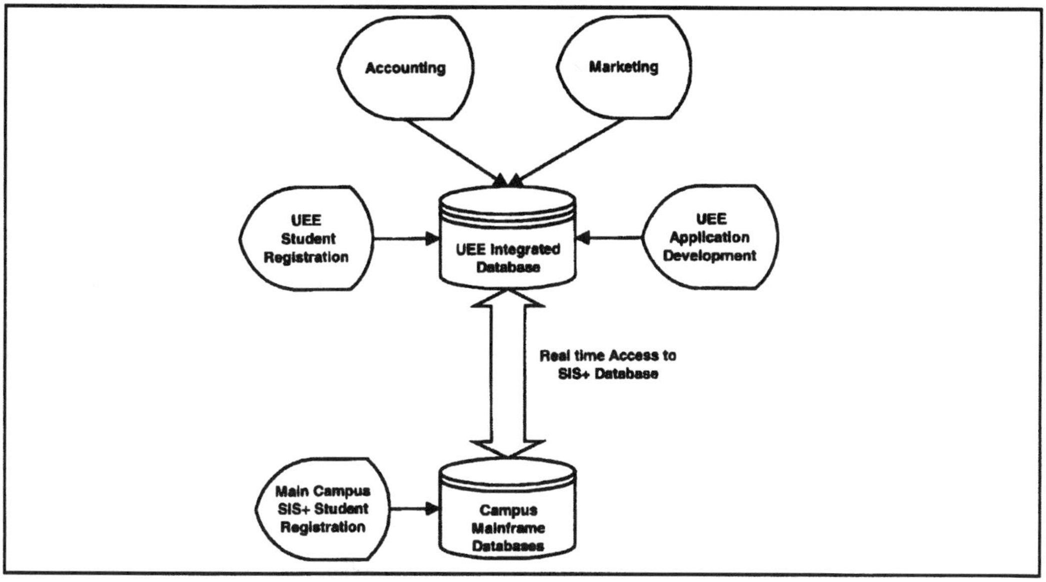

An extensive set of data flow diagrams was produced reflecting all of UEE's procedures as well as the relationship of UEE subsystems to those of the main campus systems. After careful examination of all the information, the UEE IS director decided that the UEE operation required complete reengineering, focusing on the development of an integrated client-server database management system. While a decision like this may seem to be a dramatic one, in fact, in many ways it was obvious from the beginning that this strategy would likely be required.

Figure 2 illustrates the client-server architecture that was proposed to address UEE needs. In this proposal, UEE would create a single integrated proprietary database accessible by the each of the organization's functions. The database management system that would support student registration of both credit and non-credit students as well as the development of additional database applications specific to the operation of all the functions within UEE. It would also provide UEE with the ability to create new applications such as "call-in" or "mail-in" data capture, data mining into the UEE database, etc., as the organization's needs expanded.

One of the key differences between the proposed system and the then current operation was the requirement that a proprietary online student registration system would be required. The proposed system would capture all credit and non-credit student registrants, and append, modify, and build new records in the integrated database. In this way, student records containing all main campus data elements as well as any data elements required by UEE could be built and maintained with no interference from the main campus system.

However, given main campus' requirement to provide mission-critical information processing for applications such as online student registration, the proposed design was problematic. As Figure 2 illustrates, the proposed system required a direct, high speed connection from the UEE proprietary DBMS system to the campus' mainframe-based SIS+ application. Functionally, this meant that students, who were able to register, add, and drop courses at any of the SIS+ online registration booths on campus, should also be able to obtain the same service at UEE. This level of IS service placed a burden and responsibility onto UEE that the organization was unwilling to support.

UEE's Marketing department concluded that their data requirements were not time-critical, unlike student registration information where SIS+ data entry operators are able to add and drop students in real time based on up-to-date class loading data. Accordingly, the IS staff proposed a different architecture (Figure 3).

In the new proposal, the UEE system would not be responsible for registering students. Instead, SIS+ would continue to be the online student registration system, operating from within a GUI-based

Figure 3: Actual System Implementation

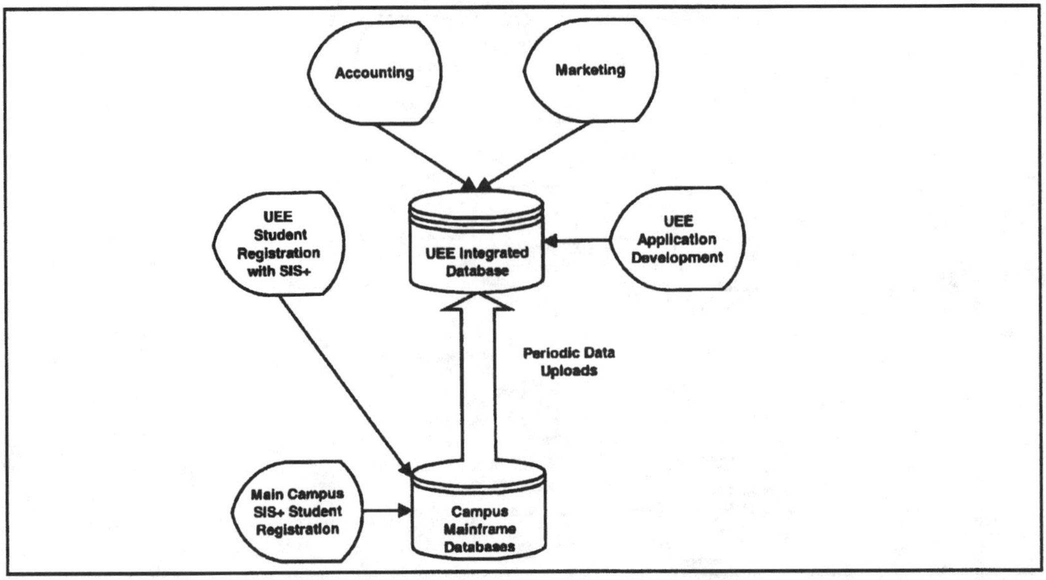

window on a UEE user's workstation, connected directly to the campus mainframe-based student database. All other UEE functions would be developed and supported directly via the UEE database management system and the UEE database. In the new system, however, student record data elements would have to come from two different sources: UEE proprietary application input, and input from the campus SIS+ database. This would require periodic uploads of records from the campus database to ensure the timeliness of UEE student records. Uploaded student records would then be joined to UEE student records to provide a more complete student record database in the UEE integrated database.

This requirement for campus uploads of data would not be problem-free. Besides Administrative Computing's limited resources, uploads of data from the mainframe could affect mainframe performance, particularly if they were required to be frequent and/or during peak system use. In order to maintain a cordial relationship with the AC, UEE decided that uploads could be performed as infrequently as once per week, but more often if possible. The reason for this latitude on the part of UEE was that it was obvious from the proposed system architecture that except for student records, all UEE-specific information would be available from within the UEE client/server database as needed. Week-old student records would be almost as accurate as more up-to-date information for market analysis purposes.

Project Status/Current Challenges

The UEE client/server hardware and software have all been installed and deployed. An IBM NetFinity multiprocessor server running Microsoft Back Office is currently supporting a SQL-Server database. Applications written in PowerBuilder and Visual Basic have been developed and are currently supporting all the Academic Programmer applications, almost all the Marketing applications, and many of the Accounting applications. During the system implementation phase of the system development, the university's Financial Affairs department implemented new campus-wide standards for financial and accounting operations. That decision set back some of the UEE schedule and precluded full deployment of the UEE system.

At the present time, AC is providing periodic uploads to the UEE database management system. For the time being, the frequency of the uploads is sufficient to support the application service necessary for UEE to move forward in its expansion plans. However, it may be necessary at some point for UEE to develop its own student registration system as was originally proposed, in order to achieve the required levels of system performance and functionality.

Table 1: Examples of Standard Report Statistical Categories and Classifications

Statistical Categories	Classifications
Headcount	Level (Freshman, Sophomore, etc.)
FTES (Full-time equivalent student)	Enrollment Type (Full-time, Part-time)
GPA (grade point average)	Enrollment Status (continuing, new, transfer, etc.)
Majors	Sex
Degrees Awarded	School
Age Distribution	Department
Ethnicity	
Citizenship	
Residence	

Figure 4: Example of a standardized report

CALIFORNIA STATE UNIVERSITY, FULLERTON 1.30.2.98

OFFICE OF ANALYTICAL STUDIES

CENSUS DATE SUMMARY OF SELECTED REGISTRATION STATISTICS:
SPRING, 1998

		HEADCOUNT		FTES	
		NUMBER	PERCENT	NUMBER	PERCENT
1. TOTAL		24137	100.0%	17600.5	100.0%
2. SEX:	MALE (M)	10098	41.8%	7317.3	41.6%
	FEMALE (F)	14039	58.2%	10283.2	58.4%
3. LEVEL:	FRESHMEN	3637	15.1%	2583.9	14.7%
	SOPHOMORES	2247	9.3%	1943.3	11.0%
	JUNIORS	5916	24.5%	4625.0	26.3%
	SENIORS	8079	33.5%	6158.4	35.0%
	GRADUATES	4258	17.6%	2289.9	13.0%
4. ENROLLMENT TYPE:					
	SIX OR LESS (PART-TIME)	6191	25.6%	2005.1	11.4%
	SEVEN TO ELEVEN (PART-TIME)	3254	13.5%	1975.3	11.2%
	TWELVE AND OVER (FULL-TIME)	14692	60.9%	13620.1	77.4%
5. ENROLLMENT STATUS:					
	CONTINUING	20732	85.9%	15756.9	89.5%
	RETURNING	242	1.0%	127.3	0.7%
	RETURNING TRANSFER	53	0.2%	35.3	0.2%
	TRANSFER	1352	5.6%	993.9	5.6%
	FIRST-TIME FRESHMEN	53	0.2%	44.5	0.3%
	FIRST-TIME GRADUATES	768	3.2%	418.9	2.4%
	OTHER/TRANSITORY	937	3.9%	223.7	1.3%

ENROLLMENT STATUS:

CONTINUING: STUDENTS WHO ALSO ENROLLED IN THE IMMEDIATE PRIOR SEMESTER AND THOSE UNDERGRADUATES WHO "STOPPED OUT" FOR ONE SEMESTER.

ALL OTHERS: CATEGORIES OF NEW STUDENTS, I.E. THEY FILED AN APPLICATION IN ORDER TO ENROLL THIS SEMESTER. ENROLLMENT STATUS DOES NOT MATCH STUDENT LEVEL FOR FIRST-TIME FRESHMEN, WHICH IS A MIX OF FIRST-TIME AND TRANSITORY.

Case 2 - Analytical Studies

Analytical Studies produces reports tend to fit into one of two categories; standardized or *ad hoc*. Standard reports are typically published biannually and consist of information relevant to each semester/academic year. *Ad hoc* reports document studies that are developed in response to specific questions asked by the academic community, and, if applicable, may evolve into standard reports. Approximately 25 standard semester reports provide profiles of statistical categories of interest broken out by various classifications (Table 1).

Other reports provide trending information over multiple academic years. Examples of ad hoc reports include retention studies for student athletes and forecasts of future faculty requirements. Figure 4 presents an example of a standardized report which summarizes selected registration statistics.

The data used to generate the reports are provided by AC in the form of flat files from the mainframe computer. These files are produced at designated intervals using COBOL programs that were written decades ago (and were revised to extract data from SIS+). The file structures change from time to time as new fields are mandated by the Chancellor's Office, the body that oversees all the universities in the California State University system. Other changes also reflect new additions/ interpretations to existing fields/code tables, such as the creation of a new major. Some changes are procedural. For instance, students were previously tracked by uniquely assigned student IDs, but currently they are tracked by their Social Security Numbers. These files are produced at different times - grade data following the completion of a semester, enrollment data after the "census" date for a semester, etc. There is considerable redundancy in these files - student names, addresses, birth dates, entrance exam scores, etc., may appear in multiple files in the same semester, and also across semesters. Course titles, department names, instructor names, and numerous other fields are replicated as well.

The following is a description of the general process which was used to produce the standard reports prior to the redesign of the system described in this case study.

"Official" files were produced using production COBOL routines and stored in designated datasets on the mainframe. Analytical Studies was notified of the availability of these files and a series of SPSS-X stored procedures was executed to identify any obvious data field changes. If necessary, AC was contacted regarding these issues and the impacted "official" files were regenerated with necessary enhancements/corrections. Additional SPSS-X procedures were then executed using the finalized versions of the "official" files to tabulate the great variety of data required for the reports. These SPSS-X procedures required their own revisions when changes in the data structure/ interpretation impacted the tabulations. Hard copies of the SPSS-X output were then used by data analysts to manually prepare spreadsheet reports. The spreadsheets were designed to perform side calculations and to produce summary lines as needed. Headers and footnotes were revised each semester. Occasionally data structure/interpretation changes would result in a spreadsheet redesign. Finally, a comprehensive review was undertaken to eliminate errors prior to publication.

While the process to produce standardized reports from Analytical Studies has been successful, it did suffer from a number of drawbacks. The entire process was labor intensive, especially the data transfer from SPSS-X hard copy output to PC spreadsheets. This data transfer was performed by student assistants who typically work for Analytical Studies for only one or two semesters. This created learning curve problems, and the sheer volume of data contributed to quality control problems. When changes were required to the SPSS-X procedures they were completed by the Director of Analytical Studies, the only employee in the organization knowledgeable about SPSS-X .

Another drawback had been the long development time typically required to generate new information and new reports. In addition, certain types of reports were just too cumbersome to be produced using the current approach. For example, one existing report presented a student ethnicity profile for the entire university. A similar report for each department and school (84 departments in 7 schools) would have been useful, but the amount of labor required under the legacy process was prohibitive.

During the Spring, 1997 semester the Director of Analytical Studies determined that the existing approach needed an overhaul. Administrative Computing was consulted, but it was quickly determined that AC did not have the manpower nor the desire to develop an entirely new application for Analytical Studies. The Director of Analytical Studies approached one of the authors to assist in a process improvement project. Following a detailed analysis phase process, the Director's improve-

Table 2: Initial Analytical Studies Process Improvement Goals/Objectives

1)	Reduce mainframe usage
2)	Eliminate SPSS-X dependency
3)	Improve validation capabilities
4)	Streamline report development
5)	Simplify report generation
6)	Expand reporting capabilities

ment goals and objectives were defined (Table 2).

Additional long-term objectives were also developed. The goal is to empower recipients of the reports (primarily Deans and Department Chairs) with the ability to generate their own reports based on "official" data. This empowerment is expected to provide parameterized queries in a decision support environment. Because of the need for standardization and the potential for data definition misinterpretation, it is expected that users will not be able to develop their own queries.

All of the initial and long-term process improvement goals and objectives could be met by developing a Client/Server application utilizing a relational database management system (RDBMS). Since the Spring 1998 university-wide initiative placed networked Windows NT Pentium II computers with Microsoft Office on the desktop of all faculty, staff and administrators, Microsoft Access was selected as the application program. "Official" (COBOL-generated) files are still

Table 3: Analytical Studies Client/Server Implementation Phases

I.	Design/Develop Relational Database
II.	Develop mainframe import, change identification and validation procedures
III.	Create "Standard" reports
IV.	Design secure Client/Server user interface
V.	Develop parameterized queries/reports for end users
VI.	Complete Client/Server application

generated by, and reside on, the mainframe, but all processing is performed by the RDBMS – thus eliminating the need for SPSS-X procedures. Using the features of Microsoft Access, the enforcement of referential integrity facilitates the data validation process, and the query/report capabilities greatly improves quality control and shortens development lead time. Table 3 presents the implementation phases of the redesign.

Project Status/Current Challenges

Implementation Phases I, II, and III of Table 3 have been completed for the standardized reports and partially completed for *ad hoc* reports. Three Microsoft Access databases were developed. The first is used for importing and validating mainframe files. The second consists of condensed data tables that contain only those fields that are relevant to report generation. The third is linked to the condensed data tables and contains the forms, queries, code modules, and associated reports.

Initially the reports were produced by exporting data into preformatted Excel spreadsheets. This approach was used because the spreadsheets already existed. While this approach reduced the initial development stage, it proved unsatisfactory because minor variations in data (new departments, majors, etc.) required major spreadsheet redesign – without any guarantee that a redesign would not be required the following semester. The reports were subsequently designed within Microsoft Access to respond dynamically to the inevitable variations in data and to eliminate the need for continual redesign. Significant improvements have been realized. Prior standardized reports often took one or two months to complete with the potential of significant data entry errors. With the new system, after a new semester's data have been validated, less than one hour is required to generate and print all standard reports without data entry or computational errors. Additional reports, such as the 90-plus page department/school student profile that in the past was desirable but impossible to produce practically, have also been completed.

Challenges that remain include the development of additional ad hoc reports, including some that are quite sophisticated (e.g. comparing graduation rates of student athletes to the general student population.) The design and development of the front end of the Client/Server DBMS is just beginning. It is anticipated that extensive front-end modifications will be considered when the prototype is available for testing.

REFERENCES

Basu, S.C., Akmanligil, M., and Kumar, A. (1995). End User Computing Issues Relating to a Global Strategy. *Proceedings of the 1995 Information Resources Management Association International Conference*, Atlanta, GA, May 21-24, 1995, pp. 305-307.

Brancheau, J.C., Janz, B.D., and Wetherbe, J.C. (1996). Key Issues in Information Systems Management: 1994-95 SIM Delphi Results. *MIS Quarterly*, 20(2).

Caron, J.R., Jarvenpaa, S. L., and Stoddard, D. B. (1994). Business Reengineering at CIGNA Corporation: Experiences and Lessons From the First Five Years. *MIS Quarterly*, 18(3).

Davenport, T. H. (1993). *Process Innovation, Reengineering Work Through Information Technology*, Harvard Business School Press, Boston, MA.

Information Associates (1991). *Student Information System Student Records User Guide*, Information Associates, Inc., Rochester, NY, 1991.

Kawalek, J., Hackney, R., and Dhillon, G. (1995). Strategic Information Systems Planning: Perspectives on the Role of the "End-User" Revisited. *Proceedings of the 1997 Information Resources Management Association International Conference*, Vancouver, B.C., Canada, 1995, pp. 439-445.

Ragavan, V. (1997). A Transition Model to Second Generation Client/Server Systems. *Proceedings of the 1997 Information Resources Management Association International Conference*, Vancouver, B.C., Canada, 1995, pp. 522-523.

Raymond, L., Bergeron, F., and Rivard, S. (1998). Determinants of Business Process Reengineering Success in Small and Large Enterprises: An Empirical Study in the Canadian Context. *Journal of Small Business Management*, 36(1).

Recardo, R. J., and Jones, D. J. (1997). A Report Card on Reengineering. *Production and Inventory Management Journal*, 36(3).

Watson, R.T., Pitt, L.F., and Kavan, C.B. (1998). Measuring Information Systems Service Quality: Lessons From Two Longitudinal Case Studies. *MIS Quarterly*, 22(1).

Wheatley, M. (1998). The Cutting Edge. *CIO Magazine*, June 1, 1998.

BIOGRAPHICAL SKETCHES

Dr. Sorel Reisman is a professor of Management Science and Information Systems in the School of Business and Economics and is Coordinator of Academic Technology for California State University, Fullerton. Dr. Reisman is a member of the Publication Board of the IEEE Computer Society, editorial board member and columnist of Perspectives in the new IEEE publication, ITPro. Dr. Reisman's publications focus on hardware, software, systems, management issues and methodologies. His recent publications include the book, Multimedia Computing: Preparing for the Twenty First Century, and papers on cost/benefits of software engineering, computer platform conversion, and mainframe to client/server migration.

Amir Dabirian is a part time lecturer of Management Science and Information Systems in the School of Business and Economics and is Head of Systems and Information Technology in the University Library for California State Unviersity, Fullerton. Mr. Dabirian received B.S. degree in Electrical Engineering from California State University, Fullerton in 1984, a M.S. degree in Electrical Engineering from California State University, Fullerton in 1988, a M.S. degree in

Computer Science from University of California in 1998 and is currently a Ph.D. student at University of California, Riverside. His research interests are self-similar network traffic and the challenges of information technology.

Dr. Roger G. Dear is a Professor of Management Science/Information Systems in the School of Business and Economics at California State University, Fullerton. Dr. Dear received his Ph.D. in Operations Research from Massachusetts Institute of Technology's Department of Electrical Engineering and Computer Science in 1976. His research interests include database management systems, computer simulation, sequencing and scheduling, and queueing theory. Dr. Dear has published numerous research papers such topics as protocols for availabiliy, reliability, and consistency of replicated files, and strategies for managing platform conversions. Dr. Dear is a member of numerous professional organizations including the Association of Information Technology Professionals.

Managerial Issues for Telecommuting

Anthony R. Hendrickson
Iowa State University

Troy J. Strader
Iowa State University

EXECUTIVE SUMMARY

In this paper the issues faced by firms in today's telecommunications environment are compared and contrasted with an actual telecommuting case study of Trade Reporting and Data Exchange, Inc. (T.R.A.D.E.), a software engineering company located in San Mateo, CA. Initial results indicate that telecommuting was successful for T.R.A.D.E because the required technology was widely available (the candidate initiated the idea and had the necessary industry and company experience) the organization could provide the flexible work arrangement while retaining a valuable employee, the employees were able to live in a geographic area of their choice, overall costs could be shared by the company and employees, the job category was an ideal fit, and existing procedures were in place for communicating and managing the geographically detached worker. As telecommunications technology evolves this arrangement will continue to challenge the firm and their employees.

BACKGROUND

Company Background

Trade Reporting and Data Exchange, Inc. (T.R.A.D.E.) is a software engineering company located in San Mateo, CA. The firm was created in 1992 to provide international corporations access to a wide variety of international trade information stored on the T.R.A.D.E'.s massive databases. T.R.A.D.E. gathers data from a variety of sources including U.S. Customs documents, Chinese Customs Administration, Dun and Bradstreet, and numerous other government trade documents. They provide general information and specific trading activities on hundreds of thousands of international buyers, distributors, and suppliers.

This data is consolidated and organized for optimal access by T.R.A.D.E. The company disseminates this information via monthly and quarterly updated CD-ROMs sent directly to the firm's subscribers. Hardcopy reports are also available. In addition, the firm offers customized reporting and monitoring services available on an ad-hoc or event triggered notification basis. The CD-ROM includes sophisticated software querying tools and wizards designed to provide subscribers a user-friendly interface to the information.

Although relatively young, the firm has grown significantly and now employs a professional staff of 50. T.R.A.D.E., Inc. is truly an international organization with offices located in Hong Kong, Taiwan, the United Kingdom, and the corporate headquarters in San Mateo. Additionally, T.R.A.D.E.

utilizes a number of licensed distributors in a number of other countries. T.R.A.D.E., Inc. was recognized as one of the ten fastest growing, private companies, in Silicon Valley in 1996. While the firm is hesitant to reveal exact figures, revenues were in the multi-million dollar range in 1996 and represents T.R.A.D.E'.s significant share in the business intelligence market.

Telecommuter Background

The specific employee we will chronicle is Dave Tucker. Tucker has worked for T.R.A.D.E. for nearly five years as a software engineer, after working for a major hardware manufacturer in Silicon Valley for nearly seven years. His responsibilities at T.R.A.D.E. include developing advanced database queries, creating programs to interface source databases into T.R.A.D.E.'s massive database system, and creating software applications for the CD-ROMs which subscribers receive.

Tucker relocated to Silicon Valley nearly eleven years ago, after he graduated with a computer science degree from a large mid-western university. Tucker and his wife both grew up in central Iowa. Now with two young children (ages 3 and 5), Tucker and his wife desired to return to central Iowa to be closer to family and friends and to provide their young children with childhood experiences similar to their own. In the spring of 1996, Tucker proposed a telecommuting arrangement to T.R.A.D.E.

T.R.A.D.E. decided to pilot test the arrangement. Tucker identified a college professor in Ames, Iowa who was interested in exchanging homes, for the summer, with someone near Silicon Valley. T.R.A.D.E. agreed to allow Tucker to telecommute for the summer and both agreed to evaluate the arrangement at the end of the summer trial. After the summer trial Tucker and T.R.A.D.E. agreed that the telecommuting alternative had substantial merit and decided to enter the arrangement on a permanent basis. What follows is an enumeration of the issues related to telecommuting and a chronicle of how Tucker and T.R.A.D.E. addressed each issue. These issues include telecommuter selection, organizational issues, social aspects, costs, and productivity measurement.

SETTING THE STAGE

Telecommuting presents organizations and their employees with a number of opportunities and a number of potential problems. Generally speaking, the benefits of telecommuting include more hours worked per day, more work accomplished per hour, and less employee stress due to commuting and/or work/personal conflicts (such as a sick child) (Curran and Williams, 1997; Race, 1993; Schellenbarger, 1993; Townsend, DeMarie, and Hendrickson, 1998). The organization must cope with employee work schedules being more flexible, in that many telecommuters may or may not keep traditional work hours. Additionally, immediate access to employee skills and services is somewhat diminished.

As shown in Figure 1, successful telecommuting requires special attention be given to: enabling technology, selecting the right job category and the right individual for the arrangement, organiza-

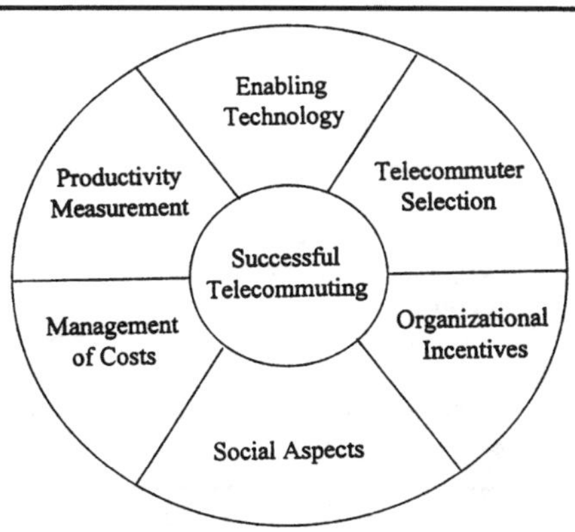

Figure 1. Telecommuting Issues

tional incentives, social impacts of dispersed workers, management of overall costs, and productivity measurement of geographically dispersed work teams. Each of these issues will be examined and compared to the specific T.R.A.D.E. example. While the specific facts encountered by Tucker represent his personal experiences with telecommuting, the issues examined apply to any telecommuting arrangement. Individuals will undoubtedly have varying degrees of problems and success with any given issue depending upon their personal background, skill level and motivations, but the categories of concern should be universal to all telecommuters and their organizations.

Enabling Technology

A number of hardware, software and telecommunication technologies are needed to support telecommuting in the software engineering industry given its requirements for distributed software development and intraorganizational communication. The specifications of the components located with Tucker, as well as some components that Tucker utilizes that are located at T.R.A.D.E.'s headquarters, are described throughout this section. Each of the items was integrated into a complete information infrastructure that supported their telecommuting needs. The components are grouped into four categories including telecommuter hardware and software located at Tucker's home office that support Web application development and intraorganizational communication, Web and database servers located at T.R.A.D.E., and the telecommunications components used for communication between Tucker and T.R.A.D.E. These components are described in Table 1.

Telecommuting Information Infrastructure Component Category	
1. Telecommuter Hardware	*Desktop PC* • Gateway 2000 P6-200, 96MB RAM, 5GB hard drive • modem with voice capabilities, Soundblaster card • SCSI DAT tape backup, 8X CD drive • UPS power supply • running Windows NT 4.0 workstation, 10MB Ethernet *Laptop PC* • Gateway 2000 Solo 2200 P5-166, 32MB RAM, 2GB hard drive, 12X CD drive • running Windows 95, 10MB Ethernet PCMCIA *ISDN router* • Webram Entre, 2 POTS lines, 4 Ethernet ports *Test Server* • Gateway 2000 P5-133, 32MB RAM, 2GB hard drive • running Windows NT 4.0 Server and Web services
2. Telecommuter Software	*Web Browser* • Internet Explorer 4.0 *Web Application Development Software* • Microsoft Visual Interdev, Frontpage 97, Visual C++, and Visual Basic *Programming Languages* • C++, HTML, VBScript, JavaScript, and Active Server Pages
3. Corporate Servers	*Database Server* • 512MB memory, 80+GB disk drive • running Oracle 8 on a HP-UX 10.0 server *Web Server* • IIS 4.0 running on an ALR Revolution Quad 4 (4 P6-200 CPUs), 256MB memory, 16MB disk drive
4. Telecommunications	• ISDN line with two voice lines (one of which was used for fax) • Hotfax MessageCenter was used to provide voice mail, fax, and dial in capabilities • US Robotics Sportser Voice 33.6K internal modem • NetINS RemoteLAN ISDN service (provided a remote LAN with five static IP addresses) • NetMeeting was used for some voice over the Internet • No video conferencing products were used.

Table 1. Telecommuting Information Infrastructure Components

The router is a unique item because it will handle both static and dynamic IP addresses. It was used to set up a remote LAN so that Tucker could have static IP addresses so that a firewall could be configured in California to allow his IP addresses through. The router also handles dynamic IP addresses given by ISPs each time a connection is made while allowing routing from multiple machines on the local LAN through the dynamic IP. The remaining items are widely available from a number of hardware and software providers.

Most of Tucker's work involved software engineering, system testing, and general trouble-shooting tasks related to the development and support of software applications used in the firm's products and services. Thus, a variety of telecommunications technology was utilized to facilitate Tucker's collaborative efforts with others at T.R.A.D.E. Specifically, Hotfax Message Center was used for electronic mail, fax, dial-in capabilities and some limited voice transmissions. Most of the voice interaction was accomplished over an ISDN connection which provided two voice channels, one of which was occasionally used for fax. In some instances it was necessary to utilize both ISDN channels at once in order to provide adequate bandwidth for some collaborative troubleshooting. In those instances, NetMeeting was also used for voice over the Internet via a dial-up connection.

T.R.A.D.E. and Tucker found NetMeeting to be very useful in these circumstances. The telecommuting arrangement did not have any video conferencing applications. Neither T.R.A.D.E. nor Tucker found this problematic since most of the interactions Tucker had were with associates whom he had previously met and had some ongoing working relationship. This is an important point since these voice connections do not provide visual support for collaborative interaction. Body language, eye contact, and physical mannerisms are filtered from the interaction with only voice connectivity (Townsend, DeMarie, and Hendrickson, 1998).

Upon reflection, Tucker expressed the need to integrate more collaborative tools and video conferencing into his telecommuting toolset. Several products were under consideration. The Internet products, such as CU-See-ME are inexpensive and provide some basic level of video interaction. Additionally, these products are relatively easy to install and are cost effective to install in a wide group of users. However, bandwidth limitations can make this solution inadequate.

Intel's Proshare units were an option that appeared very appealing. This product not only integrates the hardware (video camera) and software into one package, but the product provides a very sophisticated collaboration tool. Virtually any personal computer application can be shared between users with this product. The major drawback of this product, however, is price. Each unit is more than $1,000, and the system is proprietary. Thus each person wishing to interact with the telecommuter must also have this system. Due to the fact that Tucker supported a wide range of developers, with little or no ability to anticipate who might need to interact with Tucker on any given day, T.R.A.D.E. felt the investment in this technology to provide video connection with one telecommuter was too expensive. However, since the success of this arrangement was encouraging for future telecommuting arrangements, the firm did not rule out the possibility of this technology for developers in the future.

Similarly, collaborative technologies such as Lotus Notes was considered. This package is popular in the support of geographically dispersed collaborative teams. Notes provides collaborative tools beyond simple e-mail. Applications and documents can be shared and maintained by groups instead of individuals. Many organizations are using Notes to connect their entire professional staffs (Townsend, DeMarie, and Hendrickson, 1998). Again, the firm saw some benefit to this software, but the expense in terms of economic resources, training, and time could not be justified to support a limited number of telecommuters.

Obviously, there are an infinite number of combinations of hardware, software and telecommunication components available, but this case does identify one set of information technologies that enabled successful telecommuting. It is apparent that information technology and telecommunications services are widely available which allow telecommuting to be a feasible work arrangement for software engineering companies. But the technology is a necessary, but insufficient, condition for success. The remaining issues we discuss move beyond information technology issues to the more complex managerial and organizational behavior issues.

CASE DESCRIPTION

Telecommuter Selection

The ultimate success of telecommuting may well depend upon the initial selection of appropriate candidates for this alternative. Generally, individuals who have significant experience in a job, individuals who are above average performers, and persons with a broad range of job assignments will adjust better to a telecommuting arrangement than their less experienced, lower performing counterparts (Boyett and Conn, 1992; Fish, et. al., 1992; Kraut, 1988; Mokhtarian and Salomon, 1994; Townsend, DeMarie, and Hendrickson, 1998). More specifically, successful telecommuting candidates will have worked for the organization long enough to acquire some affinity for the corporate culture and nature of expectations in this specific work environment. Additionally, workers who have demonstrated superior technical skills (especially with remote working tools, such as a computer, modems, fax, and videoconferencing) will be more successful (Fish, et. al., 1992; Kraut, 1988; Mokhtarian and Salomon, 1994). Finally, some consideration for the individual's suitability for telecommuting may depend upon his/her personality profile and acceptance of a more ambiguous work setting.

In our example case, the firm did not initiate the telecommuting arrangement but Tucker's skills and experience are insightful. Tucker had significant experience in both the software development industry and with T.R.A.D.E. specifically. He had spent 11 years as a software engineer, four of those years with T.R.A.D.E. Thus, he had a firm understanding of his job tasks and the specific policies and goals of T.R.A.D.E. Additionally, Tucker was an extremely high achiever. When Tucker first proposed the telecommuting arrangement, T.R.A.D.E. management stated that they would have never considered such an arrangement had Tucker not been an exemplary employee whose services they wished to maintain, even if it meant altering the firm's standard operating practices. Finally, with respect to technical skills, Tucker pointed out that he felt the relationship functioned well due to his experience with the technology and his ability to provide his own technical support.

Organizational Incentives

For the organization the benefits of telecommuting are numerous. Incorporating remote workers into the organization can provide the catalyst for organization flexibility, allowing the firm to expand and contract to adjust to market conditions. As demand increases the organization can expand without the corresponding increases in physical facilities. Thus, telecommuting offers the organization the opportunity to expand with limited cost increases, or in some cases to reduce overhead by eliminating office facilities (furniture, electricity, and environmental conditioning) and space requirements (Mokhtarian, 1988; Mokhtarian 1991a; Nijkamp and Salomon, 1989; U.S. Department of Transportation, 1993).

As highly skilled human capital becomes a scarce commodity in the brain power industries of the 21st century, telecommuting offers the organization a means of attracting and retaining these limited resources (Thurow, 1996). Telecommuting may provide access to skilled employees who do not desire to reside in a particular geographic area, or who are limited due to other personal commitments (such as elderly or childcare responsibilities). Telecommuting also allows organizations to utilize workers in disadvantaged rural or urban areas, as well as incorporate the skills of workers who have left the workforce temporarily due to illness or to spend time with small children (Curran and Williams, 1997).

Although the advantages of telecommuting are numerous, there are some drawbacks to this alternative. Generally, managers are not trained to deal with remote workers, especially in terms of how to manage their activities and how to motivate them. Many managers are uncomfortable with the perceived loss of control created by telecommuting. However, most of the problems are more perceived than manifest. Good management skills in terms of delegation and performance measurement can alleviate most of the disadvantages (Katz and Kahn, 1979; Katz and Tushman, 1978).

Many of these issues were considered by T.R.A.D.E. in dealing with Tucker's request to telecommute. The firm was aware that the task of replacing Tucker may be daunting. Not only would they need to identify an individual with similar skills, but in all probability, they would face a significant task in attracting the candidate away from their present employer. Additionally, a large amount of time and resources would be spent bringing the new employee up to speed in terms of corporate culture and client expectations. Thus, given Tucker's desire to relocate, yet remain as an employee, the situation presented an opportunity for T.R.A.D.E. to retain the skills they require and provide a valued employee the flexibility he sought. Although T.R.A.D.E. would recapture Tucker's physical space, they didn't realize any significant cost savings related to physical facilities since this telecommuting arrangement was not an organization-wide arrangement. Managerial problems were not considered as crucial, primarily due to Tucker's solid history with the firm and the firm's confidence in his personal integrity and ability.

Social Aspects

The personal social benefits are the main attractiveness to most employees. Telecommuters are often unconstrained by traditional office schedules and work environments. They typically have the freedom to set more flexible work schedules to coincide with personal activities such as child, elderly, or disabled care responsibilities. The inherent lack of commuting lowers the stress, time, and financial cost of this activity. Remote workers recapture valuable time spent in daily commutes, along with reducing or eliminating the need for transportation (bus, auto, fuel, etc.) to commute (Mokhtarian, 1988; Mokhtarian, 1990; Mokhtarian 1991a; Nijkamp and Salomon, 1989; U.S. Department of Transportation, 1993).

The benefits are not without real, and potentially substantial, disadvantages (Egido, 1990). Because telecommuters typically work out of their home environments, many develop feelings of isolation. They often feel out of touch with their co-workers and feel that they miss out on informal social interactions that occur naturally in the work environment (Curran and Williams, 1997; Kraut, Egido, and Galegher, 1990; Kraut and Streeter, 1994). The lack of interaction often extends beyond co-worker interaction. Many telecommuters feel disassociated with other business professionals. Relationships that occurred and developed naturally in traditional business settings now must be recreated, often with substantial effort, in the telecommuting environment.

Working at home can cause other problems. Some telecommuters experience a sort of cabin fever due to the lack of external interaction on a routine basis. Additionally, working at home can cause strain on marital and family relationships. The homogeneity of the work and home environment offers limited stimulus variety, and can ultimately be counterproductive. The yearning for peace and quiet in order to accomplish work tasks may soon be perceived as too peaceful and quiet (Curran and Williams, 1997; Race, 1993; Shellenbarger, 1993).

Some telecommuters develop a sense of disconnect between themselves and their organizations. They may feel their skills are being underutilized and that they are out of the loop in terms of office politics and organizational gossip. While the elimination or minimization of office politics is usually cited as a productive benefit of remote work by employers and employees alike, the reality is that humans are social creatures who require some minimum level of social interaction. Sometimes, negative interaction is better than no interaction. The loss of social interaction can be mitigated if special attention is given to substituting the negative interactions with positive reinforcement of the employee's role in the organization (Hiltz, 1993; Katz and Kahn, 1979; Katz and Tushman, 1978).

Tucker experienced more of this than he initially thought possible. While T.R.A.D.E. viewed Tuckers absence as somewhat negative, Tucker initially thought the telecommuting arrangement was ideal. Tucker's superiors (Vice President of Engineering, President, and CEO) all expressed reservations with the arrangement due to Tucker's more limited interaction in informal decision making. When in the office, his superiors readily would seek out his input and advice on a myriad of business issues. Now that Tucker was not readily available (in person, although he was always accessible at his remote location), his superiors felt they were not utilizing all of his skills.

Tucker often saw this as time consuming and detrimental to his specific task agenda. However, as time passed, Tucker began to feel his political power and informal influence began to wane. He often found himself less informed on critical corporate issues and felt he was not contributing as much to the organization as he had previously. Tucker admits that he now feels somewhat less connected to T.R.A.D.E. His personal identity now revolves less around his role at T.R.A.D.E. and more around his generic software engineering skills. In an effort to recreate some of the social interactions, Tucker has joined a number of professional and business organizations, such as the Lions Club, and the local Chamber of Commerce. The success of this substitution is yet unknown, but Tucker is skeptical concerning his ability to fully recreate the same interpersonal/professional sociology he once enjoyed.

Management of Costs

The tangible costs and benefits of telecommuting are fairly straightforward and easy to address. For the organization, the costs include installation and monthly charges for long distance services to facilitate remote access to the employee which may include voice, data, and video communication. If periodic travel is required, then the cost of this travel will also need to be factored into the equation. Travel costs will not only include transportation costs but hotel and living expenses incurred during stays at the organization's site. These costs will be offset by savings from reduced costs of space (office and parking), furniture, and facilities (heating, air conditioning, and other space related benefits) (Mokhtarian, 1988; Mokhtarian 1991a; Nijkamp and Salomon, 1989; U.S. Department of Energy, 1994; U.S. Department of Transportation, 1993).

For the individual, the tangible benefits include reduced costs of transportation and the potential for reduced expenditures from unreimbursed activities such as some business lunches. Depending upon the telecommuting arrangement and personal responsibilities, there may be some potential elderly or childcare savings due to the remote worker's presence in the home environment.

In the arrangement between Tucker and T.R.A.D.E., Tucker originally proposed that he would pay the expenses of the arrangement out of his pocket. Thus, he pays for the Integrated Services Digital Network (ISDN) telecommunications lines into his home, the special software that enables him to emulate a users personal computer desktop environment on his personal computer in his home, and the associated travel expenses for the two to three day trips he makes to San Mateo every six weeks. An interesting aspect of this arrangement is that Tucker's superiors at T.R.A.D.E. expressed dissatisfaction with this unilateral arrangement. Since Tucker was incurring the costs, his T.R.A.D.E. superiors did not feel comfortable asking him to return to San Mateo for special projects outside of the normal six-week visits. With a year of experience with this arrangement, the costs are fairly well established. Thus, in the future the firm and Tucker will likely share the costs in some manner.

Productivity Measurement

But, how do we know what you are doing? is a sentiment often expressed by managers to proposals of telecommuting from employees. Metrics of productivity and performance are critical regardless of the work environment. However, less than optimal performance evaluation can be masked in traditional work environments, but vague and ambiguous management approaches fail miserably in telecommuting environments. As discussed in the telecommuter selection section, telecommuting is not for everyone and every job function. Some jobs simply require too much interaction to allow for efficient and effective skill utilization from a remote site.

This problem notwithstanding, many job functions and individuals can readily adapt to a telecommuting arrangement if good management techniques are practiced. Overall in a telecommuting environment, the emphasis shifts from control of processes to control of results. Work tasks must be planned well, delegated specifically, and be accompanied with timetables and deadlines. It is also important to provide a mechanism to periodically assess progress and issue performance feedback (Curran and Williams, 1997). Obviously, these practices are easier to accomplish with job functions that lend themselves to segmented, specific tasks, but these techniques represent superior management practices regardless of the work environment. In traditional work environments, many of these activities can be accomplished informally. With the employee and supervisor both present, progress assessment and performance feedback is relatively easy and often very informal. Planning is often overlooked since managers can easily modify employee assignments during task execution.

Tucker and T.R.A.D.E. found this area to be the least problematic to operationalize. This may be attributable to Tucker's job function. As Curran and Williams (1997) point out, the five job categories most suitable to remote working are professionals, professional support, field workers, IT specialists, and clerical support. Depending on the specific definition, Tuckers job could be considered in any one of three of the five: professional, professional support, and IT specialist. Already a well managed organization, T.R.A.D.E. incorporated most of these management techniques in its operations prior to Tucker's proposal for telecommuting.

CURRENT CHALLENGES/PROBLEMS FACING THE ORGANIZATION

Telecommuting offers substantial benefits for organizations today and into the 21st century. As the demand and competition for high-quality knowledge workers increases, this alternative will allow firms to attract and retain employees with scarce resources. At the same time, workers are demanding more flexibility as alternatives to traditional work relationships. The case study presented throughout this paper is an example of successful telecommuting. It was successful because each of the six sets of issues: enabling technology, telecommuter selection, organizational advantages/disadvantages, social aspects, costs, and productivity measurement, were addressed by both Tucker and T.R.A.D.E.

The technology needed to enable distributed software development and intraorganizational communication is widely available. Hardware, software and telecommunications capabilities continue to improve while their cost continues to decline. At the same time, telecommunication capabilities continue to evolve and these capabilities alter the way in which employees can experience the world. This continues to challenge T.R.A.D.E. and Tucker alike. However, it is apparent that technology is a necessary but insufficient condition for telecommuting success. Most issues do not involve technology implementation, but instead involve more complex managerial and organizational behavior issues.

Tucker initiated the arrangement which showed that he saw telecommuting as beneficial to him, while he also was an ideal candidate for telecommuting because he had significant experience in software development, significant experience with his company, was an exemplary employee, and had experience in doing his own technical support. The organization saw that the advantages of telecommuting for Tucker outweighed the disadvantages. Any inconveniences the arrangement may cause are outweighed by the costs that would be incurred to replace Tucker's skills with a new employee from a tight job market.

Tucker and T.R.A.D.E. saw some disadvantages arise from the lack of direct interaction with co-workers, but the ability to live in a geographic area that better matched his families needs outweighed these problems. Cost sharing arrangements can be adjusted as the telecommuting arrangement between Tucker and T.R.A.D.E. evolves. And finally, there was a fit between Tucker's job category, one of professional, professional support, and/or IT specialist that enabled his productivity to be easily measured through existing procedures. As these factors evolve, T.R.A.D.E. and Tucker must determine how the telecommuting arrangement will evolve.

While originally somewhat skeptical, the firm's experience with Tucker's telecommuting arrangement has been a positive one. T.R.A.D.E. is now more enthusiastic about entertaining similar work arrangements for other employees and contractors. The disadvantages are predictable and can be minimized. The benefits for the firm in retaining and attracting highly skilled personnel in an extremely competitive job market continues to grow. To date no additional arrangements have been undertaken, but several are being considered.

The overall conclusion of this case is that successful telecommuting can take place in today's IT environment when experienced, self-motivated, employees are involved. There are incentives for the organization to provide this flexibility, and there are procedures in place for communicating and managing remotely instead of in a traditional environment.

In the final analysis, telecommuting alters the existing social pattern of work as we have known it for the last one hundred plus years, i.e., workers leave their homes each day to toil at job sites some distance from their homes, and return each evening to rest and renew the process again. This excerpt from the U.S. Department of Energys Office of Policy, Planning, and Program Evaluation (1994) best sums up the future of telecommuting.

> *The social effects of telecommuting will occur and be judged in the larger context of advances in telecommunications as well as other changes that effectively restructure workplace interactions. It may be difficult to distinguish the effects that are specific to telecommuting. Currently, telecommuting is not widespread, and its social effects are relatively localized. Individual telecommuters, their work organizations, their families, and to a lesser extent, their communities currently are affected. If telecommuting becomes a widespread working strategy, the effects both positive and negative will become more varied and their scale will become larger, affecting individuals, the workplace, families, communities, and the Nation.*
>
> *The following issues may become particularly important: separation between home and work; increasing fractionalization of the workforce and society at large; and changes in neighborhood and community interactions. Telecommuting from home blurs existing boundaries between home and work. Employees' abilities to separate (or escape) from their work and employers abilities to invade their employees' homes may change the nature of home and work lives. Within the workplace, three major classes of personnel may develop: those who voluntarily telecommute and derive the benefits of autonomy, flexibility, and freedom from interruptions; those for whom telecommuting is required and who may not experience the advantages of the privileged class of telecommuters or have access to the benefit packages associated with long-term employment in an office setting; and those for whom telecommuting is not an option. Widespread telecommuting may enhance the safety and spirit of neighborhood and community life, though these effects may not be distributed consistently throughout a region or the country because of varying composition of communities may be associated with different levels of telecommuting. Finally, widespread telecommuting may change patterns of residential and commercial development, in part because of proximity of the workplace to the labor force would diminish in importance.*

REFERENCES

Abel, M.J. (1990). Experiences in an exploratory distributed organization. *Intellectual Teamwork: Social and Technological Foundations of Cooperative Work.* J. Galegher, R. Kraut, and C. Egido, eds., Hillsdale, NJ: Lawrence Erlbaum Associates, 489-510..

Boyett, J. H. and Conn, H. P. (1992). *Workplace 2000: The Revolution Reshaping American Business*, New York, NY: Plume Publishing.

Curran, K. and Williams, G. (1997). *Manual of Remote Working*, Gower Publishing Limited, Hampshire, England.

Egido, C. (1990). Teleconferencing as a Technology to Support Cooperative Work: Its Possibilities and Limitations. *Intellectual Teamwork: Social and Technological Foundations of Cooperative Work.* J. Galegher, R. Kraut, and C. Egido, eds., pgs. 351-371. Lawrence Erlbaum Associates, Hillsdale, NJ.

Finholt, T. and Huff, C. (1994). *Social Issues in Computing: Putting Computing in Its Place.* McGraw-Hill, New York, NY.

Fish, R., Kraut, R., Root, R., and Rice, R. (1992). Evaluating video as a technology for informal communication. *Communications of the ACM* 36(1):48-61.

Hiltz, S. (1993). Correlates of learning in a virtual classroom. *International Journal of Man-Machine Studies* 39:71-98.

Katz, D., and Kahn, R.(1979). *The Social Psychology of Organizations.* Second edition. John Wiley & Sons, New York, NY.

Katz, D., and Tushman, M. (1978). Communication patterns, project performance, and task characteristics: An empirical evaluation in an R&D setting. *Organizational Behavior and Human Performance* 23:139-162.

Kraut, R. (1988). Homework: What is it and who does it? Chapter 2 in *The New Era of Home-based Work*, K. Christensen, ed. Westview Press, Boulder, CO.

Kraut, R. and Streeter, L. 1994). Coordination in large scale software development. *Communications of the ACM* 38(3):69-81.

Kraut, R., Egido, C., and Galegher, J. (1990). Patterns of contact and communication in scientific research collaboration. *Intellectual Teamwork: Social and Technological Foundations of Cooperative Work.* J. Galegher, R. Kraut, and C. Egido, eds., Hillsdale, NJ.: Lawrence Erlbaum Associates, 489-510.

Mokhtarian, P. (1988). An empirical evaluation of the travel impacts of teleconferencing. *Transportation Research A* 22A(4):283-289.

Mokhtarian, P. (1990). A typology of relationships between telecommunications and transportation. *Transportation Research A* 24A(3):231-242.

Mokhtarian, P. (1991a). Defining telecommuting. *Transportation Research Record* 1305:273-281.

Mokhtarian, P. (1991b). Telecommuting and travel: State of the practice, state of the art. *Transportation* 18(4):319-342.

Mokhtarian, P. and Salomon, I. (1994). Modeling the choice of telecommuting: Setting the context. *Environment and Planning A* 26(4):749-766.

Nijkamp, P. and Salomon, I. (1989). Future spatial impacts of telecommunications. *Transportation Planning and Technology* 13(4):275-287.

Nilles, J. (1991). Telecommuting and urban sprawl: Mitigator or inciter? *Transportation* 18(4):411-432.

Race, T. (1993). Testing the telecommute. *New York Times*, August 6, p. F11.

Shellenbarger, S. 1993. Some thrive, but many wilt working at home. *Wall Street Journal*, December 14, p. B1.

Thurow, L. (1996). *The Future of Capitalism.* William Morrow and Company, Inc., New York, NY.

Townsend, A. M., DeMarie, S. M., and Hendrickson, A. R. 1998. Virtual Teams: Technology and the Workplace of the Future. *Academy of Management Executive*, Vol.12, No. 3, August 1998.

U.S. Department of Energy, Office of Policy, Planning, and Program Evaluation, 1994. *Energy, Emissions, and Social Consequences of Telecommuting*, Washington, D.C.: U.S. Government Printing Office.

U.S. Department of Transportation, Office of the Secretary. (1993). *Transportation Implications of Telecommuting*. Washington, D.C.: U.S. Government Printing Office.

Whitaker, S. and Geelhoed, E. 1993. Shared workspaces. How do they work and when are they useful? *International Journal of Man-Machine Studies* 39:813-842.

BIOGRAPHICAL SKETCHES

Anthony R. Hendrickson is an Assistant Professor of Management Information Systems (MIS) in the Department of Logistics, Operations and MIS, Iowa State University. He received the Ph.D. degree in Business Administration (Computer Information Systems and Quantitative Analysis) from the University of Arkansas in 1994. His research interests include virtual organizations, psychometric measurement, and object-orientation.

Troy J. Strader is an Assistant Professor of Management Information Systems (MIS) in the Department of Logistics, Operations and MIS, Iowa State University. He received the Ph.D. degree in Business Administration (Information Systems) from the University of Illinois at Urbana-Champaign in 1997. His research interests include electronic commerce, strategic impacts of information systems, and information economics.

The Planned and Materialized Implementation of an Information System

Pekka Reijonen
University of Turku/Laboris, Finland

Jukka Heikkilä
Helsinki School of Economics, Finland

EXECUTIVE SUMMARY

The object of this case study is a marketing and sales information system in two local offices of a regional telephone company. A unified, advanced client/server system was needed due to the merging of three companies into a bigger regional company, keener competition, and the growing complexity of the services provided. The system is tailor-made to meet the needs of the industry and it was developed by a software vendor in close cooperation with the nation-wide alliance of regional telephone companies.

This study illustrates the difficulties in simultaneously aligning an organization and implementing a new information system. Views on the skills and competence needed in using the system vary, and lead to the negligence of education and training. The consequent lack of skills and knowledge of some users, especially of those not using the system regularly, create profound problems in the whole work process and in productivity as the first, obvious work practices become the dominant mode of operation bypassing the desired integrated workflow. The findings are discussed and reflected to concepts of institutionalization, positive reinforcement, and productivity paradox. This case emphasizes the importance of the organizational implementation and adaptation process which ought to begin after the implementation of the technical system.

BACKGROUND

It was only in the beginning of the 1990s when the telecommunications sector was deregulated in Finland. Long distance lines (i.e., crossing the local telecommunication areas), which were previously operated by a state-owned company, opened to competition. Deregulation took place simultaneously with the emergence of radio-based telecommunications, most notably with the introduction of the wireless analog NMT and later digital GSM and DCS networks. Radio-based telecommunication is also open to competition. The mobile phone has grasped a significant share of the phone traffic (there are more than 2.3 million mobile phone subscribers in a country of five million inhabitants and 2.8 million subscriber lines). As a consequence, the rates of hard-wired long distance calls have fallen by 80 percent and the rates of local calls by 50 percent since deregulation. Deregulation has affected the regional teleoperators the least, but as the former licensing of telecommunication areas is deregulated, too, local operators have started to merge.

Currently there are two main telecommunications operators in Finland: Sonera plc (formerly known as the state-owned Telecom Finland) and the Finnet consortium, which is owned by 46

Table 1. Breakdown of the teleoperators' market in Finland (Telecommunications Statistics, 1998).

Regional phone calls	3.3 billion FIM
Long distance calls	4.1 billion FIM
International calls	1.3 billion FIM
Data communications	1.8 billion FIM
Equipment	4.7 billion FIM

regional telephone companies. There is a third player, owned by Swedish Telia, but as its market share is less than 2 percent, long-distance and radio-based telecommunication markets have practically a duopoly. Our case company belongs to the Finnet consortium, the market share of which is about 50 percent of the total telecommunications turnover, 16 billion FIM (about 3 billion USD), and the market is expected to grow to 22 billion FIM by the year 2000.

The nation-wide Finnet consortium was established by the regional telephone operators in order to provide seamless long-distance calls for their subscribers and to keep up with the pace of fast rapidly-developing technology. The consortium developed a digital SDH-based backbone network, primarily on optical cabling, for long-distance and mobile-voice calls and data communication.

The regional telephone company, called here Areal Phone Ltd. (AP), implemented a new information system (IS) about one year before this case study. The implementation was not, actually, a free choice, rather the company was driven to changes. The reason for the new IS was simple: AP is the result of a merger of three mutual, local telephone companies.[1] The three companies each had their own tailor-made operative information systems, so there was a need to unify the operations and ISs.

Table 2. Areal Phone Ltd. in figures (1997).

Turnover	145 million FIM (28 million $US)
Investments	69 million FIM (13 million $US)
Gross margin	43 %
Revenue	11 million FIM (2 million $US)
Balance sheet	320 million FIM (62 million $US)
Telephone subscribers (wired)	75 293
Cable television subscribers	25 827
Employees	240
Inhabitants in the territory	152 500
Area of the territory	3236 square kilometers (2000 mile^2)
Municipalities in the territory	23

The reasons for the merger stem from the previously mentioned changes in the telephone operators' business environment. On the one hand, deregulation has driven small local companies to seek economies of scale by merging. On the other hand, it is considered difficult for small companies to keep up with the rapid development of digital telephone technology, where different data services are becoming more and more important (for example, the percentage of Internet users in Finland is one of the highest in the world).

At present AP serves about 74 000 wire-based telephone subscribers[2] and 26,000 cable television (CATV) subscribers (see Table 2). The company functions in 23 municipalities, where there are 152,000 inhabitants. The private customer base consists mainly of urban households but a significant number of customers live in the countryside. This means that maintaining the infrastructure is a demanding and expensive responsibility of the telephone company. The digitalization of the telephone network that started from switches has continued at the subscriber end and the number of ISDN installations has been growing constantly. For example, in 1997 the turnover of AP data services grew over 30 percent. Data services are especially important in enterprises, which is the other important customer sector of AP. The possibilities of AP offering data and other add-on services are excellent because their network is totally digitized.

The main business of AP is still to maintain the telephone network and operating calls. The future trend is to increase the complexity of value-added services both for the private customers (e.g., CATV–network renewal for data communications and xDSL connections) and enterprises (e.g., running outsourced switchboards and high-speed data communications). However, as the hype for an information society continues, telephone operators are seen as vital players; hence, the interest for mergers continues: in 1997, AP acquired the telephone company of a town situated nearby. In 1997, AP also introduced their own radio-based DCS/GSM 1800 city phone network.[3]

SETTING THE STAGE

The new information system at AP was seen as a means to cope with the competition and to improve marketing, sales, installation, and maintenance of the increasingly complex value-added services. As we browsed through the documentation, it became clear, however, that the project had concentrated only on defining data structures and little attention was paid to actual business concerns and work process (re)organization.

The selection process of the new IS for AP had taken about one year. The turn-key information system chosen had not been used earlier in any of the three separate companies[4]. The coalition worked in close co-operation with a Finnish representative of a large international software vendor. The new IS for operations control was to cover practically all the functions of the company (see Figure 1).

The broad domain was not news to any of the companies, because their former ISs covered also the majority of operations. As the new IS was different from any of the three former ISs, the data

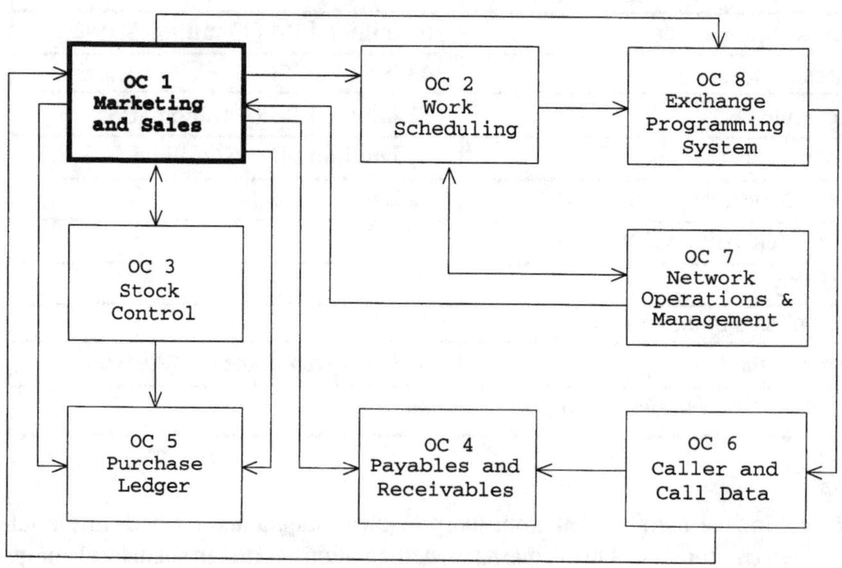

Figure 1. The subsystems of the OC system (Operations Control) and their interconnections. Subsystem #1 ('Marketing and Sales') was the main object of the deployment evaluation project.

conversion from the old ISs was anticipated to become one of the most demanding tasks of the implementation process.

The new IS was implemented on an Oracle database using Oracle Forms and Report Writer development tools. The most computing intensive routines were programmed in C++. The system runs on a UNIX-server[5] and the workstations (PCs) are connected to the server via an Ethernet-connection. The OC system is used with a terminal program, so the architecture actually corresponds to a mainframe system. Office programs, however, are used directly from the server.

The initial problems with the capacity of the server were due to the greater-than-expected number of concurrent connections. The number of connections grew high as the IS did not support incomplete tasks (i.e., if a task was interrupted all the previous input was lost and the task had to be restarted from the very beginning). Hence the users opened several concurrent terminal sessions in order to manage their work fluently in a customer service situation. At the time of the study, this capacity bottleneck had been removed and the response times of the system reached an acceptable level.

However, there arose additional, 'fuzzy' problems that led the CIO of AP to contact our research unit at the university to evaluate the situation at the marketing and sales subsystem. The perception of the problem by management was vague; it was based on casual user complaints about the poor usability of the system. According to management, the problems concerned mainly the private customer sector and were caused primarily by deficient features of the IS: the IS did not support business processes anymore because it had been designed for selling hard-wired connections instead of add-on services. Therefore, management decided to set up a project for streamlining the interface in order to improve the functionality of the IS by "reducing the number of screens by half." Our evaluation was planned to complement the requirements analysis of the forthcoming renewal project. In addition, management wanted to compare the IS usage and usability problems to earlier evaluations made by our research unit.

The subsystem studied (see Figure 1) was labeled "Marketing and sales," but actually "Customer and product" would better describe the system. This is because the marketing part of the subsystem was actually abandoned—mostly for its lack of usability and functionality—after a short trial period. The subsystem is linked to all other subsystems (see Figure 1) because it is used for maintaining customer records (e.g., names and addresses, and long-term contracts).

The goal of the evaluation was to describe end-users' work tasks, work processes, and the use situation of the IS, so that more informed decisions about further development of the organization and the IS could be made. The study was carried out at all three regional offices of AP, but we use two offices of comparable size and departments to illustrate our case.[6]

CASE DESCRIPTION

In both offices sales function of AP is organized according to their clientele, into Private and Enterprise departments. There were 15 end-users in the Private subscribers department and 11 in the Enterprise department. Management named these two groups as the target groups of the evaluation. It soon turned out, however, that about an equal number of end-users are using the same subsystems outside the marketing and sales department, and their IS usage is directly related to the work processes of the personnel in marketing and sales.

The average age of the users is about 37 years in both departments. In the Private department, nearly all users are women, but in the Enterprise department the proportion of women is only about 20 percent. The end-users are competent long-term employees and are accustomed to the organizational culture: mean employment time is 9 years in the Private department and 14 years in the Enterprise department.

The users are also experienced computer users (on average, 8-10 years of usage), even though most users have experience only in the systems used in the current work place. In the Private department, the IS is the main tool of its end-users – on average, 87 percent of the total computer usage, which is equivalent to about 74 percent of the working hours spent using the new information system. In the Enterprise department, the average usage time is 48 percent (about 25 percent of the working hours) and the variance in usage time is considerably higher. The situation in which the IS is used is also different: in the Private department, most of the usage (40 percent of the working hours) took place in customer service situations, whereas computer usage in the Enterprise department is more back-office –like; in only four percent of the working hours was the computer used while dealing with customers.

Before the IS implementation, the program vendor first trained and educated a few principal users who in turn trained their peers. The average training time was eight days in the Private department and five days in the Enterprise department —both figures can be considered impressive in light of our earlier research (Laboris, 1997)—despite the fact that in the Enterprise department the training is unevenly distributed and over half of the users had received less than four days of training. The reported reason for missing the training courses was urgent duties but, on the basis of interviews, a lack of motivation seemed to be the major reason.

Technological Concerns

The new IS has a text-based interface managed with the keyboard. This solution was decided on in the cooperative planning association long before the implementation of the system. The decision was justified by the requirement that "the appearance of the interface must be the same in all kinds of workstations like PCs, X-terminals, and ASCII-terminals." Even though consistency and uniformity are important aspects in interface design, the grounds for this decision are not relevant to AP because all the workstations are PCs. Contrary to its initial idea, the decision eventually limited the possible range of interface options (e.g. it eliminated the possibility to implement a graphical user interface) and the development tools.

As a result, the system is somewhat difficult to interact with – or, actually, there is little interaction. The system consists of 90 separate screens, and the coupling of the individual screens to the process of completing a specific task is weak. For the end-users this means that they have to memorize the screens and in which order they are used for each specific work task. This makes the system hard to learn and much practice and repetition is required in order to master the system fluently. This is confirmed by the fact that users' estimates of the time required to learn to use the system to the extent that one can work efficiently vary from two months to two years (the average was nine months). The estimates of the average learning time differ in the two departments: the average estimate of the learning time is four months in the Private and 13 months in the Enterprise department. As most of the learning takes place during system usage, it is obvious that the perceived learning time is correlated with the amount of system usage ($r = 0.59$, $p = 0.01$, $n = 18$). At the time of the study (one year after its implementation), users rated their current ability at 7.7 (on a scale ranging from 4 to 10) in the Private department and at 6.8 in the Enterprise department.

Users are unanimous that the system has been, and is, difficult to learn and to use. There are too many screens and there is little help built into the system to find the right screens in the right order. Obviously, the IS supports occasional users poorly, but with a sufficient amount of practice, the user eventually can learn it by heart. One clerk who has been using the system for 5-6 hours a day for one year expressed this by saying:

> *"That task is so routine nowadays, we have done them so much, that one does not need to think much anymore."*

But the system usage is obligatory for less often or irregularly performed tasks, too. In these cases, it has become a common practice to write the necessary information on a piece of paper and key the data in the IS later (with help from co-workers or support personnel)[7].

The overall evaluation of the system by an EUCS-measure demonstrates that all features of the system are rated lower in the Enterprise department than in the Private department (Figure 2).

The user opinions give grounds for two possible conclusions: the system really is not well suited to the work in the Enterprise department, and/or the computer self-efficacy of the users is low, i.e. the users feel that they cannot manage their computer supported work tasks properly (c.f. Compeau & Higgins, 1995).

Figure 2: End User Computer Satisfaction (EUCS) scores of the two groups. The differences in the means between the groups are statistically significant (t-test, $p < 0.05$) except for the factor 'Timeliness'.

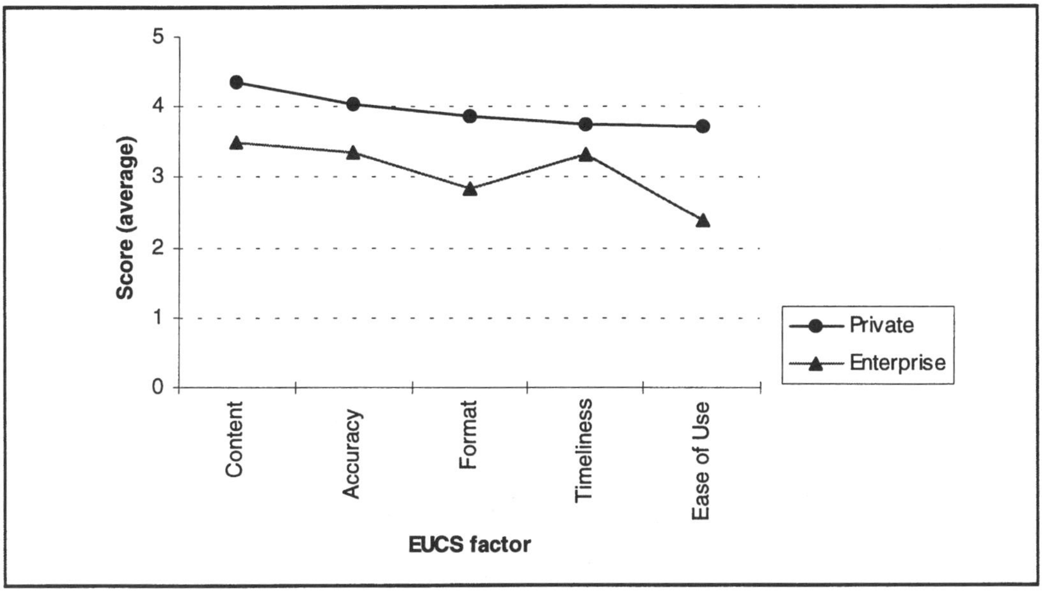

Organizational Concerns

The consequence of the poor alignment of the system functions and the actual work flow is that less-frequent users did not learn to utilize the system properly. Because some user groups (e.g. supervisors of the installation function, technicians and some of the sales managers in the Enterprise department) received little education in the system usage and on the ideal planned work flows, they developed apparently successful, but manual routines, to work around the system's limitations.

Management launched a re-design project to streamline the process in the Private department, but actually the problems were more severe in the Enterprise department. In the private sector, the service procedure was rather simple: customers were either face-to-face or on the telephone with the clerk who used the IS to carry out the service (see Figure 2). However, the system did not fully meet the requirements of marketing, selling, installing, and maintaining more complex services such as corporate ISDN-subscriptions. The willingness and skills to learn and use the system were also lower in the Enterprise department.

In the Private department, the implementation was carried out according to the plan: at the time of the study, most end-users have reached such a skill level that the speed of the work process was acceptable in both offices. The work is interrupted by breakdowns only occasionally (about half of the users have problems weekly, and 57 percent of the problems are due to the lack of skills and knowledge, mostly a problem for newly enrolled clerks). In other words, there are few severe problems in the use of the IS. Also the work processes closely follow the planned workflow. Although the system is clumsy for some individual tasks[8], and some operations are not institutionalized[9] yet, the users have learned to manage their work with the IS and to resolve the few problems with assistance from their fellow workers. In other words, private customer service either in direct contact face-to-face or by phone proceeds rather fluently. Direct customer contact also increases the clerks' motivation for developing their skills in system usage because, by being able to use the IS fluently, they could avoid embarrassing social situations which might arise if they had to say to a customer: "Well, actually, I cannot perform my work tasks."

In the Enterprise department, there has emerged an unplanned supplemental work role (substitute IS user in Figure 3) because the sales managers were unwilling to learn and use the IS.

Figure 3. The use of the IS in the sales processes (from order to invoicing): end-user roles, IS usage, and interactions in the work processes. The solid lines represent the planned interactions and work flows. The dashed lines represent the situation which was institutionalized in the other office.

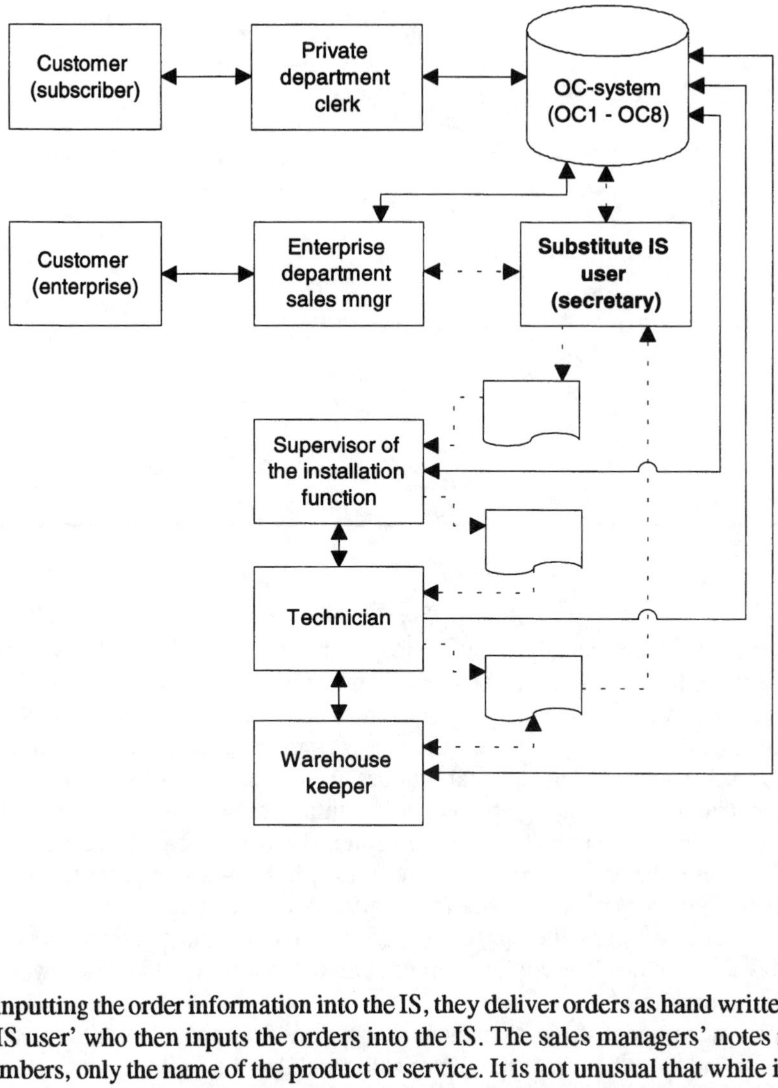

Instead of inputting the order information into the IS, they deliver orders as hand written notes to 'the substitute IS user' who then inputs the orders into the IS. The sales managers' notes rarely include product numbers, only the name of the product or service. It is not unusual that while inputting data, extra contacts between the sales manager and 'the substitute IS user' are needed because the product name[10] the sales manager uses is not found in the IS.

In the other office, technicians and their supervisors also ceased to use the IS after a few weeks of initial use. Actually, they ceased to use the IS as they were told to stop using it. This is because they had received neither enough training nor enough practice to master the IS as occasional users. For example, when the technicians were to record in the IS the working hours spent and materials consumed for a certain order, they often could not find the right order nor the right customer number. Instead, they keyed in the data in the first found occurrence of this specific customer— or created a new customer number[11]. This led to a situation where some orders were not invoiced, the incorrect department of an enterprise was invoiced, and the database of the IS was eroding because of the duplicates. Except for extra work, this also caused customer complaints and in the present competitive environment, flawless customer service, quick response, and a good reputation are critical.

In order to correct the situation, the IT usage of these user groups was replaced with pencil and paper: after inputting the work order, the substitute IS user printed out the order, and data of the later phases of the work process were documented on it. The notes made on the paper document were then inputted into the IS by the substitute IS user as the final step of the process after the installation was completed[12]. This meant new procedures and a lot of extra work in finding missing data, since the data on the paper-based documents were only occasionally complete. As a consequence, the invoicing process was delayed dramatically, in some cases up to several months, despite the constant overtime work of the substitute IS user.

A closer look at the episodes and encounters of the process that led to the ban of usage in the Enterprise department revealed that the key problem was the end-users' inability to cope with the IS. The incapability of technicians and supervisors at the other office resulted in chaos, temporarily solved by the use of paper instead of the IS. However, this arrangement became permanent. For the sales managers, it was also a question of their power in the organization: a substitute user was assigned to the 'less valuable' task of learning to use the IS and keying in the data.

Current Challenges

Even though the situation might have been avoided by managing the organizational implementation process 'better', there are two interrelated facts, which have contributed to the evolvement of the problems.

First, the IS is not user-friendly with its character-based windowed user interface, and it does not have any support for learning or remembering the flow of operations. Its functionality for some tasks is questionable as it was originally developed for a different business environment. The learning time is long (but not extraordinary) and as a consequence, some user groups are not willing to adopt the system. And we are afraid that there will be more problems of this kind. This is because the way the IS was built seems to have become an industry standard: a shared relational database is normalized to atomistic tables, which are updated by using screens and procedures (half) automatically created from the columns of the table with an application generator. This easily leads to a great number of separate screens. But, as the starting point for development is a semantic data model, the design is seldom compatible with the work procedures of the organization. It is also difficult to learn to navigate in a bunch of separate, seemingly unrelated screens.

The second problem is the inertia caused by the IS in a rapidly changing environment. For example, the IS has been designed so that there can only be one subscriber telephone number per customer ID. This means problems with the rest of the possible numbers of an ISDN line. Or, when a customer enterprise wants to regroup invoices according to its new organizational structure, there are no mechanisms for handling the update. It is inevitable that information systems freeze the current mode of operation and all changes in operations almost always call for reprogramming of ISs. And in the case of the IS, the advantage of sharing the costs of development in the first place has become an obstacle for further development. The software is owned by the consortium, the members of which have to be convinced and agree upon the need for changing the software.

To us the case clearly illustrates the common dilemma in IS implementation: an information system that is somewhat awkward to use, slightly incompatible in work processes, and difficult to change (but otherwise technically functional) can be implemented in one organization whereas in another, almost similar organization, the IS is abandoned.

DISCUSSION

The design, production, and implementation phases of a computer-based information system are commonly considered the key activities in its life cycle – in money, time and other resources. Skillful software development is about the technical maneuvers in conducting the phases. However, the value of the IS, how technologically progressive it might be, can not be realized until it is effectively implemented and deployed. As pointed out by Kling and Allen (1996), effective use of information technology does not depend on technology alone: organizational issues often play an important role. In order to emphasize the importance of non-technical factors, they use the term 'organizational implementation': "Organizational implementation means making a computer system accessible to those who could or should use it, and integrating its use into the routine work practices" (ibid. p. 269). In the world of outsourced development, this is more important than ever.

Organizational implementation necessitates, for example, that users are willing to adopt the IS and that they have the necessary skills and knowledge (both domain and technology/tools) needed in the deployment of the CBIS. In research, the need for education and training in computer-related skills is commonly recognized and its importance emphasized (Nelson et al., 1995; Clement, 1994). In practice, however, the training issues are often neglected and it has even been argued that one of the reasons for less successful CBIS implementations is the lack of user training (Mitev, 1996).

One reason for the relative ignorance of training and education issues is that the deficiency of competence usually becomes apparent only in the beginning of the usage phase of a new CBIS. If the usage of the CBIS is obligatory, users learn—sooner or later—to deploy the system while doing their daily work. When users have learned to manage their work to the extent that there are no apparent problems, the problems disappear. And, as there are no problems, there is no need to evaluate the efficiency and effectiveness of the actual implementation and work practices! Or, only the technical functionality, such as response times, is checked.

The time required to learn to use a new CBIS depends on many factors, for example, the structure and appearance of the CBIS, the amount of use (i.e. the practice obtained), and the individual learning capacity. Learning times of operative ISs, such as the OC-system, are rather long with huge variation; in most cases the learning period is several months. The learning process can be characterized as a trial-and-error diffusion process where a single user or a group of users solve the problems encountered when integrating their work practices into the features of the CBIS. It is not a uniform process, rather there are multiple simultaneous processes (Heikkilä, 1995). In this kind of learning environment people tend to accept the first functional solution and stick to it as long as it works. Its effectiveness is never questioned, nor are alternative solutions actively sought: the functional solution will do. A typical feature in this learning period is that users search for help mainly from their peers (in 80-90% of the problems) whereas manuals, help-desks and other types of support arrangements play only a secondary role (Heikkilä, 1995; Laboris, 1997). It is worth noting that managers are seldom in advising roles because they rarely can give support in CBIS-related problems. As a consequence, managers often only have a vague idea of the real CBIS deployment situation. Therefore, the effectiveness of the applied procedures depends mainly on the initially invented workable solution which becomes the best practice in that environment.

Implementation of a new CBIS generally also changes the division of labor and incentive structure which cause changes in the work tasks and motivation of individuals. After implementation users must learn this new division of labor and potentially some new work tasks as well as the functions of the new CBIS. The division of labor is planned and institutionalized rather carefully, therefore, some kind of 'role book' is thus already available when a CBIS is designed. The process of dealing out the roles to the actors is also part of the management of the implementation. However, management, control, and evaluation after implementation are often shallow and based more on beliefs, attitudes, and opinions than technological or economical data (Kumar, 1990; Kling & Iacono, 1984). Iivari (1986) has concluded the following about the intra-organizational implementation of multi-user ISs: there are actually two coexisting sets of activities in implementing an IS, or innovation in general; the technically-oriented rational-constructive tasks and a political process of resolving conflicts of interests and changing the organization.

In practice, many of the problems, which are caused by the changes in the division of labor and encountered when deploying a CBIS, are solved with the same kind of ad hoc principle as other work-related problems. As management's concerns are in restructuring the industry, the most important thing is to get the work done by the doers, no matter how it is done. And, in the same way as an individual repeats an action after a positive reward, an organization gives legitimate status rather quickly to an arrangement which seems to be functional (see positive reinforcement trap; Argyris, 1985). Finally, the division of labor becomes institutionalized and legitimated (Berger & Luckman, 1966). At this stage, the way of acting feels to be the right one and the actor becomes confident: "This is how things are done." In his book on organizational culture, Schein (1992) has described the process as follows: "When a solution to a problem works repeatedly, it comes to be taken for granted. What was once a hypothesis, supported only by a hunch or a value, comes gradually to be treated as a reality" (ibid., p. 21).

The insufficient training and the consequent emerging ad hoc work practices that get institutionalized with the IS can easily lead to inefficient and unchallenged operations, as our case study clearly demonstrates. There is growing evidence of the high hidden costs of using computers in the workplace: about 2/3 of the total costs seem to come from the invisible opportunity costs in the usage phase of the computer technology (Heikkilä, 1995; Gartner Group, 1995; van Hillegersberg & Korthals Altes, 1998). In other words, the out-of-pocket costs from the IS department budget represent only one third of the total costs. Two thirds of the costs arise from the users' working hours spent in tackling computer-related problems, advising other users, learning new computer skills, and taking part in various CBIS planning, development, and coordinating taskforces. And the number of the hours of education, training and peer-hands-on- learning on applying ISs in work settings has not kept up with the increase in IS-based activities.

How much of this hassle can be removed by training is still unclear, but the fact is that it is one of the key actions to diminish the time wasted. And what has been found so far, is that the traditional teaching courses are deficient in meeting the needs of users applying ISs in their work practices. More innovative ways of learning are needed, such as peer tutoring, support activities, and workshops for the users of ISs (Reijonen, 1998).

The problems highlighted in this case are not typical in just one kind of business sector or type of information system, but a more general trend which we have noticed in several organizations. Hence, the emphasis is on the level of the general findings and the case is used to illustrate how problems can be manifested in a particular situation.

ENDNOTES

[1] At a first glance the mutual telephone companies were geographical (natural) monopolies. But as the mutual companies were owned by the subscribers, who also elected representatives to the board of directors, the monopoly power was never actually exploited to its full extent. Nevertheless, this led to a price discrimination between leased lines and 'owned' lines, which is currently banned by the European Union regulation.

[2] The subscribers of the conventional mobile phones (NMT and GSM) are served via a separate, nation-wide company, Radiolinja, owned by the Finnet consortium.

[3] As Finnet operates GSM networks via Radiolinja, the regional telephone companies are most interested in getting their share of the radio-based telecommunications by providing geographically limited mobile services for private and business customers at favorable prices.

[4] Preliminary discussions about the development of the system had actually started several years earlier at Finnet. One of the companies that merged with AP took part in the requirements analysis. The first installations took place only a couple of years before this study, and nobody at AP was experienced with the system.

[5] UNIX environment and client/server architecture were selected because of its openness (it is possible to select between different software vendors) and versatility (technically easy to extend).

[6] The data were gathered with a survey questionnaire (26 end-users), by interviewing end-users, managers and IS-personnel (16 interviews), from documentation (e.g. documents of the CBIS development project, end-user manuals), by observing the use situation, and by using the training version of the CBIS.

[7] EUCS is End User Computing Satisfaction, see Doll & Torkzadeh, 1988.

[8] For example, it takes almost half an hour to register and schedule installation for a new telephone subscriber.

[9] For instance, some Private department clerks preferred to call the warehouse keeper for the price of a product instead of checking it themselves from the database.

[10] Both the product numbers and product names were changed when the new IS was implemented.

The new numbering/naming system has its benefits because it is becoming a standard in the Finnet-group.

[11] The design of the IS did not make the situation easier because there were no practical means for managing enterprise level customer data, so, for example, the departments of a large enterprise occurred as separate customers with unique customer numbers in the database.

[12] The use of paper-based documents also meant that the IS no longer was a real time system, so answering customer inquiries of the state of an order always required contacting supervisors or technicians.

ACKNOWLEDGMENTS

This study was financed partially by the grants from The Finnish Work Environment Fund and The Academy of Finland.

REFERENCES

Argyris, C. (1985). *Strategy, change and defensive routines.* Boston, Pitman.

Berger, P. L., & Luckman, T. (1966). *The social construction of reality.* Middlesex, Penguin Books.

Brynjolfsson E., & Yang S. (1997). Information technology and productivity: A review of the literature. *Advances in Computers,* 43, 179-214.

Clement, A. (1994). Computing at work: Empowering action by 'low level users'. *Communications of the ACM,* 37(1), 53 - 63.

Compeau, D. R., & Higgins, C. A. (1995). Computer self-efficacy: development of a measure and initial test. *MIS Quarterly* 19(2), 189 - 211.

Doll, W. J., & Torkzadeh, G. (1988). The measurement of end-user computing satisfaction. *MIS Quarterly,* 12(2), 259 - 274.

Gartner Group (1995). *Desktop computing: Management strategies to control the rapidly escalating cost of ownership.* Gartner Group.

Heikkilä, J. (1995). *Diffusion of a learning intensive technology into organisations: The case of PC-technology.* Doctoral Thesis, Helsinki School of Economics and Business Administration A-104.

van Hillegersberg, J., & Korthals Altes, P. (1998). Managing IT-infrastructures: A search for hidden costs. In Walter R.J. Baets (ed.) *Proceedings of the 6th European Conference on Information Systems,* Aix-en-Provence, France, June 4-6, 1998, 1655 - 1662.

Iivari J. (1986). An innovation research perspective on information system implementation. *International Journal of Information Management,* 6, 123-144.

Kling, R., & Allen, J. P. (1996). Can computer science solve organizational problems? The case for organizational informatics. In Rob Kling (ed.) *Computerization and Controversy* (2nd edition), New York, Academic Press, 261 - 276.

Kling, R., & Iacono, S. (1984). The control of information systems developments after implementation. *Communications of the ACM,* 27(12), 1218 - 1226.

Kumar, K. (1990). Post implementation evaluation of computer-based information systems: Current practices. Communications of the ACM, 33(2),. 203 - 212.

Laboris (1997). *A collection of customer reports on the evaluation of the deployment of information technology in organizations* (1993 - 1997). Laboris, Laboratory for Information Systems Research, University of Turku, Finland. (Only in Finnish).

Mitev, N. N. (1996). More than a failure? The computerized reservation systems at French Railways. *Information Technology & People,* 9(4), 8 - 19.

Nelson, R. R., Whitener, E. M., & Philcox, H. H. (1995). The assessment of end-user training needs. *Communications of the ACM,* 38(7), 27 - 39.

Reijonen, P. (1998). End-user training and support: A comparison of two approaches. In Walter R.J. Baets (ed.) *Proceedings of the 6th European Conference on Information Systems,* Aix-en-Provence, France, June 4-6, 1998, 660 - 672.

Schein, E. H. (1992). *Organizational culture and leadership.* (2nd edition), San Francisco, Jossey-

Bass Publishers.
Telecommunications Statistics (1998). The Finnish Ministry of Transport and Communications, http://www.vn.fi/lm/telecom/stats/index.htm

BIOGRAPHICAL SKETCHES

Jukka Heikkilä holds a Ph.D. in Information Systems from the Helsinki School of Economics. He is a Junior Research Fellow of the Academy of Finland at the Helsinki School of Economics. His current research focuses on the problem of adopting, implementing, and integrating innovative technologies to support business processes. He is also a member of the Information Systems Laboratory Laboris at the University of Turku specializing in the evaluation of the utilization problems of implemented ISs in organizations. He has published in the Information & Management, European Journal of Information Systems, Journal of Systems Management, Journal of Global Information Management, and Scandinavian Journal of Information Systems.

Pekka Reijonen received his M.Soc.Sc. in Psychology from the University of Turku. He is a researcher at the Information Systems Laboratory Laboris at the University of Turku. His research interests are usability and deployment issues of information systems in organizations and he currently is working on a project defining end-users' skills and knowledge requirements and their evaluation. His earlier work has appeared in the British Journal of Psychology, Education & Computing, Interacting with Computers, and information systems conferences.

Implementing An Integrated Software Product at Northern Steel

Annie Guénette
Nadine LeBlanc
Henri Barki
École des Hautes Études Commerciales, Canada

EXECUTIVE SUMMARY

This case describes the implementation of the payroll and human resources modules of an integrated software product in a large manufacturing organization. The firm is located in a large metropolitan city and system implementation took place following a major organizational restructuring (from a public to a private enterprise) and downsizing (from 10,000 to 2,000 employees) effort. The extensive maintenance required by the existing legacy systems and the high cost of modifying them to address the year 2000 problem motivated the company to acquire an integrated software product from a vendor, and adapt it to the organization. Implementing the software took longer than scheduled and was 35% over budget. Some of the problems encountered include conflicts between the accounting and human resources departments, technical difficulties in building interfaces to existing systems, inadequate staffing of the project team, the IT director who left during the project, and a poorly functioning steering committee.

BACKGROUND

The Steel Industry

In 1997, the North American steel industry consisted of 96 companies (Standard & Poor's, 1997). Reflecting the particularities of the steel making process (see Exhibit 1), this industry is principally made up of two types of producers: integrated mills and minimills. Integrated producers such as USX, Bethlehem, and Dofasco make use of expensive plants and equipment to produce from two to four million tons of various steel products per year. In contrast, minimills such as Nucor, Birmingham Steel, and Co-Steel Lasco produce 400,000 to over two millions tons per year by melting recycled ferrous scrap in electric arc furnaces to make a limited number of commodity carbon steel products. Originally small-scale plants serving local markets for structural steel products, North American minimills have become major players in the 1990s and compete with integrated producers in most product areas (Standard & Poor's, 1997).

A wide gap in labor productivity remains between minimills and integrated producers. For example, in 1996 Nucor's 6,600 workers produced 8.4 million tons of raw steel, or 1,273 tons per employee. LTV, the United States' third-largest integrated steel maker, produced 8.8 million tons of raw steel with a work force of 14,000; this equaled 629 tons per worker.

Minimills have a labor cost advantage because they typically employ nonunion labor, whose compensation is often directly linked to production and profits. Integrated companies employ union labor which is more expensive over the course of the business cycle.

To survive in the long term, integrated producers need a technological breakthrough that will make them more competitive with minimills by reducing their labor and capital costs. Despite aggressive cost reduction and more flexible labor practices, integrated steel making remains more capital- and labor-intensive than minimill steel making. (Standard & Poor's, 1997).

The trends experienced by all players of the industry throughout the 1990s include the globalization of major customers (e.g., automotive, appliances), continued pressure from offshore excess capacity (e.g., Russia, Asia), continued pressure for environmental improvement, rapid and accelerating technological change, and the challenges of alternate materials in traditional markets (e.g., autos, bridge construction). Many firms have responded to these pressures by reinventing steel making through corporate repositioning (e.g., increased customer focus, cost reduction, organizational restructuring), by making large investments in technology and employee skills, and through increased collaborative research efforts (e.g., thin strip casting, process modeling, electronic sensors). (Bain, 1992)

Northern Steel

Established in the northeast as a public company over 30 years ago, Northern Steel was a money losing operation throughout the latter part of the 1980s and in the early 1990s. Employing a workforce of over 10,000 employees, it was sold to an international conglomerate and became a private enterprise in 1994. The new ownership immediately started a process of downsizing the company to improve productivity and increase its market share. Over the next couple of years, numerous facilities and divisions were either sold or closed, and the size of the workforce was reduced to around 2,000. By 1997, a much leaner and customer-oriented Northern Steel had only six facilities that manufactured and distributed around 1.5 million tons of steel products in North America (including sheets, slabs, weld pipes, billets, bars, rods, wires, and pellets). In 1997, Northern Steel's annual revenues were $900M, and more importantly, it had become profitable once again. However, the organizational culture developed over the past decades as a heavily unionized public company proved more difficult to change. The apathy of a slow moving bureaucracy and a nine to five mentality could still be seen rearing its head relatively frequently despite the privatization and the downsizing, and the strong pressures they entailed to improve productivity and to cut costs in all areas.

SETTING THE STAGE

IT at Northern Steel

Since the early 1970s Northern Steel had computerized its main transaction systems including its payroll and general ledger with most programs written in COBOL. With the introduction of microcomputing in the 1980s, many user departments started developing their own applications. Some of these were quite substantial. For example, one accounts payable application which handled over $50M had been developed by a user in the accounting department.

During the mainframe years, the IT department had a budget of around $3M and had a staff of 30. By early 1990s, problems in finding and hiring personnel having the requisite skills to maintain the legacy systems, and the constant difficulties experienced in servicing the old mainframe,

The Steel Making Process (Bain, 1992).

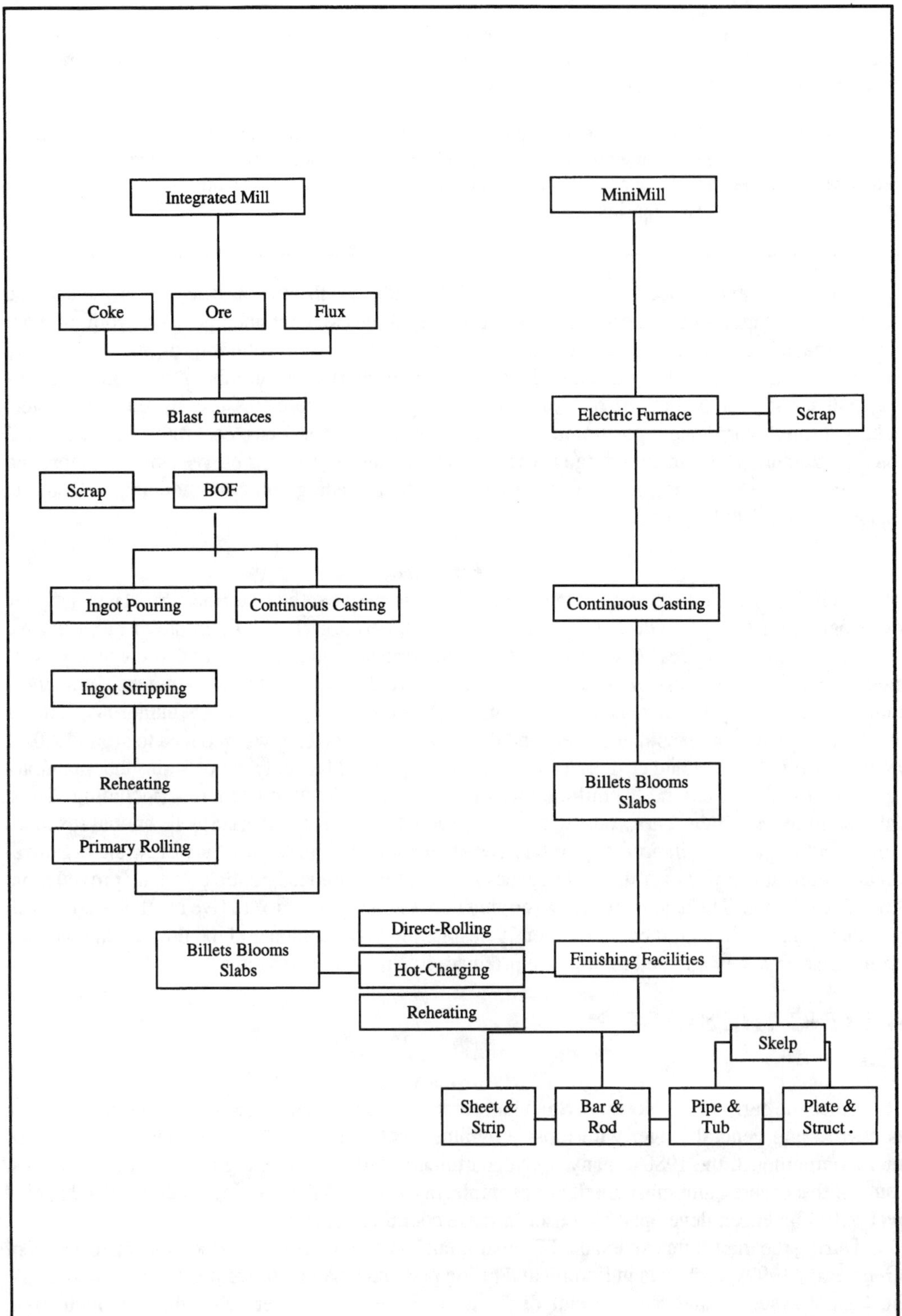

convinced Northern Steel to scrap its outdated hardware and to outsource the operation and maintenance of its mainframe systems to a large consulting company. The widespread practice of end-user development had also resulted in a decentralized IT structure. Thus, by 1990, Northern Steel's IT department was reduced to a staff of only five professionals who were responsible for telecommunications and network management, establishing company-wide software and hardware standards, and for approving their purchase. The remaining IT responsibilities were decentralized to the different manufacturing facilities and functional areas, with many former IT personnel working in, and reporting to, user departments. For example, the staff responsible for payroll systems worked in the accounting department and reported directly to the company comptroller. At Northern Steel, IT was frequently seen as a cost item. Many senior executives felt that, in general, the less Northern Steel spent on IT the better off the company would be.

The Payroll System

In 1994, an IT strategic plan was prepared and it budgeted $25M to the updating of all legacy systems and their migration towards newer and more accessible information technologies. Among these, revamping the legacy payroll systems was given top priority in view of several considerations. The accounting department, responsible for Northern Steel's payroll, was quite anxious to update the legacy payroll system and applied continuous pressure to prioritize its implementation. The extensive modifications the payroll system needed in order to deal with the Y2K problem added considerable urgency to the problem. Moreover, the existing payroll programs mostly contained spaghetti code written by programmers many of whom had long left the organization. As a result, the payroll system was already costly to maintain and difficult to manage. Frequently, changing one of its parts introduced bugs to its other modules. Thus, making all the Y2K changes to the legacy system was, at best, a scary proposition. Finally, the fact that the employees of Northern Steel were represented by several unions each with a different collective bargaining agreement meant that the payroll system was already complex, and had to be modified every time Northern Steel signed a new collective bargaining agreement.

The idea to replace the existing payroll system was initiated in 1980, and a proposal was obtained from Integrated Business Solutions (IBS), a software development/consulting company. However, Northern Steel's lack of resources at that time had prevented the then public company from pursuing the issue further. Following the privatization of Northern Steel, and according to the 1994 strategic plan, alternative solutions to replacing the legacy payroll system were examined. Initially, outsourcing the payroll was considered. The available products promised improved payroll accuracy and up to 80% reduction in payroll preparation time and offered a wide range of options including tax filing and deposit services, automatic check signing and insertion into envelopes, unemployment compensation management features, and a variety of management reports. While such an alternative appeared satisfactory from a purely payroll perspective, it would do little to improve the existing human resource (HR) systems, many of which were also tightly coupled to the payroll system.

At this time, the director of the HR department suggested the purchase of an integrated software product to replace the existing payroll and HR systems. In his dealings with a large bank, he had observed Integrated Business Solutions' (IBS) software product being used, and had been quite favorably impressed by the bank's positive evaluation. Since Northern Steel lacked the necessary resources for in-house development, adopting an integrated software product to replace both the payroll and HR systems simultaneously, while initially more expensive and challenging, appeared as a better long term solution. Given the general consensus about the need to scrap the legacy payroll system and adopt an integrated software product, it was decided to prepare a request for proposal (RFP) and to send it to suppliers.

In June 1995 the IT department prepared a relatively short RFP document that consisted of 14 points. These provided broadly described requirements (e.g., the modules needed such as payroll, workman's security, employee skills, medical etc.; the need for compatibility with the planned organizational platforms such as Unix and Oracle, and a client/server architecture; the possibility of modifying the product so as to fit Northern Steel's business processes) and guidelines concerning what information the suppliers needed to provide (e.g., the estimated cost to Northern Steel of modifications that would be needed, the cost of maintenance services, the availability of support from the supplier). Suppliers of integrated software products were contacted and asked to submit offers. The evaluation of the submitted proposals was completed by the end of September 1995.

Whereas other competitors offered software that integrated several business processes, including HR and payroll, IBS sold a "best of breed" product that consisted only of an integrated payroll and HR system but its offer included considerable customer support and service. With strong backing from the HR department, the Selection Committee chose IBS's offer as being the most suitable for Northern Steel. However, this was not a unanimous decision since the accounting department did not agree that IBS's software was the one that most suited their needs. Despite opposition from accounting, the Selection Committee voted to go ahead with IBS and an estimated project schedule and budget was prepared. The budgeted amount included the necessary funds to purchase IBS's payroll and HR modules, including its employee benefits and compensation modules, and to finance a project for their implementation. IBS's pension module was excluded from the purchase in view of the fact that Northern Steel had relatively recently implemented a new pension system. The necessary interfaces between the existing pension system and IBS would have to be built during the project. The project schedule and budget were submitted to senior management for approval and the allocation of the requisite funds.

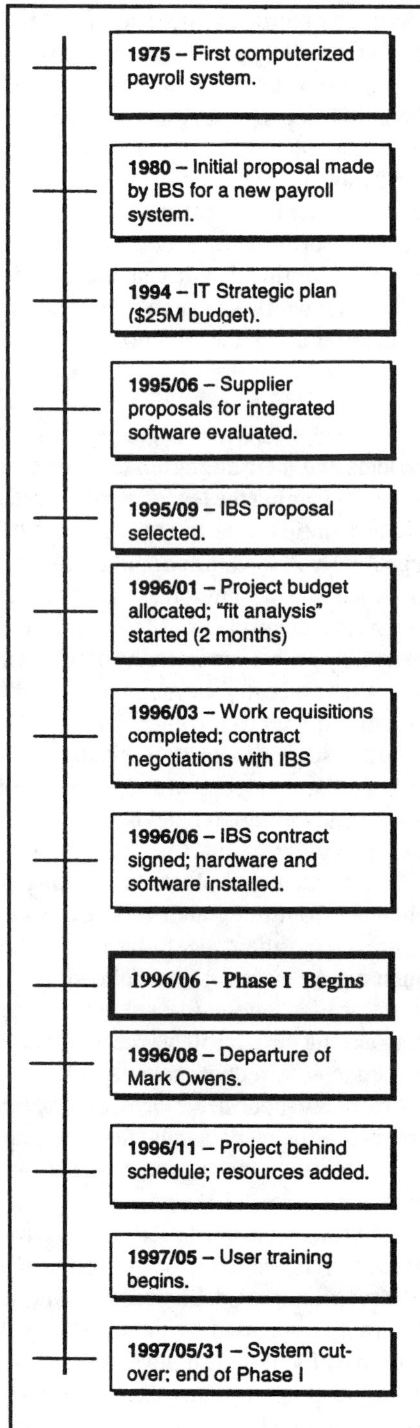

CASE DESCRIPTION

Senior management approved the Selection Committee's decision, and allocated the requested funds in January 1996. Based on the Committee's recommendation, Steve Collins was assigned as project leader. Working in the accounting department and reporting to the comptroller, Steve had been with Northern Steel since 1980 and was well-known across the company. Over the years, he had developed several accounting applications as an end-user and had gained experience as a project leader by managing several projects in accounting and other departments. Steve was quite excited about the IBS project. While this was going to be Northern Steel's first experience with the implementation of an integrated software product and Steve's first attempt at managing a large project team, he did not foresee too many problems: "With a package, all we have to do is implement it, and it should not take too long."

To oversee the project, a Steering Committee was formed. Its members included the IT Director, Mark Owens, an experienced and well-respected professional, and senior executives both from the accounting and HR departments. In addition, a five-member project team including two IT department staff and one representative each from accounting and HR was also created. The selection of each was left to their respective departments.

The next two months were spent in a period of "fit analysis." During this time, a User Committee consisting of ten HR employees, two accounting employees and one IT staff was also created. This committee's mandate was to examine the IBS software and to recommend a list of modifications to be made to IBS so as to satisfy Northern Steel's payroll and HR information requirements. The modifications and additional functionalities requested by the User Committee were communicated to the project team in an informal manner, and at times, verbally. Examples of the type of changes requested from the IBS software included changing the paycheck format, incorporating the different overtime rate calculations, calculating payroll based on hourly salaries, and preparation of new reports. Based on the requests they received from the User Committee, the project team prepared a written work requisition and subsequently evaluated their feasibility and decided which IBS functionalities to keep and which ones to modify according to the work requisitions.

All too frequently, the modifications and supplemental functionalities requested by the User Committee appeared more like a wish list to the project team. According to Steve, users had difficulty understanding that certain hard to reconcile differences existed between the way things were done at Northern Steel and the way the IBS software was designed. As a result, the project team ended up discarding approximately half of the requests received from the User Committee.

By March 1996 a list of work requisitions was finally arrived at and submitted to the legal department for approval. These were approved in June 1996 and a contract for a two-phase project signed with IBS. Phase I of the project, scheduled for completion in February 1997, was started immediately following the signing of the contract. The objective of this phase was to implement the payroll module of IBS as well as parts of its HR system such as its insurance, accidents, and HR administration modules. IBS started to install the hardware and software environment (the network, the server, work stations, Unix and Oracle) and to train Northern Steel's IT staff in their use.

During the first six months of the project, little progress was made. All user members of the project team continued to work full time in their original jobs and so, could not devote much time to the IBS project. Moreover, the prevalent organizational culture meant that many users had little motivation to stay after hours for the sake of the project. In addition, replacing the old systems with IBS meant that a number of jobs would disappear from the HR department, creating some anxiety and lack of cooperation among the HR personnel. Another blow came in August 1996 when Mark Owens left Northern Steel. Up until then he had been one of the project champions and had been instrumental in resolving several conflicts that had arisen between accounting and HR. His departure also created a certain void in the Steering Committee. Its meetings were gradually held less frequently and eventually disappeared.

An added difficulty which slowed progress stemmed from the technical challenges encountered in building interfaces between IBS and the existing pension system. Moreover, during this period Northern Steel was in the process of negotiating a collective bargaining agreement with one of its large unions. Preoccupied with this process, many senior executives were unwilling to spend too much time on the IBS project. For example, when Steve tried to discuss the project with the director of the HR department, the response he got was "Listen, I am trying to negotiate salaries for the next five years and we're talking over a billion dollars here, so don't bug me."

By November 1996 the project had fallen behind schedule by several months and Steve was quite concerned with the slow rate of progress. He finally convinced senior management that remedial action was needed and this resulted in the addition of two more members to the project team, one from accounting and one from HR. In addition, all team members were now assigned to the project on a full time basis. These changes improved the situation considerably, and the project started to advance at a more rapid pace.

The next six months were spent with making the necessary modifications to the IBS software and developing its database. While this activity was IBS's responsibility, all data being converted had to be validated by the users concerned. Their input was necessary since they knew all the compensation and benefits rules and procedures that were being incorporated into hundreds of tables in the new database. However, this process took longer than anticipated. The old payroll and HR systems consisted of numerous poorly structured files some of which contained over 200 fields.

Cleaning them up and making sure all the data they contained would be conserved in Oracle's relational tables was in itself a challenging task. According to Steve, an added difficulty came from the user representatives in the project team who had problems grasping elementary database concepts and needed a lot of help with the design.

Moreover, the user members of the team were frequently unable to decide without consulting other users. During these consultations it became clear that many of the work requisitions discarded earlier in the project were actually needed. For example, at Northern Steel, salaries were calculated on a daily basis but the IBS software determined paycheck amounts by first converting these into hourly rates which were then summed up over two-week periods. This process sometimes resulted in rounding errors of one cent. However, during consultations it was discovered that not only any such deviations were totally unacceptable to the users (and to the unions) but that paychecks had to be exact to three decimal points. This meant that there was no choice but to make the necessary changes to the IBS software to satisfy this particular requirement. Numerous such requirements re-surfaced as Northern Steel's business processes were once more passed under scrutiny. As a result, many previously discarded work requisitions had to be re-integrated into the modifications already made to the IBS software. During this period, Steve frequently felt frustrated by the lack of flexibility exhibited by the user departments who approached the whole process with a "Today we are at this point and we won't go back" attitude. In addition, the project was being delayed further and since all budgeted funds had been spent, additional funding had to be requested from senior management. Steve felt the pressure mounting.

Nailing down all the requirements, completing the database design, writing and testing more than the thirty data conversion programs needed to populate the IBS tables took the project well into May 1997. At this time, preparations to cut-over from the old systems to IBS were undertaken. The accounting and HR personnel were trained over the first three weeks in May through a 3-day training program that was repeated to groups of users. Those who participated during the project were trained first, and in turn, they trained the other users.

While running the old and the new system in parallel was considered as an option, it was discarded due to the approximately twenty additional personnel and the associated expenditures it would have required. All the data from the old systems had to be converted and transferred to the IBS databases rapidly and without error. This requirement was mainly due to the unions who were quite inflexible with regards to potential delays or errors in the issuing of worker paychecks. Despite firm promises that all errors would be eventually corrected, they refused to accept any delays or mistakes in the paychecks. This meant that the all conversion and cut-over activities would have to be completed over a weekend, putting additional pressure on the project team.

The cut-over took place during the last weekend in May. Tests were run first with one employee, then with 50, and finally with all employees. During these tests, both the old system and IBS had to be run in parallel and this turned out to be more complicated than anticipated. In addition, some of the conversion programs were found to contain several errors, all of which had to be corrected on the fly. It all made for a very stressful Sunday. However, by late Sunday night all modules were working properly and IBS was up and running.

After a four-month delay from its scheduled completion date and at a cost that was 35% over budget, the IBS system was put into use at the end of May 1997. During the year that followed it has functioned without any major problems. The users have no major complaints regarding functionality or ease of use.

CURRENT CHALLENGES

In August 1998, Steve was wondering whether or not to start the second phase of the IBS project. During Phase II, scheduled to take six months, the remaining HR modules of IBS which include the medical, workman's security, and employee skills systems were to be implemented. These modules were considerably more complicated than those installed in Phase I and Steve was not yet sure that enough time had elapsed to allow everyone to catch their breath and to get used to the IBS software.

ENDNOTE

[1] In order to maintain confidentiality, the name and geographical location of the companies concerned as well as the names of the individuals involved in the case have been disguised.

REFERENCES

Bain, T., *Banking the Furnace: Restructuring of the Steel Industry in Eight Countries*, W.E. Upjohn Institute for Employment Research, (1992).

Standard & Poor's, *Metal: Industrial Survey*, (July 24, 1997)

BIOGRAPHICAL SKETCHES

Annie Guénette has an undergraduate degree in Information Systems from the École des Hautes Études Commerciales in Montréal where she is also completing her MSc in Information Systems. Her MSc thesis examines integrated software products in order to compare their flexibility in supporting business process changes.

Nadine LeBlanc has an undergraduate degree in Information Systems from the École des Hautes Études Commerciales in Montréal where she is also completing her M.Sc. in Information Systems. Her MSc thesis compares three learning approaches in the use of multimedia in education.

Henri Barki is Professor in the Information Technologies department and Director of Research at the École des Hautes Études Commerciales in Montréal. His research interests focus on software development project management and journals where his publications have appeared include Information Systems Research, Information & Management, INFOR, Management Science, and MIS Quarterly.

The Institutionalisation of User Participation for Systems Development in Telecom Éireann

Tom Butler
Telecom Éireann, Ireland

Brian Fitzgerald
Executive Systems Research Centre, University College Cork

EXECUTIVE SUMMARY

It was in 1984 that Telecom Éireann first introduced institutional mechanisms which facilitated employee participation in the formulation and execution of corporate strategy. However, almost ten years elapsed before the full benefits of user participation were realized in the development and implementation of organizational information systems. Two systems development projects that are perhaps exemplars of the manner in which user participation was and still is effected in Telecom Éireann, and which offer unique insights into this multi-faceted phenomenon, are described herein. This case study not only illustrates why user participation is important for systems development in organizations, it also provides evidence that user participation is insufficient for success in systems development if appropriate attention is not given to change management issues associated with the implementation of developed systems. The lessons learned by Telecom Éireann in addressing such issues helped it to evolve its participative policies into a partnership approach to organizational change that helped ensure the success of its strategy of IT-enabled organizational transformation.

BACKGROUND

As the Republic of Ireland's major telecommunications utility, Telecom Éireann provides universal telecommunication service and enjoys a monopoly in many areas of its business. Being a state-owned company, Telecom Éireann's majority shareholder is the Irish Government, which retains a 65% stake in the organization. The remaining shareholders include Telecom's employees, who hold a 15% stake, and two European telecommunication's operators—KPN (PTT Telecom BV) of Holland and Telia (AB) of Sweden—who jointly own 20% of the company. KPN and Telia also possess options to purchase a further 15% of Telecom's equity. Telecom Éireann entered into a strategic alliance with these companies in January of 1997 in a deal that saw KPN and Telia offer Telecom access to a global telecommunications platform that would enable it to deliver improvements in both the quality and price of its products and services. Despite the many competitive challenges it faces, Telecom's future seems bright, but this was not always so. In its first set of published accounts in 1985, Telecom Éireann reported a loss of IR£65 million on a turnover of IR£389 million with a debt of under IR£1 billion. Some fourteen years on, however, in the published accounts for the year ending 1997/98, the company announced a profit of IR£155 million on a turnover of IR£1.35 billion, with a debt of IR£172 million. In the intervening years, Telecom reduced its cost base by cutting staffing levels from over 18,000 to 11,000. From the very beginning,

Telecom's management realized that it had to divest itself of the bureaucratic image and practices of its previous incarnation as a civil service department within the Irish Government. The then CEO was faced with the considerable task of changing the culture, structure, and business processes of the organization—this change was effected within the framework of industrial democracy. It was within this framework that the company and staff labor unions entered into a policy of participation on all issues of major concern. To help operationalize its policies, the company instituted a profit sharing scheme for all employees based on the achievement of financial and operational targets; it also changed its organizational structure to decentralize decision making in relation to operational matters to regional functional units.

Telecom Éireann's annual report for the year 1994/95 describes several key forces for change in the telecommunications industry in Ireland viz. technology, competition, regulation, globalization of the telecommunications market, and customer choice. In order to meet the challenges posed by such forces, the company's newly appointed CEO felt that Telecom had to radically improve its ability to meet customer needs by (a) targeting sales, marketing and service delivery capabilities at segmented customer markets; (b) providing customized, competitively-priced packages to business customers; and, finally, (c) minimizing its operating costs through streamlining customer contact, service delivery, billing, and repair operations. The use of information technology (IT) was considered by the CEO to be the pivotal factor in enabling this strategy. Some of these initiatives were not novel as measures were already established to rectify certain weaknesses in the company's operations; for example, the need for both of the information systems (IS) described in the case study was articulated in early 1994. These systems were proposed in order to help minimize operational costs and streamline telephone service delivery and repair processes.

Under the leadership of its new CEO, Telecom underwent a radical transformation in the four years up to 1998: for example, the company's organizational business processes and structure were changed significantly (see Exhibit 1, Appendix A, for the current organizational chart) and its IT infrastructure was altered to facilitate and support such change. Although the participative policies and practices that existed within the organization in 1994 were sufficient for the degree and pace of organizational change undertaken in the 1980s, the CEO deemed such policies and practices to be ill-suited for the radical measures he had in mind. Existing participative structures and processes were, therefore, enhanced and augmented to ensure that both management and staff were committed to the transformation of their operational roles and work practices in the achievement of corporate strategy. One of the most significant measures undertaken by the company was to replace the existing profit sharing bonus scheme with an employee share ownership plan (ESOP) that saw staff acquire some 15% of the company's shares in exchange for which they would agree, among other things, to cooperate and participate fully in the IT-enabled transformation of the organization. It must be noted that it was the labour unions that first tabled and championed this initiative in the face of notable opposition from the government of the day; however, the ESOP deal had the unequivocal support of the CEO and his management team. This change in the organization's approach to user participation in the IT-enabled transformation of the organization did not occur overnight, the structure and process of user participation for information systems development within Telecom Éireann had been in the process of change since 1994.

SETTING THE STAGE

While Telecom Éireann's policy on employee participation was applied with success in many of the company's operational areas in the first ten years of business, and the industrial unrest seen in the 1970s became a thing of the past, there was one area of the company's operations in which this policy had little effect—that is, in the development and implementation of corporate information systems. In the first ten years of its existence the company did not have a coherent policy on the use of information technology to informate and transform its business processes. Certainly, several large-scale corporate systems, e.g. transaction processing, financial and management information systems, were introduced or upgraded; however, there was little in the way of integration taking place due to the non-alignment of business and IS strategies—indeed, it must be said that until 1994 the company did not possess an IT strategy as such. Neither was there any attempt to institute change into company operations based on the introduction of information systems: change, when it did occur, was incidental and of little overall consequence. Users participated in the development of corporate information systems, but given the prevailing approach to systems development and to participation by users in this process, the benefits associated with user participation in the introduction of IT were never fully realized.

In the mid-to-late 1980s, personal computers (PCs) and networking technologies were in widespread use throughout the organization; these technologies were not employed to automate or informate business processes in conjunction with corporate IS, rather their main use was in providing office automation and personal productivity tools to end-users. Nevertheless, as a consequence of the introduction of such technologies, administrative and operational managers and staff began to fill the information vacuum by developing their own PC-based solutions to informate their activities and hence work smarter. However, because of the uncoordinated ad-hoc approach to the development of these systems, coupled with an overall lack of cross-functional integration and knowledge transfers between members of the business community, the systems developed and implemented by end-users remained ineffective in a macro-level context. The major advantage of this trend in end-user computing was to heighten an awareness of the benefits that could accrue from the use of IT in supporting business processes and organizational decision making. The experiences that end-users had in this area proved to be crucial in later years as a cadre of IT-literate managers and operational staff was available to participate in the development, implementation, and use of corporate information systems.

The major reason for the aforementioned uncoordinated approach to the development and use of IT was that, up until 1994, the organization's IS function was a relatively obscure department within the personnel directorate. As such, the Information Systems Department (ISD), as it was then called, was merely seen as a servant to business directorates and, because of this power asymmetry, it lacked the political muscle to institute its own policies for systems planning, development and implementation. Up until the early 1990s, political issues and the scope of the IS function's development activities colored its relationships with the broad community of users in the organization. From about 1990 on this situation began to change as information systems with a wider organizational scope and impact began to be developed and implemented. As a result, the ISD began to establish good working relationships with the business community. All this had a positive impact on the type and degree of participation by users in the systems development process. However, because of the aforementioned lack of political clout, the IS function could not acquire the type and degree of participation it desired to ensure success in all its development endeavours.

In 1993, Telecom's board of directors concluded that the company would have to be in a position to respond to increasing competitive pressures or face disastrous consequences. In order to effect the necessary change, a new CEO was selected and appointed in 1994. This individual proposed a number of innovative and radical plans to help transform the organization. Telecom's new CEO maintained that the success of the company's business activities and the attainment of performance targets was (and still is) very dependent on the quality of the support and services offered by its IS function. Accordingly, in 1995 he changed the IS function's position within the company structure and elevated it to directorate status. This change in the IS function's standing was prompted by the recognition of the pivotal role IT would play in the achievement of the company's strategic objectives. In reality, however, this change in the IS function's status had already taken place in 1994 when the CEO first assumed the reins of control and informally invested a strategic role in the function. As a consequence of this change, the power asymmetry that previously existed between the IS department and its business clients was effectively mitigated. Hence, many of the negative effects of 'political' influence and infighting among company directorates and functional units, especially in relation to the focus and prioritization of systems development activities, and the quality of user representation on development projects, were overcome. This change in status, coupled with other related events, allowed the IS function to effectively manage its relations with business units and associated user communities within the organization. In enabling developers to manage better their relationships with users; it also greatly enhanced the quality of user participation in systems development within the organization. The IS function's experiences with the development of the two operational support subsystems described in this case study bears witness to this.

Thus, the newly formed IT Directorate (ITD), or Group IT as it is now known, was a centralized functional unit whose chief responsibility is the development, integration, maintenance, and support of all corporate IS. Based in Dublin, it has a staff of over 240 spread among its eight divisions. In keeping with its new status and strategy, the directorate was restructured internally in 1995 as part of an ongoing endeavor to create a *"customer-centric culture"* using an account management strategy; this change in orientation was supported by the introduction of an intra-directorate *"customerfirst"* (organizational term for what is Total Quality Management, TQM) quality program (see Appendix Exhibit 2, for the ITD's organizational chart as of 1995). The IT director viewed this change in culture as being critical to the success of the directorate's various activities—particularly, in its efforts to maximize the benefits of participation by users in systems development. It was the first serious attempt to align the directorate's IT strategy, infrastructure, and processes with business strategy, infrastructure and processes. This change pre-empted the organizational restructuring begun under the CEO's Organizing-to-Compete strategy (OTC) to some degree. The implementation of the OTC in 1997 saw the number of divisions in the ITD increase from six to eight in order to closely align the ITD's sub-functions with those of the newly restructured core business directorates (see Appendix Exhibit 3). The new divisions include the Financial and Administrative Systems, Sales and Marketing Systems, Billing Systems, and Operations systems sub-functions. The remaining sub-functions include the operational support system (OSS), IT Strategy, IT Architecture, and the Data Center divisions.

A JOINT STRATEGIC CONSULTATIVE PROCESS

Telecom Éireann's participative approach to the implementation of organizational policy and decisions was recently underlined when the company reiterated its policy in this area viz. *"The process of consultation with unions in regard to all the implications for staff of technological change, is one to which the company remains fully committed."* To give effect to this policy, the company has instituted several joint bodies; for example, the Computer Liaison Committee (CLC), whose members are drawn from the company's management team as well as the labor unions, deals exclusively with issues surrounding the introduction of information systems within the organization. Two other technology-related forums of note here are the Joint Technology Committee (JTC) and the Joint Working Party (JWP). However, in order to institute the radical transformation planned for the five years from 1997 to 2002, a more innovative approach had to be adopted by the company. In 1995 the CEO put together an umbrella forum called the Joint Strategic Consultative Group (JSCG) to introduce and implement a new approach to the company's partnership approach to change. A framework agreement for the transformation of the company was drawn up by management in consultation with the unions at this forum. Of particular emphasis here was the continued recognition by all concerned of the need for a high degree of user involvement in expediting IT-based solutions across the enterprise.

The Structure and Process of User Participation within Telecom Éireann

One of the areas where Telecom Éireann's participative approach in the formulation and implementation of its business policies is particularly evident is in the development of its corporate information systems. In order to operationalize this approach, users at all levels, from top management to operational staff, participate in the development, testing and implementation of corporate IS. Accordingly, each systems development project within Telecom has a designated business owner or project sponsor. For large projects, a development steering group (DSG) is formed from the constituencies of interest within the organization; managers from the relevant business areas and the ITD normally comprise these groups. Two project managers jointly manage each systems development project: a user project manager drawn from the business constituency, and a development project manager drawn from the ITD. The latter manages the physical development of the system;

Figure 1 The Structure of User Participation and Involvement in Systems Development in Telecom Éireann

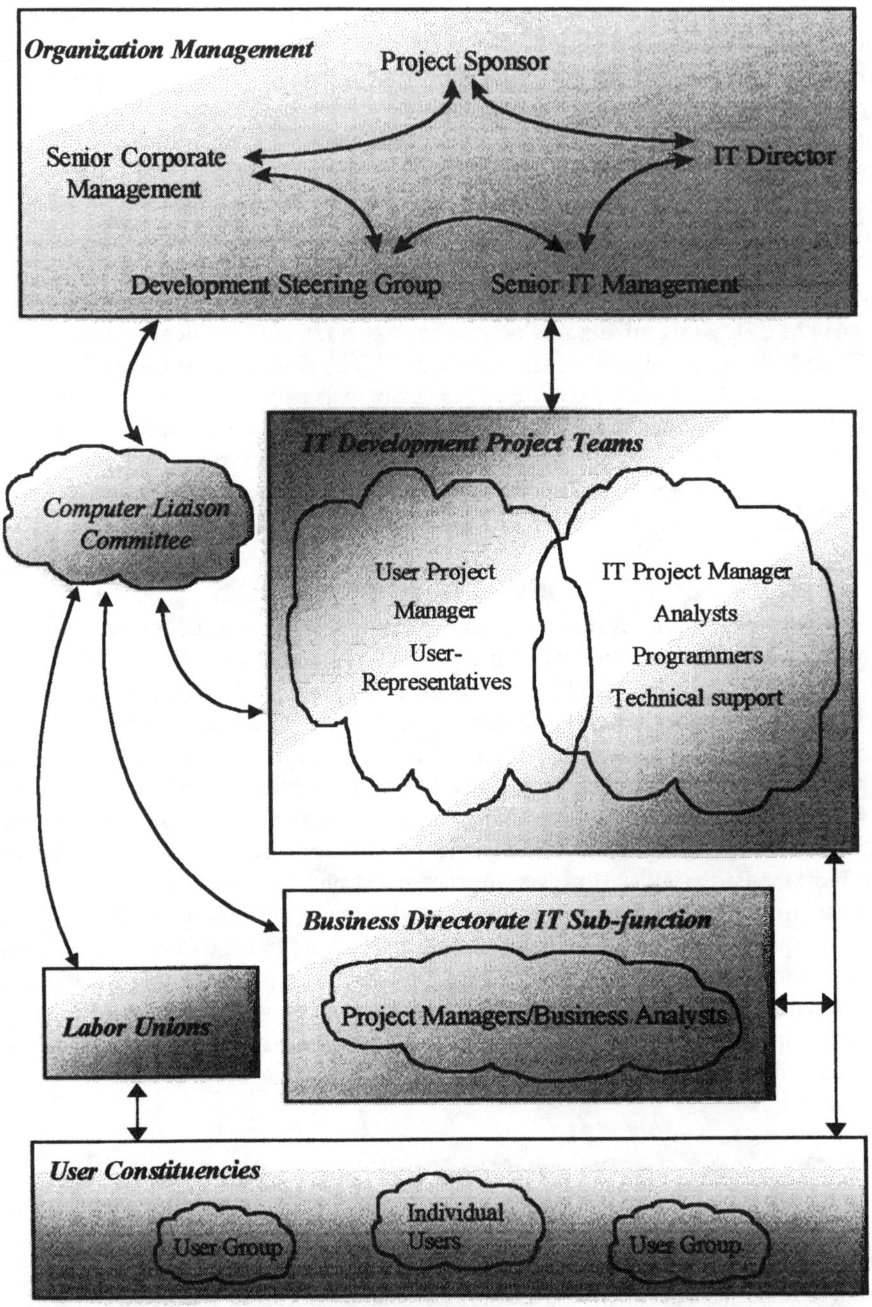

Legend: The double-headed arrows indicate the paths of communication and influence that exist between the different constituencies/entities/social actors.

the former manages business user input into the project in areas such as the provision and management of user representatives, user groups, user test teams, and infrastructural resources etc. Development teams normally consist of one or more user representatives from interested business constituencies and a team of developers from the IT Directorate. User representatives actively participate in most development activities, apart from programming and the technical aspects of systems development. Although key users are interviewed to elicit system requirements, user groups are also formed to provide the development team with a core group of users for further requirements analysis and to verify and ensure that the system, as developed, will meet these requirements. User participation in Telecom is characterized by both the industrial democratic and participative approaches to systems development. Hence, a combination of participatory design (PD) and joint application design (JAD) are utilized in systems development.

Figure 1 illustrates the structure that operationalizes the company's participative policies in the area of systems development. Apart from the IT sub-function within business directorates, which is a relatively new element in this structure and which was introduced in 1998, the figure describes the different constituencies within the organization who were involved or who participated in systems development in 1994/95. This structure, and the processes that underpin it, is generally considered to have had a positive effect on user attitudes toward and behavior in development activities. As one of the ITD's project managers pointed out, the participatory mechanisms employed within the organization *"provide users with opportunities to express their 'world views'...help resolve political conflicts, and help negate potential 'them-and-us' situations arising between developer and user."*

CASE DESCRIPTION

As previously indicated, the restructuring of the company's IS function by the newly appointed IT director gave new meaning and effect to the concept of user participation for systems development within the organization. Empowered by its new status as a business unit within the company, the IT directorate's managers and systems developers could now secure and encourage suitable business users to actively participate not only in traditional analysis and design activities, but, also, throughout the development process, where required. The establishment of a *"quality-based culture"* by the IT director also meant that the already positive attitude developers had toward users was enhanced by the *"customer-focused"* orientation of the director's quality initiative. On the other hand, users were now more aware of the power and benefits of IT, and were confident in the ability of their labour unions to maintain tenure of employment and conditions of service for those effected by the introduction of new IT systems. This meant that users who actively participated in the development of new systems, which radically altered their operational roles and working conditions, were not, as one user put it, *"turkeys voting for Christmas."* This was the general climate in which the development of the information systems now described took place.

The Generic Appointment System (GAS) and the Geographic Information System (GIS) development projects were two of the first undertaken by the newly formed IT Directorate in 1995. The genesis of these projects predates the introduction of the company's overarching IT superstructure—the operational support system (OSS) which lies at the core of Telecom's business and IT strategies. That said, the GAS and GIS constitute two vital elements of this superstructure and, as such, provide technology platforms that help implement the company's strategies and enable its new business processes.

The Generic Appointment System grew out of a business need in one key area of the company's operations—its telephone repair service. Business managers across the organization recognized the need to make efficient the manner in which repair service workloads were managed, and associated service appointments made with customers. One of the goals to be achieved by introducing this new system was the elimination of unproductive visits by operational staff to customer premises when customers were absent. The GAS would also assist supervisors in their task of allocating workloads to their repair teams, which consist of operational staff. Both groups therefore had a keen interest in the development and implementation of this system as it impacted on some of their basic functions. The GAS also supports the operation of the company's ten fault-handling and repair centers.

The Geographic Information System was developed to provide a graphical database of the telephone network in the general Dublin area. Heretofore, the planning and drawing office functions manually recorded network-related details using paper-based records and maps. The business manager responsible for this project recognized that there would be significant improvements, in terms of economic and operational efficiencies, to be gained in using a GIS in this area of the company's operations. However, the implementation of the GIS meant that a radical change had to take place in one of Telecom Éireann's operational business processes. Accordingly, the development of the GIS posed significant challenges to the business sponsor, users and developers alike. On the one hand, there was the issue of change management associated with the radical change in work practices/roles of the users in operational units who then performed telephone network mapping, planning, and record handling duties. On the other, there was the challenge of developing a highly complex and sophisticated information system within a proprietary application development environment.

Table 1 presents an analysis of both the GIS and GAS projects in terms of the salient issues and impact of user participation on the development of these systems. It can be seen that the characteristics, impact, and positive outcomes associated with user participation varied little across these projects. The following sub-sections provide a more detailed exposition of several salient aspects of participative development in the case.

GAS and GIS Project Characteristics

A development team that consisted of a user project manager, a development project manager, two analysts, the CASE vendor consultant, one programmer and a user representative carried out the development of the GAS; three user groups and several individual users formed the bulk of participating users from the constituencies of interest. A computer aided systems engineering (CASE) enabled rapid application development (RAD) approach saw development take place within a three month time period; that said, the implementation of the first phase of the GAS took a further six months. As a distributed IS, the GAS is comprised of 8 relational databases that serve up to 180 windows-based PC terminals in fault-handling centers around Ireland, and a further 400 in operational depots nationwide. The GAS project came in on time and budget.

The GIS development team consisted of a user project manager, a development project manager, two analysts, three programmers, two user representatives, and a team of ten end-users whose primary role was to input graphical data and carry out test functions. User groups were also drawn from the two constituencies of interest—the drawing and planning functions. Consultants from the software vendor also participated in the development process. The GIS was built around a proprietary graphical database engine that serves up to 40 high-end workstations. The first phase of the GIS development took almost two years to complete. The implementation and rollout of the first phase took a further year. The project failed to meet the scheduled completion date, and also exceeded budget.

The Content and Context of User Participation in the GAS and GIS

Although adhering to the basic structure and process of participatory systems development, the content and context of user participation in the GAS and GIS development processes differed in significant ways. Developers on the GAS project, for example, needed to understand how staff in the fault handling centers accepted and tested faults on customer lines and equipment and then dispatched them to repair teams and, since fault-handling staff were the initial point of contact for customers, the manner and circumstances in which fault-related service appointments were made with customers. Customer service provision and repair teams are responsible for the repair of customer lines and equipment and the provision of new lines and customer premises equipment—both of these activities involved the making of appointments with customers to gain access to their premises to either install or repair equipment. If involved in the repair process, technicians would receive customer appointments from the fault-handling staff or, alternatively, as with the provision of customer service, they would make appointments themselves. The customer team leader was very much left out of the loop in the control and execution of these business processes, except in regard to allocation of certain types of work and the routine issue of customer service orders to team members. Business managers felt

Table 1: A Comparative Analysis of Issues related to User Participation in the GIS and GAS Projects

1. General Dimensions	Description	GAS	GIS
Type of system under development	Operational Support Sub-System.	√	√
Participation vs. involvement	Users immediate to the project team(s) participated, while the majority of users were mainly involved in the development process.	√	√
Type of user participation	Consultative, representative, consensual.	√	√
Degree of user participation	Participation by advice (indirect) ranging to participation by strong control (direct).	√	√
Content and extent of participation	Pan-lifecycle for user representatives, that is they were present throughout the development process. Individual users and user groups participated formally at key points in the development process—e.g. analysis and design, testing and implementation —and informally throughout.	√	√
Formality and influence of participation	User representatives were co-opted into the development project team. Development steering groups were formed for management users, and user groups were formed to participate in requirements analysis, design verification and testing. Significant user influence was exerted, especially through labour unions and joint management/union forum.	√	√
Organizational perspective on participation.	While user participation in both projects was characterised by the industrial democratic form of participation, there also was an element of participative management in evidence. A combination of participatory design (PD) and joint application design (JAD) were utilised for systems development.	√	√
Users participating	User project manager, user representatives, user groups, and individual users employed. Joint staff/management bodies were also involved.	√	√
Location of development project team	On-site at the business client's offices.	√	√
2. Contingencies			
2.1 Organizational Variables			
Organizational policy on systems development	Organizational policy on participative development fully implemented.	√	√
Influence of organizational culture on team subcultures.	Shared organizational culture ensured that the team subculture was receptive to user participation in systems development.	√	√
Time for development	Although there was a very tight project time schedule, it did not impact negatively on the degree of user participation.	√	√
Financial resources available	Budgetary resources did not affect the degree or quality of user participation.	√	√
Top management commitment	A high degree of support existed from organization and IS function management. A high degree of top management support existed in the first phase, but this waned in subsequent phases. There was also a lack of support from senior IS function management.	√	√

2.2 Project-Related Factors			
Initiator of the project	Business Management.	√	√
Project complexity	Moderately complex project, several functional groups were involved, functional boundaries crossed.	√	
	Highly complex project, functional boundaries crossed.		√
Degree of task-structure	Low-to-Medium-level task complexity, moderately defined business process.	√	√
Development Technology available	CASE workbench (IEF) that fully supported prototyping, significant impact on the quality of user participation: user representative trained in CASE tools.	√	
	Proprietary development tools, SSADM employed in analysis and design, user representative trained in SSADM.		√
Expected change brought about by the system	High degree of change for one user constituency. New business process supported.	√	
	Radical change to user role-related activities in two user constituencies.		√
2.3 User-Related Factors			
User perceptions of organizational climate	Users felt that a favourable development climate existed.	√	
	Users were of the opinion that the organizational climate was negative; however, they felt that a favourable development climate existed.		√
Willingness to participate	Users eager to participate.	√	√
Ability to participate	The use of dual project development teams (user and developer), greatly facilitated user participation.	√	√
User characteristics & attitudes	Very positive attitudes by users. User computer literacy a problem. Shared organizational culture of benefit in accommodating different 'world views'.	√	√
3. Factors within the participation process that impact on the degree and effectiveness of user participation			
User/analyst relationships	Very good. Relationships were enhanced by the existence of a common organizational culture and favourable development climate in project teams.	√	√
Influence & power relationships	Several institutionalised checks and balances existed which countered any power asymmetries or political opportunism that may have arisen. This was due to the implementation of organizational policy by all the constituencies involved in systems development. Positive management attitude toward and acceptance of user input.	√	√
Communication	High degree of user/analyst communication.	√	√
	Greatly enhanced by on-site development, training the user representative in IS development method and tools, and the prototyping approach adopted.	√	
	Some improvement in communication brought about by user training in SSADM and on-site development.		√
4. Variables moderating the participation-success link			
Perceived control	The type and degree of user participation gave users a sense of ownership and control over the system as developed, despite eventual reservations over the systems utility.	√	
	Change management difficulties dominated and coloured user attitudes toward the system.		√
Desired level of participation	Good fit between desired & actual levels of participation by users.	√	√
Perceived importance and relevance of system to users	Medium to high degree of relevance as evidenced by the change management and industrial relations difficulties.	√	√

that team leaders needed to exercise more control over the activities of the technicians in their teams, and the new appointments system would help effect this. Operational users from these three constituencies therefore participated in the development process. Business managers participated in development activities either directly as part of the development steering group or indirectly through the user project manager. Then, of course, the influence of the labour unions had also to be factored in, as any change in work practices would ultimately require their agreement. In summary, developers needed to understand existing operational processes and also work with users and their managers in order to create a more efficient customer appointment process in order to map it unto the proposed information system.

The user requirements for the GIS were equally complex, but in a different way. Here, developers had to capture in great detail the geography and content of the Dublin metropolitan telephone cable network infrastructure. They also had to understand how this information was utilized within the network planning process so that the process could be effectively enabled using IT. Company draftsmen were traditionally responsible for supplying the network planners with the raw material for the planning process—that is, detailed drawings of the network infrastructure. This work was highly skilled and labour intensive. The draftsmen's knowledge therefore had to be captured by developers and embedded within the new system. In addition, the views of business managers and labour unions had to be taken into account due to the radical nature of change to the existing business process that would be introduced by this new system.

The key actors in the elicitation of the detailed user requirements were undoubtedly the user representatives. These individuals acted as interpreters and advisors to the developers in their efforts to understand the existing business processes and the manner in which the new systems would impact on the proposed changes to these processes. In the GAS project the user representative was intimate with existing processes and work practices across the constituencies of interest. A planner and a draftsman acted as user representatives on the GIS project. User project managers on the development teams looked after the interests of managers in the associated business constituencies. In both projects, user representatives were trained in the CASE tools and development techniques employed by developers, and participated in the use of these tools and techniques alongside members of the development team. This was a significant departure from usual practice in Telecom Éireann. In addition, user representatives also took an active role in the implementation of these systems. In the GAS project, for example, the user representative had sole responsibility for the technical implementation and rollout of the new system. Users who did not participate directly on development teams did so at individual interviews and at group sessions with the systems analysts and user representatives during the requirements analysis phase. In addition, these users also participated in prototyping activities. In both projects selected users were formed into teams in order to comprehensively test the developed systems.

Issues Impacting on the Effectiveness of User Representatives and User Groups in the Development Process

The high level and quality of participation in the GIS project was commented on by one developer: *"The team greatly benefited from the presence of user representatives. I was up to speed with user needs all the time."* These sentiments were strongly endorsed by developers in the GAS project also. In regard to both project teams, participating users were fully aware of the favorable attitude that developers had towards their contribution and responded accordingly. Nevertheless, developers in the GAS and GIS projects articulated the need for more active participation by certain users as it was felt that an increased level of participation by such users could have helped mitigate many of the contentious change management issues surrounding the implementation of these systems. Although formal and informal communication mechanisms existed within the development project teams to facilitate the expression of users' views on the systems development product to developers and business management—e.g. formal project meetings, user group sessions, representations via the user representatives and user project managers etc.—many users decided to make representations directly to their labour unions in order to have particular issues raised at meetings of the Computer Liaison Committee (CLC). Why then did users resort to such tactics given the availability of more direct mechanisms? A possible reason for this was offered by a user on the GIS team who was of the opinion that such mechanisms were inappropriate because many of the problems users had with the new systems were of a 'political' nature—that is, they were related to the changes wrought by the new systems on user work-related roles, remuneration, responsibilities and conditions, the automation of certain tasks by these systems, and so on. Users felt that the project team managers did not possess the necessary wherewithal to resolve these issues locally. In both projects, the development project managers were totally against modifying agreed system features to mollify particular users—so too were senior IS function managers. Even so, it must be said that, on the whole, users were eager to participate and get involved in developing the GAS and GIS—for a small minority it was only to find ways to protect the status quo, it was clear, however, that the majority had a keen interest in shaping and influencing their future working lives.

Although the CLC acted as a formal forum for thrashing out political issues that arose as a consequence of systems development, the airing of such matters were not confined exclusively to this body. The development-related group workshops provided a quasi-formal mechanism for highlighting such issues because of the manner in which the workshops were constituted. Each workshop consisted of developers from the relevant project team and end-users from one of the user constituencies with an interest in the development process. The group workshop sessions on the GAS project tended to be used as a platform for political infighting between different user constituencies, as users from participating groups would introduce arguments to oppose or alter system features favored by users from other groups and operational areas who were not present at the individual group sessions. Users in all groups also tended to play on the known objections of absent groups in order to influence development outcomes—in terms of system features that supported planned changes to existing business processes—in their favor. Because of the degree of political conflict between the various groups, the user representative on the GAS project observed that there was a need *"to have all the user groups affected by the systems development project present at each of the workshops; this avoided the emergence of a 'them and us' situation between users, and between users and developers."* In the GIS project a similar situation existed in that there was a clear conflict of interest between the network planners and the draftsmen. But, as one developer pointed out, *"they seemed to pull together fairly well and put their differences aside while participating in the development of the GIS...their grievances were with the company, not with each other, so they left it to their respective unions to deal with."*

Choosing the 'right type of user' to participate in systems development was a problem that exercised the minds of business and IS managers and developers alike prior to the commencement of systems development on the GIS and GAS projects. IS managers and developers were eager to secure the most knowledgeable and proficient user project managers and representatives in order to make their *"lives that much easier"* in arriving at a full set of user requirements and in converting these requirements into a system that would be accepted by the business constituency. At a time when developer resources were scarce, issues like developer productivity and project life span were uppermost on the minds of IS function managers. This led one senior manager in the ITD to argue that *"if the ITD were going to commit scarce and valuable resources to a project, then business managers should do likewise."* In the GAS and GIS projects the selection of user representatives was perceived as a key issue due to the active role that they were expected to play throughout the development process, and in the subsequent testing and implementation of the developed systems. On the business side, however, it was clear that managers had to balance the need to maintain their most experienced people in key areas, with the need to ensure that the new systems would adequately capture reflect business needs. On top of this, the staff labour unions also exercised a significant say in the choice of operational user selected to participate in systems development as either members of the user groups or as user representatives. It is important to note, however, that the existence of a committed project sponsor on the business side for both the GAS and GIS projects was seenby developers and users to be decisive in all this, in that it was at this level that the final decisions were made in regard to the quality of user representative and the makeup of the user groups. In any event, it was evident that developers were quite pleased with the choice of user representatives and user group members in both projects.

The coordination and control of developer and user development-related activities was uppermost on the minds of the project managers involved. One scarce resource in any project is time, and the more time developers spend with users, the less time they spend developing. Project managers were mindful of this and set out to manage closely these activities in order to achieve the right balance. In any event, this problem was largely overcome by the fact that both user representatives and developers often shared the same tasks, especially in the requirements analysis and design phase. As expected, regular project meetings helped developers to keep abreast of each others' progress and activities; and the joint nature of such meetings provided user and development project managers with an opportunity to keep both user representatives and developers abreast of external issues such as industrial relations problems etc. This forum also provided a mechanism for user representatives to air their views on the manner in which the development of the GAS and GIS systems was proceeding and on the operational features of the emerging systems. Here, user representatives could formally convey the views and wishes of the user constituencies whom they represented. It must be noted, however, that because user representatives were expected to participate closely with developers and, indeed, often perform the same tasks as the developers on the project teams, there was, as one user put it, *"the danger of going native"* and losing not only sight of their own role and purpose but, also, credibility with the user constituencies whose views they were supposed to represent. The user representative on the GAS team therefore spoke of the need to *"take [him]self out of the immediate development environment, and get back into the field"* in order to maintain contact with his work associates.

This problem of *"going native"* also applied to the GAS and GIS user project managers, but was to some extent mitigated because the development teams were sited at the main business center where they maintained their office accommodation—this allowed user project managers to maintain close formal and informal social contact with their peers. As previously indicated, user project managers were responsible for the provision of project-related accommodation, materials and facilities on the one hand, and for the implementation and testing of the developed systems on the other. They also managed the user resource throughout the development project and looked after user training once completed. There was also a very important role to be played by them in ensuring continuing commitment by business management to development goals and objectives. In the past, development teams had experienced difficulty in obtaining the required level of user involvement in many of the above areas and welcomed the contribution that business managers could make as user project managers.

Pan-lifecycle End-User Participation and the Benefits of On-site Development in the Business Area

Pan-lifecycle participation in systems development refers to the active participation and involvement of end-users at practically all stages of the development process. As indicated previously, developers, project managers, and senior IS managers in Telecom Éireann considered this approach to system development to be vital for the success of their development endeavours. Nevertheless, a senior IS manager pointed out that the benefits which accrue from this approach to user participation really depend on the type of business process being supported by the target system, and on the level of social and technical complexity of the development process. The business process supported by the GAS, for example, had a low to medium degree of task structure—in other words, the day-to-day tasks of operational users were neither well-defined nor highly structured—and the development project possessed a moderate level of technical complexity. The GIS system was similar in that the business process it supported possessed a medium degree of task structure, but the technical complexity of development process and its product was relatively high. In addition, both projects clearly exhibited a high degree of social complexity as users from several functional areas were involved in the development of these systems. Taken together, then, these factors indicated that a high degree of user participation was required because of the need to accommodate in-depth the views of several 'competing' user constituencies. The term 'competing' is employed here because users from the different functional areas involved in systems development of the GAS and GIS felt that they were involved in what could be described as a 'zero sum game' for control over the business process.

Another major consideration in adopting this particular approach to user participation was that due to a scarcity of developers in the IT directorate, users were encouraged to become more actively involved in systems development activities, particularly at the design and implementation stages. In addition, IS managers considered that the type and degree of user participation they desired could only be achieved through the policy of on-site development at the business users' offices—indeed, both projects were housed on the same floor in one of Telecom Éireann's Dublin business centers. Prior to the development of the GIS and GAS, most systems development took place off-site, that is, within the IT directorate's own place of business. Senior IS and development project managers recognized that there were significant benefits to be gained from on-site development at business users' own accommodation. For example, IS managers thought that this policy would provide additional opportunities for informal and indirect user participation, thereby improving user/developer communication and fostering good relations at all levels. Certainly, the level of formal and informal contact that resulted from developers and users sharing the same office space confirmed this belief. As the GIS development project manager pointed out, *"day-to-day contact between all concerned facilitated the growth of a community spirit between [his] team of developers and the users involved in the project."* Having developers and users in such close proximity also helped negate some of the ill feeling toward the new systems and the concomitant change to existing business processes as users perceived such change to be coming from within, rather than being thrust upon them by an outside agency.

User Participation and the Use of Technology for Systems Development

The traditional approach to information systems development in Telecom Éireann has been centered on internal development of handcrafted, custom-built solutions using the systems development life cycle (SDLC). The traditional approach was modified for the GAS project in that IT, in the form of an integrated computer aided systems engineering (I-CASE) workbench or application development environment (ADE), was employed throughout the development process in what was a prototyping-led rapid application development approach (RAD). Some use of older CASE technology was made on the GIS project, but only in the diagramming and modeling activities in the requirements analysis phase. User representatives on both projects received training in the use of these technologies. The user representatives were at one in stating that such technologies facilitated a more active role for them as participants in the development process; however, the type of technology employed on the GAS project was deemed to be of greater benefit in this regard. Developers too commented on the improved level of user/developer communication in these projects over previous development projects where CASE technologies were not employed. In a sense, communication was improved because the CASE technology enabled a sharing of 'world views' between developer and users on the properties of the emerging system. It did this by offering a common schema or language that mediated or negated the traditional schism between technically-oriented developers and business-oriented users. This feature or benefit was even more apparent with the ability of the I-CASE tool-set to prototype the new system by allowing the user representative and members of the user groups to design elements of the system in close cooperation with developers. Thus, the systems requirements for the GAS were elicited and refined through a dialectic process that saw developers and users jointly decide on what was both technically feasible and desirable in terms of system functionality. The lesson from this experience was that technology, far from excluding users from a role in the development process, required more input and participation by users and hence helped realize many of the benefits that user participation can bring to systems development.

User Participation and Management of Change in Systems Development

Project managers, developers, and users all agreed that the issue of change management associated with the implementation of the GAS and GIS systems exerted a critical influence on the trajectory of the development process and its outcomes in terms of the difference between the planned systems features and those contained in the systems as developed and implemented. Of course, all this had a concomitant impact on the business processes to be supported by the new systems. For example, the user representative on the GAS project reported that *"staff at the fault handling centre felt that their jobs were being whittled away and that the control of the fault handling system was being shifted to the repair teams.*" This situation engendered a negative attitude towards the new system within this user constituency and strongly influenced the deliberations of the Computer Liaison Committee (CLC) in having the system's features modified somewhat in order to arrive at a suitable compromise for users in this operational area.

Even though the development project teams were embedded within the user community, and user groups were employed in the elicitation/verification of requirements, in what could be described as a fully participative development exercise, industrial relations problems arose in relation to both systems, as developed, when the time came to implement them. Although the GAS had been accepted as developed by all the constituencies of interest, the CLC over-rode decisions taken and agreed to by the user representative and the user groups. This situation occurred despite the fact that one of the CLC's members had been involved throughout the development process as a participating member of the one of the user groups. A developer on the GAS project provided an explanation for this and reported that influential users who did not participate in the development process had voiced their *"unhappiness with system features...[and that this had] prompted the CLC to say no to the implementation of the system."* Hence, prior to its implementation at a trial site, several modifications had to be made to the GAS in order to address these objections. A very similar scenario existed in relation to change management issues that arose during the life of the GIS project. Although business managers in the relevant areas were aware of the potential for significant change management problems to develop when the system was implemented, they took no action prior to the development of the GIS to address these issues. These problems related to the radical nature of the change in work-related roles, responsibilities and remuneration of one of the user constituencies involved, and although these users were satisfied with the system as developed, they were unhappy with the consequences of the system's introduction. Therefore, the absence of adequate managerial attention to the issues of change management meant that, although both systems were developed with the active cooperation of users, both projects encountered user-related obstacles and resistance at the implementation stage. This proved to be the major weakness with the company's approach to user participation for systems development, and was an issue which would have to be addressed if the CEO's ambitious IT-enabled strategy was going to achieve any measure of success.

The general characteristics of user participation in both development projects described in the case accords well with the model presented in Figure 1. Taking into account the radical changes that were planned to the company's structure and business processes, the major problem facing the company's managers was how to preserve participative practices that were productive, and which would help give effect to its strategies, and eliminate those that were counter productive. The following section discusses such issues.

CURRENT CHALLENGES/PROBLEMS FACING THE ORGANIZATION

What has been described in this case study is perhaps an exemplar of the institutionalization of user participation for systems development. In both development projects there was a happy ending in that the systems were implemented and have been successful in use. There was, however, *"a fly in the ointment,"* as one IS manager put it, in that the considerable effort applied by developers and users in the construction of two key operational support sub-systems almost came to naught due to the mismanagement of change. Despite the high level of commitment by all parties to participative development practices, and the positive influence this has had on the culture and climate of the development environment and, also, on the development trajectory of the GAS and GIS systems, when it came time to implement these systems within the organization to support the proposed new business processes, significant problems arose because of the disagreeable change that the systems would bring to the work-related roles and responsibilities of operational staff. The company's senior managers felt that business management and the labor unions should share the burden of responsibility for this state of affairs: business managers, for example, were unwilling to address change management issues head on and, instead, hoped that such issues would somehow be resolved during systems development by the user and development project managers; on the other hand, the unions were unreasonable and unrealistic in that many of their demands were overly influenced by the views of a powerful minority of their members. This was the major lesson learned by both Telecom Éireann's senior management team and labor union leaders from their experiences with user participation in projects like the GAS and GIS. Union leaders and company management recognized that, because of the harsh competitive realities of the telecommunications business in the 1990s, the issue of change management was going to be critical to the success of the company's strategic goals and objectives. Hence, as a consequence of past experience, it was something that the company and the unions took great pains to address when IT was employed to enable the transformation of the organization from 1998 onwards.

To recap then, it is clear that while the institutionalization of user participation, as described in relation to the GAS and GIS projects, was effective in purely development terms, it had, however, very little impact on the issue of change management in the implementation of these systems. Nevertheless, among the benefits of this approach to systems development were that participating users were empowered to take an active role in all development activities and non-participating users had their perspectives and interests taken into consideration; as consequence of this, users developed favorable attitudes toward the new systems and were committed to their use once the change management issues were resolved. It is also clear that the successful elicitation of what were complex requirements in the GAS and GIS were better apprehended and understood using this approach. Furthermore, the pan-lifecycle nature of participative development ensured that the end product closely matched user requirements and, hence, facilitated a high level of user satisfaction with the developed systems. It must also be noted that, given the significant challenges facing the company in developing and implementing what is a formidable portfolio of proposed systems over a very short time frame (1998-2002), Telecom Éireann is fortunate to have developed a competence in mining and applying its human resource in this area, as the experiences recorded in these two examples indicate. However, the problem facing Telecom Éireann's management and its partners, the labour unions, was: 'How could they maintain and build on the benefits of user participation and overcome what could become a fatal *'Achilles heel'* for the company—that is, organized resistance to change by its employees coupled with poor management of change by the company's business managers?'

To address the problems associated with resistance to change and transformation management, the labour unions proposed a blanket, up-front agreement to cover all future IT-enabled change to the company's business processes, whatever the consequences for the staff concerned, in exchange for a 15% stake in the company for their members. As indicated in the introductory sections, this has been agreed to and has been implemented as of mid-1998. It is unlikely then that the type of industrial relations problems which plagued the implementation of the GAS and GIS will arise in relation to the introduction of planned future systems. Still, this will not mitigate the potential for conflict between different constituencies of users. Developers will still have to be aware of and sensitive to such issues, particularly in their potential to affect the product more so than the process of systems development. As of late 1998, business users are participating in development projects, very much as in the GAS and GIS, which will see their role-related responsibilities and remuneration altered significantly. In addition, many of the affected staff will have to transfer to new positions within the organization or will have to consider voluntary redundancy. The role that the unions play has also changed: for example, the CLC still exists, but it is more of a facilitator of change, rather than an participatory forum that has the potential to be a source of problems for the systems development process and its product. Once agreement had been reached on staff cooperation in the transformation of Telecom Éireann, the challenge facing the organization's managers was how to develop and integrate the proposed IT infrastructure with a limited developer resource? In order to address this challenge, the IT director and his associates in other corporate directorates adopted an innovative participatory strategy that builds on and extends that described in the case.

At present, the IT directorate is developing what it terms as its 'Future Methods of Operation' (FMO); this approach is based on its experience in reengineering corporate business processes in conjunction with Bellcore (a US telecommunications company) and the Business Process Design directorate of Telecom Éireann. As part of its strategy, the ITD has instituted a 'buy versus build' policy that sees developers source 'off-the-shelf' or 'canned' application packages. The focus has therefore changed to the integration of custom-built, 'turnkey' solutions and 'off-the-shelf' third-party vendor applications. The practice of comprehensive user involvement in the development of organizational information systems is to be maintained and extended in new and innovative ways. It is clear, however, that quite apart from those systems that will be developed in-house in their entirety, the scope for user participation in systems will be developed externally and will be limited. This poses a challenge for the IT and business directorates to ensure that such systems meet user needs and requirements. However, potential problems here may be mitigated by a new user-centric initiative in the area of the provision of corporate IS. As part of the recent corporate restructuring a decision was taken to decentralize many systems development activities to IT sub-functions within each business directorate in the Telecom Éireann group, for example. The new IT sub-functions will consist of management and staff from user constituencies who will act as project managers, business analysts and systems implementers and who will liaise with project managers and developers from the IT directorate and with outside vendors in the planning, development and implementation of new information systems. The roles of user project managers who are seconded to the IT directorate to participate in systems development will remain as previously described in the case, however business users who will take up positions as project managers in the IT sub-functions of business directorates will have extra responsibilities viz.

- Preparing the business case for development
- Identifying user requirements
- Process mapping
- Preparing request for proposals (RFPs), evaluating responses and recommending solutions
- Developing and implementing project plans
- Communicating with the staff involved/affected by the project
- Liaison with the IT directorate and external suppliers
- Ensuring delivery of system components to agreed time scales
- Acceptance testing of system developments to specified requirements
- Co-ordination of IT training
- Managing risks and resolving implementation-related problems
- Preparing reports and presentations for key stakeholders
- Hand-over of live system.

In all cases, primary responsibility for systems development will still rest with the ITD.

In a new and innovative strategy that addresses the critical shortage of IS developers, and which provides business users with an opportunity to play a frontline role in the development of corporate information systems, satellite or decentralized IT development units are to be set up in specific geographical locations within the organization. Staff for these new units are to be recruited from the internal labor market and will consist of business users who have a predominantly technical background and whose present work areas are over-resourced. Once recruited, these individuals will complete a six-month long training program that will provide them with skills in systems analysis and programming. The IT director has stated that clear benefits will accrue from this strategy, as former business users will bring their technical and business knowledge to bear in their new roles as developers.

The foregoing then are some of the major issues being addressed by Telecom Éireann's IT directorate as it undertakes its task of planning, developing and implementing the information systems in its development portfolio. The pivotal role that participating business users play in the realization of these endeavors continues to be emphasized within the framework of industrial democracy.

BIOGRAPHICAL SKETCHES

Tom Butler is a telecommunications engineer with Telecom Éireann, Ireland's major telecommunications service provider. He holds an M.Sc. from the National University of Ireland, at Cork

(UCC), and is currently pursuing his Ph.D. at the same institution. He has been a visiting lecturer at UCC since1997. His research interests include information systems development, CASE, user participation, organisational change, and the implications of IT for the emerging knowledge-based theory of the firm. He has a particular interest in the application of the constructivist philosophies of Martin Heidegger and Hans Georg Gadamer for research on information systems. His research has been published and presented at several international conferences and he has a number of forthcoming book and journal publications.

Brian Fitzgerald is senior researcher at the Executive Systems Research Centre (ESRC) at University College Cork and is currently an associate editor for The Information Systems Journal. He holds a PhD from the University of London and he is actively involved in applied research projects in the areas of systems development approaches, foundations of the IS field and executive information systems. His work in these areas is published in various books and international journals, including The Information Systems Journal, Information and Management, INFOR, the Journal of Information Technology, and the International Journal of Information Management. He has also presented his research at international conferences and seminars in Europe and North America. Having worked in industry prior to taking up an academic position, he has more than 15 years experience in the IS field.

Appendix

Exhibit 1: Organisational Chart for Telecom Eireann (as at 1998)

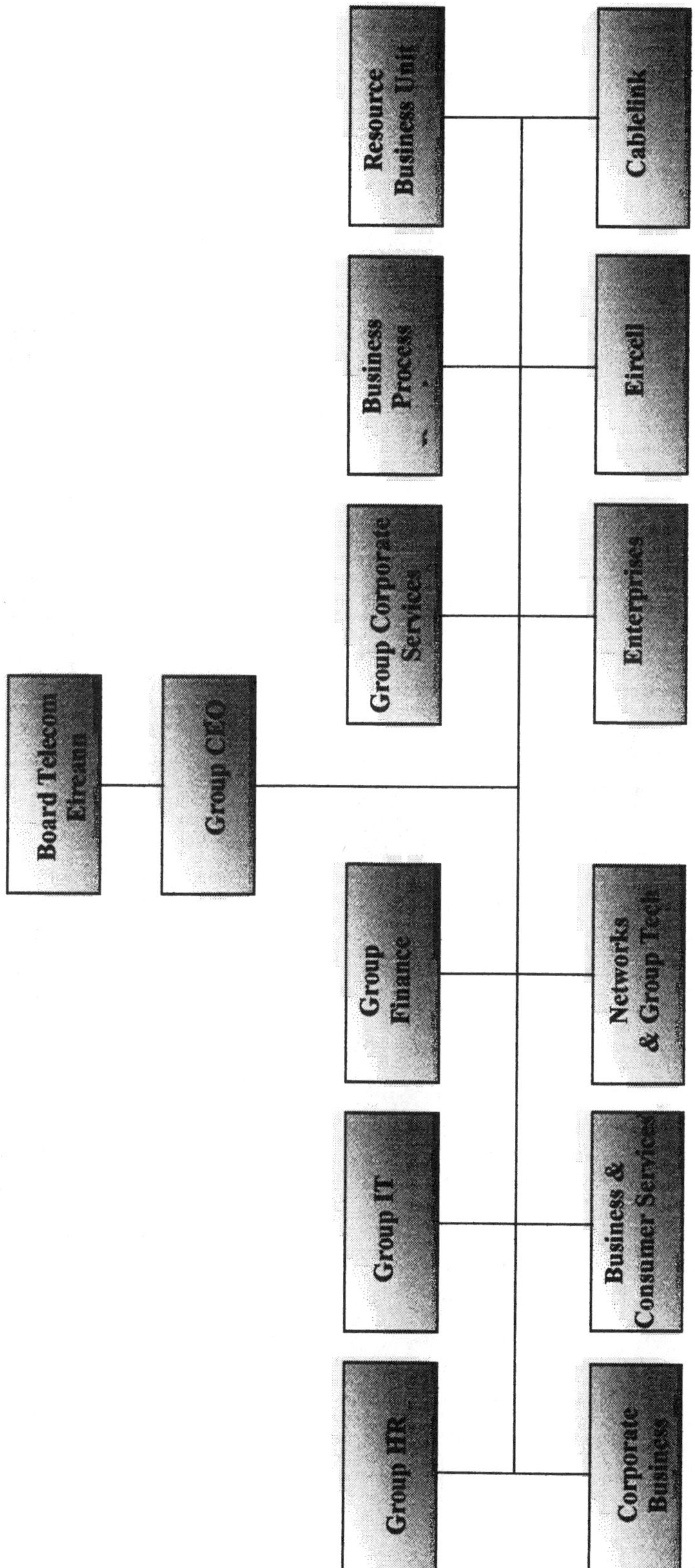

Exhibit 2: IT Directorate (as at 1995)

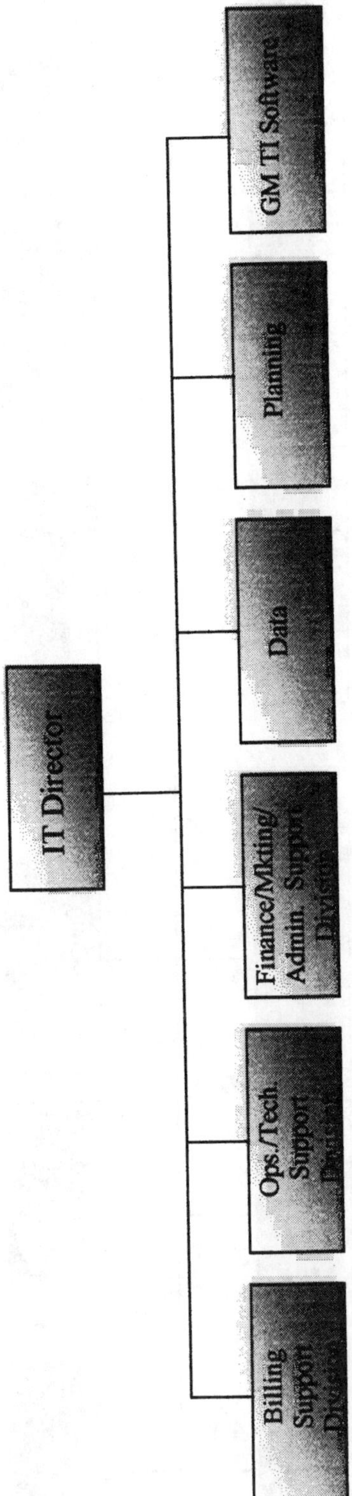

Exhibit 3: IT Directorate (as at 1997)

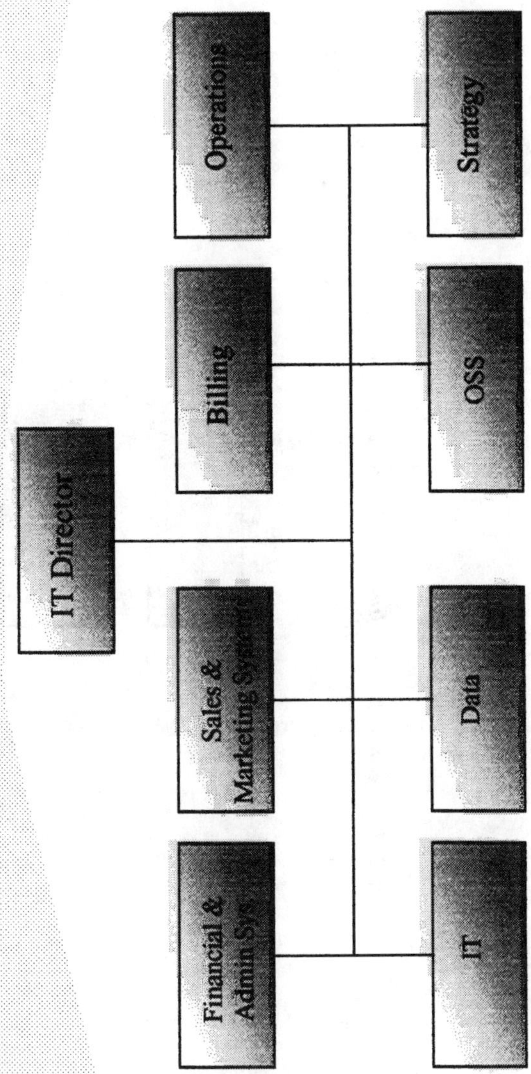

Privatizing Military Housing Through Information Technology

Guisseppi A. Forgionne
University of Maryland Baltimore County, USA

EXECUTIVE SUMMARY

The armed services must provide its personnel with acceptable housing at minimum cost within the vicinity of military installations. To achieve these housing objectives, the Department of Defense (DOD) has entered into experimental joint ventures with private developers to construct attractive housing projects on military installation property, with some of the projects reserved for military personnel. To support the analysis of the joint ventures, the DOD needed a methodology that would help officials evaluate the feasibility and cost implications of the housing projects. A decision support system, called the Housing Revitalization Support Office System (HRSOS), has been developed to provide the necessary support.

The HRSOS architecture is based on a combination of database, econometric, simulation, and decision support techniques. Its deployment can help the Department of Defense to realize significant economic and management benefits. Future enhancements, motivated by the challenges from the current system, promise to increase the power of HRSOS and to further improve the DOD's ability to manage its housing projects.

To obtain the benefits, the HRSO experience suggests that system design, development, and implementation should be a team effort through an adaptive design strategy. It also indicates that an integrated suite of software development and implementation tools, offering rapid prototyping, computer assisted software engineering, and object-oriented analysis, can promote this strategy. The strategy is likely to work well in a hybrid project-technology virtual organizational form that is established and administered by the practicing top manager.

BACKGROUND

The United States Department of Defense (DOD) consists of the armed forces (mainly the Air Force, Army, Marines, and Navy) and several support offices (most notably the Inspector General, which serves as the major fiscal oversight group for the DOD). Technically, the President of the United States is the head of the DOD. In effect, however, a political appointee, the Secretary of Defense, is the DOD's chief operating officer. There are several deputies, and many department chiefs, who assist the Secretary in performing her/his responsibility.

The main roles of the DOD are to provide a national defense, enforce foreign policy, promote national interests, and support U. S. allies. A force in excess of 2 million civilian and military employees implements strategies, policies, and procedures that fulfill these roles. Such implementation requires the establishment and maintenance of hundreds of military installations throughout the United States, Europe, Asia, and other places in the world. For security and other practical reasons, personnel (and often their families) must be acceptably housed on, or near, the installations. The Department of Defense (DOD) wants this acceptable housing to be within a one-hour commute of the military installations.

Historically, the thousands of needed housing units have been acquired, maintained, and managed through the Department of Defense with government funds. The process has become quite expensive, and military personnel and officials have not always been satisfied with the resulting accommodations. To improve the situation, the DOD recently has sought remedies from the U. S. Congress. In response, Congress passed the Military Housing Act of 1995, providing the DOD with a series of desired authorizations.

In search of improved efficiency and effectiveness, public officials have proposed and experimented with the privatization of local, state, and federal government services (Chi, 1998; Brunsdon and Crossmit, 1998; Kotlikoff, Smetters, and Walliser, 1998). Governments have contracted with private organizations to collect garbage, house prisoners, build structures, and perform other essential services (Boubakari and Cossert, 1998; Helsley and Strange, 1998; Haarmeyer and Moody, 1998). Successful privatization efforts influenced the U. S. Congress to include such an initiative in the 1995 Military Housing Act.

Partially because of government budget cutting, there has been significant movement within the armed services to privatize military housing. Top-level policy makers realize that all the services can benefit from privatization initiatives. Using the 1995 Act as authority, the DOD established an agency, called the Housing Revitalization Support Office (HRSO), to develop innovative initiatives that would help achieve the housing objective. This office has a Director, four deputy directors, and a support staff of three technical specialists, one secretary, and eight armed service representatives at its suburban Washington, D. C. headquarters. There are no true IS/IT personnel in the office.

For the past five years, HRSO's management group has planned housing initiatives, developed procedures to implement the plans, and then communicated the procedures to the housing managers at each military installation. Such policies, procedures, and actions are audited by Department of Defense, Government Accounting Office, Office of Management and Budget, and other government agencies for compliance with existing laws, regulations, and guidelines. Since audit reports can significantly influence available funding, HRSO typically is very responsive to auditor suggestions on housing initiative policy and practice.

Installation housing managers collect data pertinent to the planning process, communicate the data through various information systems to the HRSO, implement HRSO-developed procedures, and administer onpost assets. Traditionally, these installation managers have been given much discretion in exercising their responsibilities. Moreover, the HRSO has relied heavily on installation managers' input in formulating housing initiative policies, procedures, and practices.

Figure 1 gives the organizational chart relevant to HRSO housing initiative management. Currently, managers in this organization influence decisions about approximately $30 billion worth of onpost housing assets. HRSO has an annual budget of about $7 million, and a long-term allocation of approximately $10 billion to design, develop, and implement housing projects that would affect these assets.

As Figure 1 indicates, the HRSO organization is largely hierarchical in nature. For the most part, major decisions are made by the Director, with considerable input from the deputies, service representatives, and installation managers. Such input is sought after and valued by the Director. Supporting analyses are implemented and managed by the deputies and service representatives. Some, but not all, of these deputies and representatives are IS/IT literate, but none are IS/IT experts.

SETTING THE STAGE

At any military installation, the projected supply of available government housing may be insufficient, or of inadequate quality, to meet the personnel demand expected at the site. To reduce,

Figure 1: Housing Revitalization Support Office (HRSO) Organizational Chart

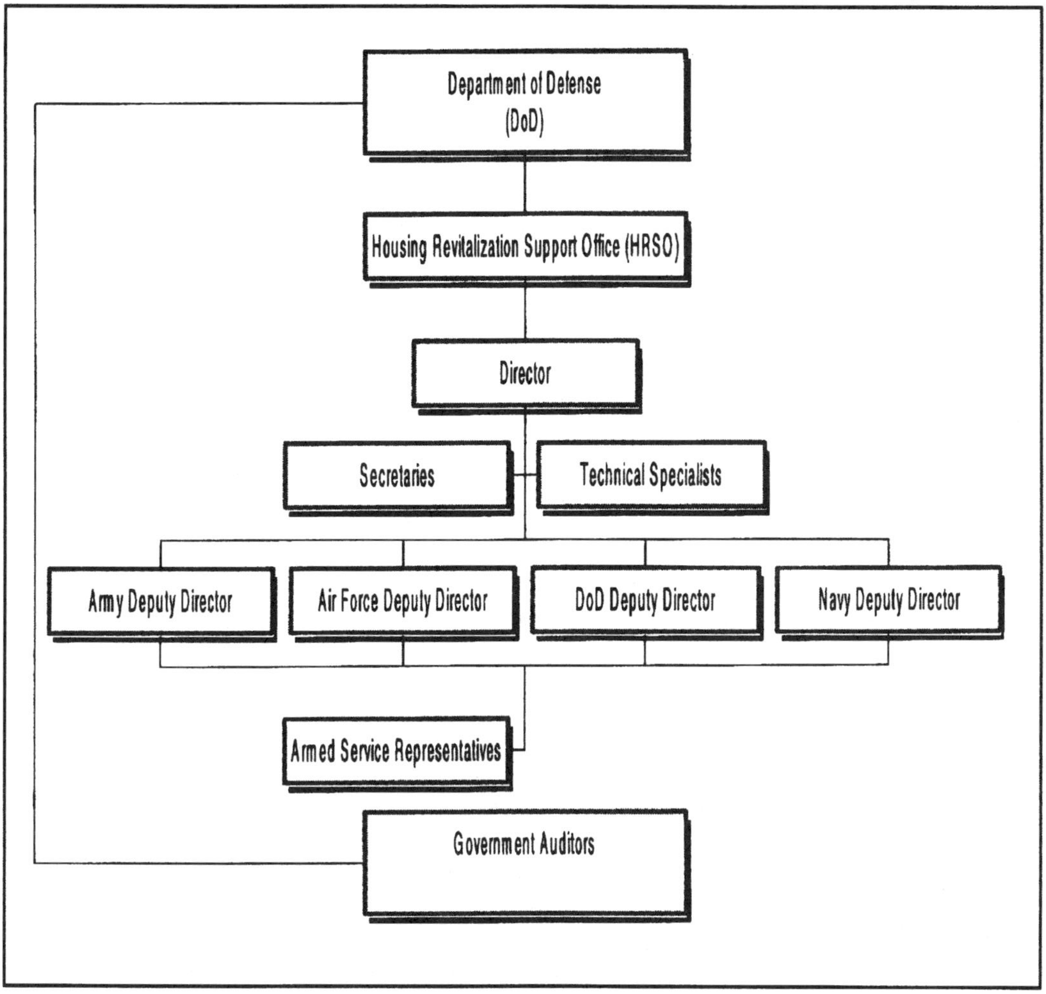

or possibly eliminate, this deficit, the DOD's Housing Revitalization Support Office (HRSO) has developed experimental programs for revitalizing the stock of military housing by attracting private capital investment. In these programs, private developers construct attractive housing projects on military installation property, and the federal government provides subsidies, loan guarantees, or other financial commitments to the developers.

The Federal Credit Reform Act of 1990 established a system of budgeting that requires agencies to: (a) measure a program's total cost and (b) set aside budget authority sufficient to cover the commitment. Furthermore, the Manual on the Federal Budget Process and the Office of Management and Budget (OMB) Scoring Model establish procedures and processes to measure the dollar amount to be set aside. As a result, any HRSO program must be supported by detailed and systematic analyses of the joint venture's financial feasibility and potential credit impact on the federal government. Such analyses involve the process shown in Figure 2.

Venture Capital Analysis

Any venture capital analysis begins with a forecast of the number of military personnel renting or owning offpost. Next, HRSO officials forecast the number of military personnel that ordinarily will be placed in onpost government-owned housing. These two forecasts become the basis for predicting the number of personnel expected to rent newly constructed or renovated onpost housing after the implementation of the venture project.

Figure 2: HRSO Venture Capital Analysis Process

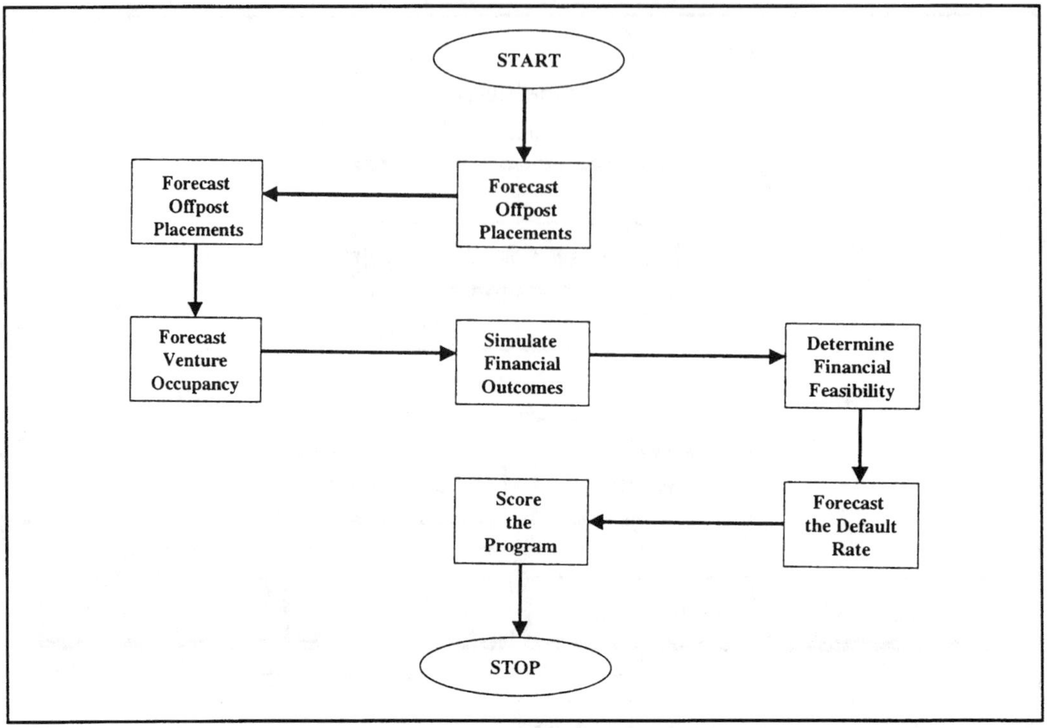

Multiplying the expected rentals by the soldiers' housing allowances gives the revenue generated from the venture. This revenue less the corresponding operating and management cost provides the annual cash flow yielded by the program. Using these flows over the housing's life, officials can simulate the internal rate of return and net present value from the investment. Such financial outcomes then determine whether or not the project is financially feasible.

Once financial feasibility has been established, HRSO must forecast the likelihood of default (the default rate) by the private developer of the onpost housing project. This default rate, in turn, will be entered into the OMB scoring model for potential credit impact on the federal government. Scoring seeks to ensure that the actual and expected costs associated with the use of each program are fully recognized in both budget and financial reporting. Every potential DOD housing revitalization program must be "scored" by the OMB model.

Process Support

The Statement of Federal Financial Accounting Standards (SFFAS) No. 2 requires the establishment of a systematic methodology, such as econometric modeling, to project default costs of each housing revitalization program. The methodology should create an audible procedure that: (a) formalizes and documents the subsidy authority performance and assumptions, and (b) defines the key relationships between the program performance, economic and other private and military housing indicators.

The HRSO has neither the staff, nor the internal technical expertise, to fully comply with the SFFAS, Federal Credit Reform Act, Manual on the Federal Budget Process, and OMB Scoring Model analysis requirements. Consequently, officials have relied on large consulting firms to support the venture capital analysis. These firms have assigned knowledgeable domain experts to collect the necessary data, generate projections, and convert the projections into financial reports and OMB scores. Such support is offered on a fee-for-service basis.

Data are collected from military information-system-generated status reports and through interviews with HRSO personnel and military installation housing managers. Projections are generated by simple extrapolation of current parameter values. Financial outcomes are simulated with a spreadsheet-based model that replicates defined relationships between expected rentals, cash flows, project investments, and returns. The project's default rate is estimated mainly through the collective opinion of the domain experts. A score is obtained by substituting the estimated default rate and other forecasted parameters into the OMB scoring model.

As part of the fee-for-service contract, the consulting firms provide the software, maintenance, and technical help needed to capture and store the collected data, project parameters, simulate financial outcomes, and generate reports. As in many organizations, then, virtually all of the IS/IT requirements are outsourced to these firms (Sridhar and Balachandran, 1997). All IS/IT tasks are initiated and managed by the consultants, with some oversight by HRSO deputies and service representatives.

Support Deficiencies

The consultant-based support is labor intensive, time consuming, and costly. Typically, a variety of senior HRSO officials, military housing officers, and private consultants take many weeks to gather the pertinent data, forecast parameters, perform the spreadsheet-based financial analyses, and report and document the venture capital analysis. During the process, the agency spends approximately $60,000 per installation in consulting fees.

There are also conceptual flaws in the consultant-based support. The extrapolation forecast assumes that current patterns will persist, and the methodology offers no explanation for the underlying pattern. In addition, the spreadsheet simulation is based on restrictive subjectively determined assumptions and an averaging process that may be unrepresentative of actual housing and financial conditions. These flaws make it difficult to provide the objectively based evaluations desired by government regulators, and they could result in erroneous project decisions.

CASE DESCRIPTION

The conceptual flaws, high cost, and potential adverse project consequences encouraged HRSO officials to investigate alternative forms of support for the venture capital analysis. Campaigning by a visionary service representative at HRSO funded an external study that researched the use of internal IS/IT tools for such analysis. This research has resulted in the development of improved forecasting and simulation models and the ***Housing Revitalization Support Office Systems (HRSOS)*** decision support system for delivering the consequent knowledge in a timely, cost effective, and transparent manner to affected officials.

HRSOS Architecture

HRSOS was developed iteratively, using the Adaptive Design Strategy (ADS), by two university researchers working in conjunction with HRSO officials, OMB support staff, and the consultants. During the development, the researchers, usually in a collective manner, reviewed the latest roll out with the participating officials. Comments and suggestions for enhancements were tested and evaluated in real time, typically in prototype form using rapid application development (RAD) toolkits available within the development software. Detailed technical changes were made by the researchers over a relatively short period of time (typically a few weeks), and the system changes were presented to the officials for review and evaluation. The evolutionary process continued until the latest version of the system was finalized. The entire process took about six months.

The system is made available through an easy-to-use computer system with the conceptual architecture shown in Figure 3. As this figure shows, the decision support system interactively processes inputs into the outputs desired by HRSO officials.

Inputs

HRSOS has a data base that captures and stores economic variables and housing characteristics for the military installations' Housing Market Areas (HMAs), relevant onpost data, and pertinent financial conditions. Economic variables (such as the poverty rate and average population age) and housing characteristics (such as average rents and housing stocks) are collected from the U.S. Census.

Figure 3: HRSOs Conceptual Architecture

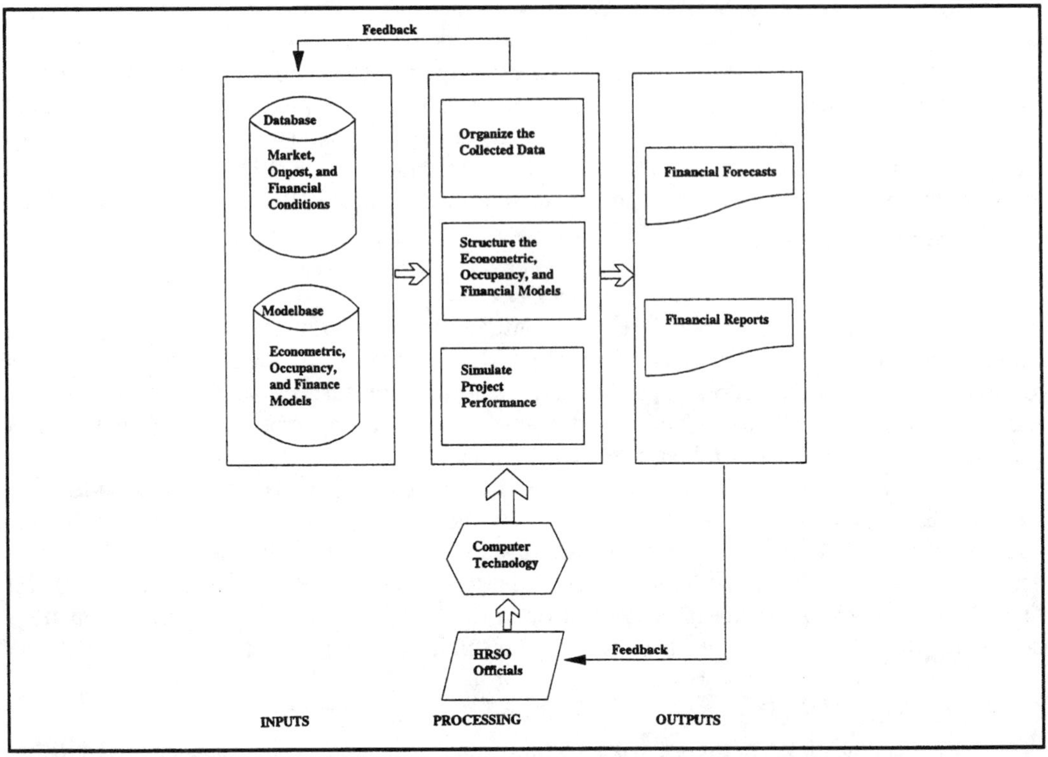

The onpost data (such as installation populations and military housing allowances) are extracted from other military information systems. Financial conditions (such as borrower characteristics and the effective interest rate) are obtained from: (1) the U. S. Department of Housing and Urban Development's (HUD's) Consolidated Single Family Default Monitoring System and (2) FHA Trends of Home Mortgage Characteristics for Section 203(b) Mortgages Insured. Economic theory, officials' judgment, previous HRSO analyses, and preliminary statistical analyses suggested the data to capture and store.

There is also a model base that contains statistical procedures, data conversion, econometric, occupancy, and finance models. The statistical procedures are used to categorize attribute data within the HMAs and to calculate summary statistics for the economic variables, housing characteristics, and financial conditions within the HMAs. Data conversion formulas are used to convert the captured economic, housing characteristic, financial condition, and military data into the variables needed for HRSOS's analyses, evaluations, and reports.

One econometric model forecasts the number of military personnel renting or owning offpost. The offpost forecasts become inputs into an occupancy model that predicts the number of military personnel that can be expected to rent newly constructed or renovated onpost assets after the venture project has been implemented. A simulation model transforms the onpost rental projections into financial outcomes. These outcomes, which include the cash flows, internal rate of return, and net present value from the venture project, predict the financial feasibility of the housing project. A second econometric model predicts the default rate that can be expected from the program (housing revitalization tool) used to finance the venture project. A score is obtained by substituting the estimated default rate into the OMB scoring model.

Processing

The decision maker (a HRSO official or staff assistant) uses computer technology to perform the financial analyses and evaluations. Currently, the system executes on an IBM-compatible

Pentium-based microcomputer with 16MB of RAM, a color graphics display, and a printer compatible with the microcomputer. It runs the SAS information delivery system through the OS/2 (or Microsoft Windows) operating system. Computer assisted software engineering was implemented through SAS/BASE, SAS/ETS, SAS/ASSIST, SAS/AF, and SAS/FSP, while object-oriented analysis (OOA) concepts were supported through SAS/FRAME. This configuration was selected because it offered a more consistent, less time-consuming, less costly, and more flexible development and implementation environment than the available alternatives.

Acting as an electronic counselor, the decision support system sequentially guides the user through an effective financial analysis and evaluation. The system delivers pertinent expertise by: (a) including audit-supportable economic analyses, (b) automating the complex financial computations, and (c) generating policy-specified detailed supporting reports. System operations, which are performed in an intuitive, timely (typical five-minute-session), and error-free fashion, liberate the user to focus on the creative aspects of venture capital management.

As indicated by the top feedback loop in Figure 3, organized data, structured econometric, occupancy, and financial models, and performance reports created during HRSOS' analyses and evaluations can be captured and stored as inputs for future processing. These captured inputs are stored as additional or revised fields and records, thereby updating the data and model bases dynamically (Wijnhoven, 1998).

Outputs

Processing automatically generates visual displays of the outputs desired by HRSO officials. Outputs include return and default rate forecasts and financial reports that provide the justified knowledge and wisdom needed to effectively plan venture capital projects. These reports are in the form of tables that display forecasted net present values, internal rates of return, and default rates for various HRSO program (mainly direct loan and loan guarantee) initiatives. The user has the option of printing or saving the reports.

As indicated by the bottom feedback loop in Figure 3, the user can utilize the outputs to guide further HRSOS processing before exiting the system (Tuttle and Stocks, 1997). Typically, the feedback will involve sensitivity analyses in which the user modifies market conditions, program initiative parameters, financial terms, or other pertinent factors and observes the effects on financial outcomes.

CURRENT CHALLENGES/PROBLEMS FACING THE ORGANIZATION

Scientific testing revealed that HRSOS' models generated accurate forecasts of market conditions, housing requirements, and financial outcomes. Based on the test results, the Housing Revitalization Support Office decided to examine the potential value of HRSOS. The system has been in an evaluation phase for the past nine months. Preliminary results from this ongoing evaluation indicate that the decision support system will have significant economic and management benefits, present key challenges, and offer important lessons for the Housing Revitalization Support Office.

Successes and Failures

By utilizing the decision support system, HRSO officials can rapidly: (a) obtain accurate, consistent, objective, and reliable housing data, (b) analyze the local housing market and its impact on military housing, (c) perform error-free financial feasibility and default rate computations and evaluations, (d) justify the analysis with detailed reports, (e) examine the sensitivity of the analysis to local market and privatized housing policy changes, and (f) flag data and information deficiencies. These enhanced capabilities can enable HRSO officials to more efficiently and effectively manage the revitalization programs under their control. The gain in efficiency and effectiveness can save the DOD a total of

$30,000 per site x 60 sites = $1,800,000

The decision support system was developed and implemented for about $60,000. These cost savings are quite timely in light of federal budget restrictions and the military's current direction of large scale force reductions and base closures (Chu, 1996).

Despite these economic and management benefits, the HRSO still is evaluating the system. During this evaluation, the decision support system serves mainly as a backup to the consultant-based methodology. Such reticence can be traced to organizational inertia, internal politics, and the system's profound challenges to the organization's culture and climate (Ione, 1997; Tolsby, 1998).

Limitations and Challenges

During implementation, system experimentation educated users about the devastating consequences of inaccurate housing projections, while the development demonstrated the irrelevance and redundancy of much pre-innovation collected data. When implemented fully, the innovation will alter the work design for, and supervision of, the venture capital analysis process. Requisite operations and computations will be simplified, automated, and made error-free. Training requirements will be reduced to a minimum. Processing efficiency will be dramatically increased. User-inspired creative market-area and housing policy experimentation will be facilitated and nurtured. Management learning will be promoted. Knowledge capture will be expedited.

In short, HRSOS's usage would substantially reshape the organizational culture. Faced with significant time pressures and limited staff, HRSO leadership may be reluctant to take on this burden at the present time. In addition, officials have developed and cultivated strong and enduring relationships with the consultants. These consultants also have important contacts and allies within the government agencies that oversee HRSO programs. For these reasons, it may be politically wise for HRSO to preserve these consultant relationships.

There is a final data challenge. At the present time, there is limited data available to measure the default rate for HRSOS's second econometric model. These data limitations precluded a meticulous application of theoretical financial models and, as in the consultant-based analysis, resulted in potentially underestimated default rates for some military installations. If the politics can be worked out, the requisite data can be obtained from other government agencies, thereby eliminating this shortcoming.

System Lessons

Improved accuracy in forecasting housing privatization requirements is important to the military and society. Underestimating requirements can alienate civilian populations competing with military personnel for housing, leave thousands of military personnel inadequately housed, lower personnel morale and family well-being, and jeopardize military preparedness. Unnecessary housing construction and renovation can waste scarce natural resources and, in the process, alienate local residents, environmentalists, military housing managers, and other interest groups.

The potential of HRSOS may convince the DOD that the privatization initiatives can be substantially enhanced with decision support systems. Consequently, the HRSOS-supported venture capital analysis can be offered as the standard for the process. With adaptations, the HRSO's decision support system may also be applicable to potential privatization efforts of other federal, state, and local public housing authorities (You, 1998).

Regardless of HRSOS' legacy, the application offers useful lessons for decision support systems development and management. These lessons involve team composition, development tools, and organizational issues.

Team Composition. The HRSOS decision support system is effectively delivering to the user, in a virtual manner, embedded econometric, management science, finance, housing, and information systems expertise specifically focused on the management problem. Any single human technical specialist typically will not: (a) be proficient with, or even aware of, all pertinent tools, or (b) possess sufficient domain knowledge to fully understand the management situation, propose trials, or interpret outcomes (King, 1998). While practicing managers will have the domain knowledge, they usually will not have the technical expertise to effectively develop and implement relevant technology (Osterlund, 1997).

The HRSO experience suggests that decision support system design, development, and implementation should be a team effort. In addition, the team should be composed of the affected managers, information system personnel, and technological specialists proficient with the tools needed to address the management problem. When the personnel have decision support system education and training, the team may consist of relatively few people. For example, the HRSO needed a team of only four people. There were two top HRSO officials (who participated on a sporadic basis), one decision support system professor (with technical expertise in various information systems, econometrics, and management science), and a finance professor (with technical expertise in housing analysis).

Development Tools. The HRSO experience suggests that decision support system design, development, and implementation should be a team effort executed through an adaptive design strategy. It also indicates that an integrated suite of software development and implementation tools, offering rapid prototyping, computer assisted software engineering, and object-oriented analysis, can promote this strategy (McGibbon, 1997). While the SAS System for Information Delivery offers one such suite, there are others available (such as the FOCUS toolkit offered by Information Builders).

Organization Issues. Forecasting by top managers for planning purposes is inherently a semi-structured (or even ill structured) problem. When initially confronted with such situations, these managers have a partial understanding of the problem elements and relationships. Typically, their understanding evolves as they acquire more information, knowledge, and wisdom about the problem (Eardley, Avison, and Powell, 1997). Decision support systems are designed to support such decision-making. Relying on the information center, or other traditional information system organization, to design and develop a decision support system will likely be ineffective (AlBanna and Osterhaus, 1998). These types of organizations are staffed by personnel with general skills, limited technological expertise, and restricted problem-specific knowledge. Development and implementation will follow a prescribed pattern designed to provide standard solutions to relatively well-understood and well-structured problems.

At the start of the HRSOS development process, the research team was assembled by the visionary HRSO official. Each expert was assigned respective data location, extraction strategy, and system development tasks by this official. System development and implementation was managed by the HRSO official. Most work was performed at the sites of the technology specialists. Distributed collaborative work tools, most notably electronic mail and audio conferencing, were used extensively to communicate work in progress and manage the effort.

This experience suggests that the HRSO's hybrid project-technology organization may work well for decision support system design, development, and implementation. The organization would be virtual rather than physical. A project team would be established and administered by the practicing top manager. Team technology specialists would be drawn from within and outside the organization to match the expertise needed for the specific project (Joyce, McGee, and Slocum, 1997). Telecommuting and distributed collaborative work would be allowed and possibly encouraged.

FURTHER READING

Adelman, L. (1992). *Evaluating Decision Support and Expert Systems.* New York: Wiley.

Holsapple, C. W., & Whinston, A. B. (1996). *Decision Support Systems: A Knowledge-Based Approach.* Minneapolis: West.

Nonaka, I., & Takeuchi, H. (1995). *The Knowledge-Creating Company: How Japanese Companies Create the Dynamics of Innovation.* New York: Oxford University Press.

Salton, G. J. (1996). *Organizational Engineering: A New Method of Creating High Performance Human Structures.* Ann Arbor: Professional Communications, Inc.

Sauter, V. (1997). *Decision Support Systems.* New York: Wiley.

Sengupta, K., & Abdel-Hamid, T. K. (1993). Alternative conceptions of feedback in dynamic decision environments: An empirical investigation. *Management Science*, 39(4), 411-428.

Sprague, R. H., & Watson, H. J. (1996). *Decision Support for Management.* Upper Saddle River: Prentice-Hall.

Stair, R. M. (1996). *Principles of Information Systems: A Managerial Approach.* Danvers: Boyd & Fraser.

REFERENCES

AlBanna, S. J., & Osterhaus, J. (1998). Meeting the software challenge: A model for IS transformation. *Information Systems Management*, 15(1), 7-15.

Boubakari, N., & Cossert, J. (1998). The financial and operating performance of newly privatized firms: Evidence from developing countries. *The Journal of Finance*, 53(3), 1081-1110.

Brunsdon, J. D., & Crossmit, C. W. (1998). Privatization of Fairbanks municipal utility system. *Government Finance Review*, 14(3), 25-27.

Chi, K. S. (1998). Privatization in state government. *Public Administration Review*, 58(4), 374-376.

Chu, D. S. C. (1996). Resizing the federal government. *OR/MS Today*, 23(4), 38-41.

Eardley, A., Avison, D., & Powell, P. (1997). Developing information systems to support flexible strategy. *Journal of Organizational Computing and Electronic Commerce*, 7(1), 57-77.

Haarmeyer, D., & Moody, A. (1998). Tapping the private sector: Approaches to managing risk in water and sanitation. *Journal of Project Finance*, 4(2), 7-23.

Helsley, R. W., & Strange, W. C. (1998). Private government. *Journal of Public Economics, 69(2), 281-304.*

Ione, A. (1997). Information, description, cognition, invention. *Informatica*, 21(3), 423-434.

Joyce, W. J., McGee, V. E., & Slocum, J. W. (1997). Designing lateral organizations: An analysis of the benefits, costs, and enablers of nonhierarchical organizational forms. *Decision Sciences*, 28(1), 1-25.

King, W. R. (1998). IT-enhanced productivity and profitability. *Information Systems Management*, 15(5), 70-73.

Kotlikoff, L. J., Smetters, K. A., & Walliser, J. (1998). Social Security: Privatization and progressivity. *The American Economic Review*, 88(2), 137-141.

McGibbon, B. (1997). High performance through team building. *Object Magazine*, 7(9), 57-59.

Osterlund, J. (1997). Informatics for technology management. *International Journal of Technology Management*, 14(2-4), 415-438.

Sridhar, S. S., & Balachandran, B. V. (1997). Incomplete information, task assignment, and managerial control systems. *Management Science*, 43(6), 764-778.

Tolsby, J. (1998). Effects of organizational culture on a large IT introduction effort: A case study of the Norwegian army's EDBLF project. *European Journal of Information Systems*, 7(2), 108-124.

Tuttle, B., & Stocks, M. H. (1997). The effects of task information and outcome feedback on individuals' insight into their decision models. *Decision Sciences*, 28(2), 421-442.

Wijnhoven, F. (1998). Designing organizational memories: Concept and method. *Journal of Organizational Computing and Electronic Commerce*, 8(1), 29-55.

You, C. (1998). Privatization – A major strategy of public service provision in China: A case study of urban housing policy & management. *International Journal of Public Administration*, 21(9), 1363-1385.

BIOGRAPHICAL SKETCH

Guisseppi A. Forgionne is Professor of Information Systems at the University of Maryland Baltimore County (UMBC). Professor Forgionne holds a B.S. in Commerce and Finance, an M. A. in Econometrics, an M. B. A., and a Ph. D. in Management Science and Econometrics. He has published 23 books and approximately 100 research articles and consulted for a variety of public and private organizations on decision support systems theory and applications. Dr. Forgionne also has served as department chair at UMBC, Mount Vernon College, and Cal Poly Pomona. He has received several national and international awards for his work.

Reinventing Business Processes Through Automation: A Case Study

David Paper
Utah State University, USA

EXECUTIVE SUMMARY

Although numerous firms have experimented with business process reengineering (BPR), many have achieved less than dramatic results (Davenport, 1993; Hammer and Champy, 1993; Kotter, 1995). To explore possible reasons for BPR underachievement, an in-depth case study was undertaken. The case focuses on events surrounding the redesign of a fundamental business process (Personal Lines Insurance) at Safeco Corporation. Redesign is led by the Information Systems and Services (ISS) department as they are in charge of the Personal Lines Systems (PLS) that support the business. Analysis of the interviews provides a detailed description of the project and uncovers possible reasons for failure of the reengineering effort.

BACKGROUND

Safeco is one of the largest diversified financial corporations in the United States with annual revenues exceeding $4 billion. It was founded, as the General Insurance Company of America, in 1923. Its operating companies employ more than 7,500 people in 100 offices. Although property and casualty insurance is its largest operation, the company engages in life and health insurance, real estate management and investment, commercial credit, surety and asset management.

Safeco draws its strength from its employees, innovative technology, and a conservative investment philosophy. Safeco recruits employees with excellent communication skills, the ability and desire to learn continuously, and the ability to adjust to changing environments and technology. Safeco recruiting was discussed with a senior systems analyst. "The three top skills needed in the IS (information systems) profession are communication, learning ability, and adaptability. Written, oral, memos, e-mail, and voice mail communications are exchanged on a daily basis. Our field changes daily, so you must enjoy and be motivated to learn on a daily basis. You also need to be able to adjust to changing environments and technologies because things are discontinued, companies are acquired, and standards go away" (D. Green, personal communication, August 19, 1997).

SETTING THE STAGE

Case study analysis of Personal Lines Systems (PLS) redesign at Safeco Corporation began with a site visit on July 26, 1997. Data were gathered from interviews, annual reports, observation, e-mail, and informal discussions. Four people were formally interviewed at Safeco including the Vice President of Information Systems and Services (ISS), the Assistant Vice President of Field

Operations, the manager of PLS reengineering, and a BPR consultant. Contact has been consistently maintained via telephone, e-mail, and fax. The researcher last had contact with a director of information systems (IS) on March 13, 1998.

Each interview lasted approximately one hour. The researcher used a set of open-ended questions, related to process improvement, to guide interview discussions. However, spontaneity was encouraged by allowing respondents to discuss any issues they considered important to the research. The list was checked periodically to make sure that all questions were addressed. Transcription was done within 2 days to reduce information loss.

The unit of analysis is the Personal Line Systems (PLS) reengineering project. PLS provides information technology (IT) support for the Personal Lines Insurance business. Personal Lines is a core business at Safeco that offers automobile, boat, fire, and homeowners insurance products to its customers. The focus of the redesign effort is to provide "aligned" automation of business processes in the Personal Lines business. Keep in mind that past automation was not congruent with business process changes and activities.

PLS was chosen for business process reengineering (BPR) due to heightened competition in the core business, increasing technology costs with less than encouraging return on investments, and an effort by ISS management to move toward an enterprise-wide information sharing paradigm. Competition in the insurance business was escalating rapidly because nontraditional businesses are now allowed to compete for insurance products. For instance, many banks were beginning to offer insurance products. To better serve the information needs of customers and decision makers, Safeco invested heavily in new technology. However, IT investments were hard to justify with traditional accounting methods and ISS was concerned that new systems development was not properly aligned with business processes. Finally, the long-term goal of ISS was to provide information sharing on an enterprise-wide basis. ISS realized that information would not be truly enterprise-wide unless systems were aligned with business objectives and met customer needs.

CASE DESCRIPTION

The BPR effort at Safeco is lead by ISS. Its mission is to facilitate the cost-effective use of IT in the company's basic business activities; activities intended to achieve long-term profitability and growth. In order to accomplish this mission, ISS management realizes that management support technologies should be tied to business objectives, which should be tied to fundamental customer needs. BPR is underway in ISS, but not across the organization. According to the vice-president of ISS, "BPR is internally-driven. It must be CEO-driven to work in the organization as a whole. Successful BPR requires cultural change. To effect cultural change, the mind set of people must change. To change the mind set of people, we need to model behavior. Without top-down support, the mind set of people won't change and BPR will probably fail" (D. Armstrong, personal communication, July 25, 1996).

The BPR effort originated with Jan Dyer, Director of PLS. "Jan (Dyer) was in the best position to lead the BPR effort. She had a keen understanding of the business and worked well with technology people that support and maintain PLS" (D. Armstrong, personal communication, July 25, 1996). Jan's first move was to identify a core business to redesign. "Automation of systems at Safeco is common and ongoing. But, these investments weren't really paying off. Customer satisfaction wasn't increasing and cycle time wasn't decreasing. I felt that the real problem was that the business processes weren't changing with automation. Personal Lines was chosen because it is a core business and top management felt that the PLS team was used to dealing with change" (J. Dyer, personal communication, July 25, 1996).

Top management felt that PLS provided an ideal platform to implement a large scale BPR project. If this project worked, it would provide positive inertia for future projects. The goal of PLS redesign was to reduce cycle time for rate changes and change the culture (rate changes are a fundamental business issue at Safeco, thereby, critical to the success of the business. Rate changes refer to the cost to the client for different levels of insurance coverage). Rate changes were taking approximately 18 weeks and needed to be reduced to less than 8 weeks to remain competitive. The other problem was the organizational structure inherent at Safeco (the structure is multi-layered and rigid which impedes communication flow and makes process improvement difficult).

Any kind of major change is difficult and BPR calls for major changes in the way a business operates. "For us to get good at BPR, we must be able to cross departments and businesses with our systems. To accomplish this, our people need to be able to question the status quo. We need to be able to work better in cross-functional teams. Finally, we need to build consensus through negotiating rather than dictating. Hopefully, accomplishing these goals will help us change to a process culture" (D. Armstrong, personal communication, July 25, 1996).

PLS reengineering was the first step toward justifying the value of systematic automation of critical information systems. If top management could be convinced of its value, it could help ISS launch BPR as an enterprise-wide effort.

PLS History

Before 1970, many claims and exchanges of information were handled manually. During the 1970's, ISS began to implement PLS to move to a fully automated system. It's goal was to standardize information processing to reduce redundancies and increase data accuracy. The outcome of automation was geared toward efficiency. Safeco management was hoping to reap tremendous cost savings in terms of reduced paperwork, reduced re-keying of data, and easier access to decision making information. Safeco found that automation didn't reduce costs. Actually, costs were rising due to dramatic increases in technology updates, technical personnel staffing, and maintenance of legacy systems. Moreover, tremendous cost savings were not being realized.

In the 1980's, ISS began to refocus automation efforts on job pieces. ISS would try to shave an hour off a project on one desk and an hour off a project on another desk. The idea was to save time on a piece-by-piece basis by job task. It was hoped that the end result would be a significant savings in time for an entire job. The major problem with this type of automation was that people expanded what they did to fit the job. As such, time saved on one piece would inevitably be transferred somewhere else. This caused even a bigger problem because it was even harder to find out where and how the time was used.

In 1987, Jan Dyer began questioning the automation strategy. She noticed that automation wasn't adding value to the business and wasn't reducing total costs. Her insight was that automation wasn't causing people to work differently. "We automated systems without teaching people to work differently. We just automated a paper process without taking into account the business need and the capability of technology to completely rethink how things could be done" (J. Dyer, personal communication, July 25, 1996). BPR appeared to be an ideal approach to rethinking the business with the power of technology. During the next few years, ISS began to explore the BPR approach by sending key executives to training seminars, encouraging reading of BPR books and magazines, and adding BPR literature to the Safeco library.

During the 1990's, ISS adjusted its automation strategy to include BPR principles and behavior modification techniques. As a first step to process redesign, ISS had to map the existing process, that is, they had to find out what people actually did along the process path. To accomplish this, interviews at every desk level were administered. Two to three people were interviewed to get a clear frame of reference. A better understanding of the existing process enabled system automation to be better aligned with the business. However, it did not provide a cross-functional view of the process.

Currently, automation is becoming more cross-functional. Consensus management is becoming an alternative to autocracy. Teamwork and team charters are replacing individual tasks. Creativity is becoming an accepted ideal. ISS is encouraging people to be more creative and more concerned with the success of the overall business. Jan Dyer is charged with changing workers' current frame of reference and pushing them to think "out-of-the-box". Workers are responsible for ownership of their projects in a team setting. Jan is working to change the way people think of their work by trying to change behaviors. As a first step, she is trying to change the reward system to award teamwork, creativity, and innovation. Until the system changes, people will not be encouraged to change their behaviors. "Many times, people are waiting to be told to do something. They need to be encouraged to think more on their own. They are not used to solving problems. We encourage this behavior by rewarding quality, individual accomplishment, and status quo thinking" (J. Dyer, personal communication, July 25, 1996).

The role of ISS is changing very rapidly. ISS is no longer just a cost center or service organization. It is a leader. ISS must understand that it serves customers across the enterprise. People are now driving IT. They expect more useful technology as IT continues to advance at an incredible rate. The major problem is that ISS is staffed with people who have not traditionally had to be responsive to external customers. Most of the people who do the work are programmers or are associated with programming tasks. "Programmers can no longer think in terms of applications. They need to satisfy both internal and external customers. They need to think in terms of customer needs since people use all of the systems. They need to look at the business process first and then see if automation will support it. ISS needs to refocus its paradigm from that of an internal to that of an external view" (J. Dyer, personal communication, July 25, 1996).

PLS Steering Team

In 1995, a steering team was formed to guide the PLS redesign project. The composition of the steering team included representation from the key partners (customers) involved, leadership positions, PLS analysts, and PLS experts. The team consisted of ten members: Anne Randall, Rich Striker, and Jim Griffiths (customers), Jan Dyer (Director of PLS), Dennis Ellermeier (Reengineering Manager), Cherly Hays (Auto, lead experience), Kory Wilson (Quality), Becky Sorenson (Customer Communication), Lynne Schoebelen (System Support), and Bette Beck (Reengineering Project Manager). The team was staffed by Deetsy Armstrong, Vice-President of ISS.

Although the CEO endorsed the formation of the steering team, he was not actively involved in any of its activities. Responsibility for success was left entirely up to ISS. It was clear from the interviews that top management was not an active participant in the BPR experience at Safeco. The reasons for this were unclear. However, the Vice-President of ISS offered some insight. "The BPR effort is not CEO-driven. It will be difficult to enact enterprise-wide change without this support. It will take time to convince the CEO of the value of BPR. Until that time, we will continue to struggle" (D. Armstrong, personal communication, July 25, 1996). Another problem is that the existing leadership is older. They grew up in a business environment that was much more stable. The life-cycle of technology development was measured in years rather than months. The idea of a flat organization was considered ridiculous.

The steering team had five responsibilities. First, they were charged with providing focus and clear direction for BPR efforts throughout the department. Second, they were supposed to make BPR happen, that is, they were the facilitators of the process and charters of the process improvement teams (PIT). Third, they were to act as coordinator and counselor to PIT. Fourth, they were charged with evaluation of PIT performance and redirection of PIT activities as necessary. Finally, they were to look at the organization from a systematic (enterprise) view to identify obstacle to BPR success. Basically, PIT were responsible for doing the work and the steering committee was responsible for facilitating PIT.

To help the steering team ease into such a dramatic change in job roles, an organizational psychologist (Peter Scontrino) was brought in to discuss what to expect from employees when undertaking a major change effort. In addition, Peter discussed what employee involvement is and is not. Suzy Maloney conducted a problem solving seminar to give the steering committee a set of tools and techniques to pass on to line workers.

The steering committee intentionally did not take ownership of the project. It wanted the teams to take ownership to increase employee involvement and buy-in. In addition, the steering committee considered itself an overseer rather than a driver of the initiative. According to the reengineering manager, this may not have been the correct stance. "On reflection, we (steering team) should have been more proactive. After all, PIT were not used to taking ownership of these type of projects. Actually, I think that the idea of ownership is great, but responsibility of success should be shared by PIT and the steering team" (D. Ellermeier, personal communication, July 25, 1996).

Culture

The legacy culture at Safeco is rigidly hierarchical. Reporting protocol requires that information be passed through the chain of command. Moreover, teamwork and process thinking is not the norm. ISS realizes that the culture must change for any permanent changes to take place in the way work is done. To combat cultural inertia, ISS formed the steering team, PIT, and invested in BPR training.

To date, the formation of PIT for the PLS and other BPR projects is the only real structural change at Safeco. The team structure is not yet mainstream. Moreover, teams tend to dissolve upon completion of projects. The reward system has not really changed. It still rewards individual performance only and doesn't really take into account creative output. Nevertheless, ISS still believes that Safeco is moving in the right direction. The fact that the PLS project is underway is evidence of the organization's commitment to change. However, it is concerned that enterprise-wide change will not be realized in the near future. "Top management is reluctant to endorse enterprise-wide structural changes. It is used to operating is a certain way. It needs more evidence of the viability of BPR before it will move" (D. Armstrong, personal communication, July 25, 1996).

PLS Reengineering

This section details the design of the project, expectations, and anticipated changes. It is organized into the five major parts of the project — automation strategy, BPR methodology, process management and support, PIT, and IT. To successfully implement the project, ISS management was cognizant of the major changes — process change, organizational structural change, and philosophical change — that would be required.

Automation Strategy

The vision that guides BPR is called the Automation Strategy for Personal Lines. The automation strategy is defined to guide the PLS redesign effort and make sure that it stays aligned with the overall company focus of providing quality and timely customer service. The strategy concentrates on developing five key areas with PLS redesign — reinventing the independent agency distribution process, being more responsive to market conditions, supporting a "one company" image, supporting multiple business models, and pushing ownership and financial accountability down to those who do the work.

Reinventing the Distribution Process.

The predominant method of distributing Safeco insurance products is through its independent agents. Since agents use IT to deliver insurance needs, redesign is focused on developing consistent input and output forms. The new interface is called single-entry multiple company interface (SEMCI). SEMCI offers a standard input/output form for all systems. The result is that agents get a consistent look and feel to the systems they use to help customers.

Being Responsive to Market Conditions.

PLS improved system responsiveness by decreasing process complexity and increasing program modularity in its legacy systems. By simplifying business processes in conjunction with system redesign, IT flexibility was increased and programs were easier to change and maintain.

Supporting a "One Company" Image.

Customers don't care how services are delivered, they just want them! Therefore, a pivotal focus of PLS redesign is to project a seamless image of the company to customers. To provide this image, redesign will need to become customer focused rather than internally focused. Redesign of the customer information database to allow 24-hour information access is the first step. To facilitate accessibility of information, redesign efforts should also include positioning point of service (POS) systems where customers really need them.

Supporting Multiple Business Models.

New ways to market and sell insurance are being developed by Safeco business partners. To support these new approaches, PLS should be flexible and responsive. The redesign effort should concentrate on modifying the existing database design to support flexible and rapid development of new products. However, database changes will not help unless the IT management structure is redesigned to respond more quickly to business needs. Structural changes should concentrate on flattening the IT organization and redeploying displaced people to other areas in the organization.

Create Ownership and Financial Accountability.

Systems span functional departments and even organizational boundaries. Hence, PLS needs a new and broader way of looking at their systems. Better communication is needed between the business and IT units. The concept of strategic investment has been introduced to address this problem. The strategic investment approach begins by identifying the systems that are considered strategic. The next step is to separate the normal management activities from funding issues. This reduces the pressure of trying to justify cross-functional systems within a single department's budget. Expense is charged by usage rather than to a particular department. This should increase ownership since users pay for their own systems.

What Went Wrong - Automation Strategy

Technology projects like SEMCI interfacing, POS, and program modularity development have been successful. However, these projects are technology-based and require little cross-functional cooperation. Cross-functional ownership and accountability initiatives haven't fared as well. People across functions are not used to working together and have no real incentive to do so. Moreover, the organizational structure and environment hasn't really changed. What is clear from the case is that the support role of top management is not active. PIT is charged with doing the work. ISS is charged with facilitating PIT. Nevertheless, top management is not actively involved. It expects ISS to make the project work. Without cross-functional support, PLS redesign will probably be just another automation project. The major goal of the project — alignment of technology changes with cross-functional business processes — will most likely fail.

Reengineering Methodology

There are three parts to the Safeco BPR methodology — investigate, redesign, and implement. "As we start this journey, we need to take stock of where we are today, decide what our destination is, and develop an itinerary to help you (the PLS reengineering team) decide how and when to join this journey" (B. Beck, personal communication, January 12, 1995).

The first step is to identify (investigate) what the project is supposed to accomplish. An investigation team (with a representative from each unit) is formed to identify core processes, gather ideas and feedback, participate in brainstorming sessions, and develop cost/benefit analysis. Final approval is made by the vice president of ISS.

The second step is to redesign poor processes identified in step one. Redesign involves taking stock of what is happening today, brainstorming alternatives, and selecting the most viable option(s). Solutions are then presented to the department and business partners to get help in deciding which is the best one. The teams responsible for redesign are called process improvement teams (PIT). These teams are chosen by management based on expertise along the process path. The overriding mission of the PIT is to redesign processes to do more, faster, at a higher level of quality, with the same number of people.

The third step is to implement the best solution. Prototyping is an effective method for testing out ideas prior to full-scale implementation. "We do a good job with prototyping. For instance, we prototyped a small rate change and a small project in the context of redesigned processes to test for effectiveness. Prototyping helps us get the bugs out of a system before we put it into production" (D. Ellermeier, personal communication, July 26, 1996). The PIT needs to understand that the scope of the project must address the entire business process. Optimizing particular parts of the process may not be beneficial to the process as a whole. Although the PIT is responsible for implementation, explicit and active vice-president support and sponsorship is critical.

The BPR project is conducted along the model of a regular project. A dedicated project leader directs the efforts of the PIT. Management oversight includes sponsorship at the vice- president level or higher, significant guidance at the director level, and detailed guidance at lower levels of management. Using the regular project methodology for BPR is useful because everyone is comfortable with how it works. Radical change is important, but development and implementation procedures should not be.

What Went Wrong - Reengineering Methodology

ISS realizes that its methodology has problems. The redesign stage of the methodology could be more systematic. "Our methodology is not as systematic as we would like, but we continue to work toward a better way of approaching BPR problems. Our up-front planning is pretty good, but the redesign stage could be enriched" (D. Ellermeier, personal communication, July 26, 1996).

Another major problem not addressed by management is the lack of BPR training for the PIT. The PIT can read material on BPR housed in the Safeco library and learn by doing. However, formal training of the methodology is not in the budget.

Process Management and Support

The mechanisms in place to encourage change include a PIT structure for attacking business process problems, a steering team to guide teams, an initiative to change the existing reward structure, and vice-president support. The BPR team structure consists of a steering team and PITs. The steering team sets milestones for BPR projects, selects project priorities, and facilitates PIT efforts. The PIT does the work. Vice-president support is leveraged to reduce or remove political obstacles to clear the way for change. Vice-president support is visible and active. The vice president of ISS takes pride in their department's role in PLS reengineering. "Our department (ISS) leads the reengineering effort for the entire company. We are proud of this, but we hope that upper management will take a more active role in enterprise-wide BPR. We think this will happen, but we must first show the company how much value can be gained from our PLS efforts" (D. Armstrong, personal communication, July 25, 1996).

The steering team and vice-president are involved in major decisions, but the goal is to push decisions down to those who understand and do the work. Although decisions are made at a lower level, the PIT should report proposed changes to the steering team for feedback. The steering team is in place to facilitate the decision-making process, not direct it.

The process manager (PIT leader) makes sure that redesign efforts get done in a timely fashion. He or she works with a system manager to make sure that business changes can work with system changes. The process manager directs the resources to projects. The system manager is responsible for system integrity and individual systems development to support the project. The process manager has equal or greater authority than the system manager during the time the resources are dedicated to the project. In other words, he can stop the system manager from jerking the resources away from the project.

Although the process manager is the team leader, he or she does not dictate team policy. The team works together, *as a team*, to make decisions. However, with autonomy comes responsibility. The PIT is responsible for measuring and meeting target outcomes, that is, they are responsible for the success of the project. The support structure is in place to provide resources and organizational support.

The culture is changing in several ways to facilitate teamwork. System managers are working together to select the best project manager candidate for a given project. In addition, team selection begins before a new project is started. Finally, buy-in from the vice-president level is sought early on in the process. However, some cultural issues still need to be addressed. The existing reward system is still based on individual performance. Unless a shift is made to rewarding team effort, team work will not flourish. Physical separation of team members continues to be a potential problem, but relocation is not a reality at this time. System and project managers will continue to have conflicts because each has a natural tendency to look out for their own best interests. The support structure (vice president and steering teams) must continue to address these issues and intercede, where necessary, to resolve conflicts.

What Went Wrong - Process Management and Support

The PIT structures are relatively flat, but the environment where process changes are made is not. Dramatic process change across functional lines will be difficult until the organizational structure changes. As successful BPR projects are finished, the company should better understand how to recreate its structure. In the meantime, the focus will be on developing more of a process perspective among all employees. System managers need to continue to work with project managers to identify the best candidates for projects. Employee and system managers need to work together to identify growth opportunities. Another problem area is the reward structure. Since the new reward structure is still in its infancy, it will take time to modify it to foster team effectiveness.

Real change has been very slow due to lack of enterprise-wide support. Since ISS is spearheading the BPR effort instead of the CEO, cross-functional obstacles to change are still in place. Although it is relatively easy to achieve cycle time reductions in individual activities, it is extremely difficult to achieve overall (real) cycle time reductions across entire cross-functional processes. Given the current stance of top management toward BPR, real cycle time reduction will most likely be minimal.

PIT

PLS has excellent people, but they might not be deploying them in the most effective manner. Hence, the team staffing strategy is focusing on building teams that will work on cross-functional process problems rather than departmental or functional ones. Training to cultivate a process-oriented mentality is a priority. Courses are being offered to help managers and team members excel in a team-oriented environment. Examples of courses include *Leading and Facilitating Effective Teams and Meetings* (provides practical techniques in creating a successful team and getting the most from meetings), *Getting to Yes* (focuses on negotiating and conflict resolution skills), and *The Team Approach to Quality* (shows how each person on the project team plays a critical role in creating a high quality product).

An effective tool to get team buy-in is the PIT team charter. The charter is a formal contract with team members. It acts as a rallying point to get everyone on the same page. The charter includes sponsorship, objectives, ownership, and resources. Each team has a sponsor. The sponsor is a person (or persons) to whom the team is responsible or reports. The sponsor acts as liaison to management. The charter statement is a document that includes general objectives and a brief description of the process problem. The document passes ownership of the process problem to the PIT when it agrees to the charter. The charter also describes how the project links with BPR objectives. The resources needed by the team should be outlined in the charter. Resources may include funding, team training, support staff, materials, and access to documentation and data. In addition, reporting guidelines, milestones, project completion dates, deliverables, and celebration guidelines should be a part of the charter.

PLS has articulated a strategy for building successful teams. Team member selection is at the heart of the strategy. When identifying team members, the following should be considered:

- Ideal team size is 6-8 people. absolute maximum is 10.
- Members should be people who work closely with some aspect of the process
- Candidate should have appropriate technical knowledge to be a meaningful contributor
- Candidate should have understanding of his work unit and be able to describe its work flow
- Candidate should understand business impacts of project and business strategy of department
- Team composition should include people along the process path (upstream and downstream)
- Team representation from groups likely to be affected by process changes is desirable

- Team member is allocated resource to project, not intrusion on their "real job"
- Candidate will have *time* for team meetings and assignments
- Diverse make-up of personality styles (drivers, analyticals, amiables, expressives) is desirable
- Team member should be open to feedback from others, interested in innovation, focused on team goals over individual ones, willing to express point of view, willing to take risks, able to meet deadlines, listens to others, and willing to think "out-of-the-box"

Although team composition is very important, it takes a special kind of manager to foster teamwork. Managers need to be role models as well as coaches. If managers want real process improvement, they have to lead change. They have to show others that they are willing to take risks, think "out-of-the-box", and continuously learn and strive toward process improvement.

What Went Wrong - PIT

The team charter transfers ownership of the project to the team. It also lays out the guidelines for reporting, resources, and authority. It is an excellent first step toward creating real empowerment. ISS appears to understand that its people do the real work by encouraging teams to develop their own team charters. However, the existing environment is still rigid and functional. Political problems continue to surface as the charter takes the team across functional lines. "Business people across functional lines want Safeco to be excellent. They want to improve customer satisfaction. They also want quality information from PLS. However, they don't want to change unless they are convinced that it will help the business without hurting their position in the organization. Active CEO involvement would greatly help, but is not forthcoming at this time" (D. Ellermeier, personal communication, July 26, 1996).

IT

Technology is often the enabler. Technology provides opportunities to revamp labor intensive work flows (policy assembly vs. laser statements), rethink third party communication (paper vs. EDI), and refocus on core business processes" (B. Beck, personal communication, January 12, 1995). PLS reengineering is lead by ISS because IT is at the heart of all Safeco businesses. Rate changes, document processing, claims, communications, and information flow depend on IT. PLS reengineering won't work without it. IT is really the enabler of other BPR enablers. IT is used to facilitate the BPR methodology, help people solve problems, and help information flow in a flatter organizational structure.

Process improvement is not new at Safeco. ISS has been automating its systems for many years to cope with changes in the marketplace. Many of the existing PLS processes are already automated with IT. What is new is the idea of BPR. Automation of a poorly designed business process merely speeds up bad quality. However, redesign of an existing business process, with IT, can greatly speed up a well-designed process and improve quality at the same time. Expected performance improvements include reduced cycle time, increased quality, and improved customer satisfaction with the same number (or fewer) of people. However, these improvements may not be possible without major changes to the structure of its information systems and the structure of the organization.

PLS reengineering has caused ISS to rethink its approach to IT. Safeco technology is predominantly mainframe-based. However, the mainframe is not necessarily a problem. The mainframe can still be used for processing large volumes of information while newer technologies can be used as front- and back-ends to the mainframe. In addition, newer technologies are being developed to work with the Internet. The key is rethinking and redesigning business operations using IT as the predominant facilitator. A potential danger is relying too much on IT to solve problems. Technology is not a panacea. It is a tool to facilitate BPR.

Workflow automation is the glue that ties the PLS reengineering objectives — planning and prioritization, teamwork, and system redesign — together and workflow automation depends on IT to make it work. To automate redesigned work flows for future flexibility, the future of technology at Safeco must be considered. For instance, the future LAN environment, the internal controls requirements, and the future of the mainframe environment at Safeco should be integrated with all redesigned business processes.

ISS has in place a BPR methodology, a team structure, and is changing its organizational environment to facilitate workflow automation. It needs to continue to work closely with business managers and business partners to integrate technology into its business processes. Once existing business processes are mapped, they need to be redesigned with the enabling power of IT in mind. To make this work, the PIT needs to have representatives from both the business process and ISS.

What Went Wrong - IT

Automation with technology is a priority at Safeco. Safeco also realizes that redesign and automation go hand-in-hand. The major problem is that the BPR effort is spearheaded by ISS. This is not a weaknesses in and of itself. However, the implications are that top management is not really an active supporter of the BPR effort and definitely not a visible one. Although IT is the lifeblood of information flow, top management has the only real power to remove cross-functional roadblocks and make permanent changes to the organizational structure.

CURRENT CHALLENGES/PROBLEMS FACING THE ORGANIZATION

The lack of active and visible top management support permeates the case study. It appears to be the critical flaw in the Safeco BPR plan. The basic parts are in place. PLS has a rigorous automation strategy, reengineering methodology, team charter, and a PIT structure. It also embraces IT as a primary facilitator. However, only functional improvement can be achieved without enterprise-wide support.

The organizational structure is not flat and is not very flexible. The PIT structure is a great start, but the limitations of the work environment will most likely impede any real cross-functional success. Within the PIT structure, people have real empowerment. Outside the structure, empowerment is not the norm. Problems arise when trying to enact permanent cross-functional change because some people involved along the process path are never involved.

Process improvement through automation is very common. However, process redesign in conjunction with IT automation is relatively new. Alignment of IT with process change is difficult because the organizational structure is not flattening as quickly as automation is taking place.

By their own admission, they are not there yet. Safeco realizes that its organizational structure cannot change overnight. A better understanding of the positive impacts and consequences of BPR must be understood before major change can occur. All Safeco employees must understand why changes are needed for BPR to work well.

The researcher recently discovered that the PLS reengineering project is no longer active. "The PLS project fizzled out. We have turned our attention to other projects" (C. Purcell, personal communication, March 11, 1998). The failure of PLS should not be a surprise because of the limitations brought out in the case. The researcher's major concern is: as Safeco moves to other projects, are they really addressing the root problem?

Although the case uncovered several potential problems, the company may not be able to address them all. In this case, top management support is the critical issue. Two other important issues are the current organizational structure and the reward system. Safeco is focusing its attention on other projects it considers more important at this time. Keep in mind that Safeco is undergoing tremendous growing pains and it will take time before the organizational structure can change. In addition, it seems that top management is reluctant to actively commit to BPR.

Safeco appears to be moving in the right direction in several areas. It has established an excellent team charter to pass ownership to team members, identify reporting responsibilities and sponsors, and grant authority to make decisions. It also has an excellent automation strategy which details overall objectives where redesign with IT can be used to improve existing business operations. Directives are set up to help clarify the roles, responsibilities, and authority of process managers and system managers. The directives call for process managers and system managers to work together to solve real business problems rather than wasting time over political turf battles. The company is also very aware that automation without redesign may not work well since it has been automating for many years with less than optimal results. Finally, the attitude at Safeco is very positive. Safeco employees really care about their company and want to make it better. Safeco management appears to be working very hard to create an excellent environment for change.

REFERENCES

Davenport, T.H. (1993) *Process Innovation: Reengineering Work Through Information Technology.* Harvard Business Press, Boston, Massachusetts.

Hammer, M. & Champy, J. (1993) *Reengineering the Corporation.* Harper Collins Books, New York, NY.

Kotter, J.P. (1995) Leading Change: Why Transformation Efforts Fail. *Harvard Business Review*, March-April, 59-67.

Paper, D. (1998). Creating Value for the Customer at Barnett Bank: A Qualitative Study. *International Journal of Applied Quality Management*, 1, 1, 45-58.

Paper, D. and Dickinson, S. (1997). A Comprehensive Process Improvement Methodology: Experiences at Caterpillar's Mossville Engine Center, in *Cases on Information Technology Management in Modern Organizations*, eds. Khosrowpour, M. and Liebowitz, J., Chapter 9, Hershey, PA: Idea Group Publishing.

Peters, T. (1992) *Liberation Management: Necessary Disorganization for the Nanosecond Nineties.* Fawcett Columbine, New York.

BIOGRAPHICAL SKETCH

Dr. David Paper is an assistant professor at Utah State University in the Business Information Systems and Education Department. His academic credentials include a bachelor of arts in computer science from Southern Illinois University, a master of business from Arizona State University, and a doctorate of business from Southern Illinois University. His research focus is on business process reengineering (BPR) and total quality management (TQM). To augment his academic knowledge of BPR and TQM, he conducts longitudinal case study research with companies such as Caterpillar, Honeywell, Safeco Insurance, Barnett Bank, Moore BCS, and Fannie Mae. He is also a certified evaluator for the Shingo Prize for Excellence in Manufacturing. He has several refereed publications appearing in journals such as Long Range Planning, Accounting Management and Information Technologies, Annals of Cases of Information Technology Applications and Management in Organizations, Journal of Computer Information Systems, Information Strategy: The Executive's Journal, and Journal of High Technology Management Research. His industry experience includes Texas Instruments, DLS, Inc., the Phoenix Small Business Administration, and Wides Village Real Estate Company.

Building an Internet-based Workflow System: The Case of Lawrence Livermore National Laboratories' Zephyr Project

Judith Gebauer
University of California, Berkeley, USA

Heike Schad
University of St. Gallen, Switzerland

EXECUTIVE SUMMARY

Lawrence Livermore National Laboratories' Zephyr System demonstrates how emerging technologies can help streamline procurement processes and improve the coordination between participants in engineering projects by enabling new ways of collaboration. The project also shows the success of a highly pragmatic approach that intentionally covered only standard situations, rather than additionally automating the exceptions. Assigning purchasing responsibilities to the end user reduces the involvement of the purchasing department in operational activities. This streamlined the process, resulting in major time savings, cost reductions, and improved quality. Left with less day-to-day purchasing operations, the purchasing department has more time for strategic tasks such as selecting and pre-qualifying new suppliers, negotiating contracts, or implementing new procurement systems. The case reinforces the notion that the use of information technologies can result in major benefits when aligned with process re-organization efforts.

BACKGROUND

In 1949, San Francisco saw the christening of the California Zephyr train that subsequently connected Oakland and Chicago until 1970. The route of the California Zephyr covered 2,525 miles and took an average of 2 1/2 days to complete, without changing trains. The innovative stainless-steel streamliner helped speed up the time to travel between the U.S. West Coast and the industrial areas in the North and East.

Similarly, the eponymously named Lawrence Livermore National Laboratories' (LLNL) Zephyr project helps accelerate business processes from concept through delivery and payment of engineering prototypes by connecting LLNL's engineering, procurement, and suppliers. Initiated in 1994, the project is conducted by LLNL's Engineering Directorate in partnership with Procurement, Administrative Information Systems, and Computations. It shows the benefits of moving an inter-organizational process from a traditional paper-based path to a Web- and e-mail based distributed workflow system. A closer look at the system reveals several factors that were critical to its success such as accompanying organizational changes.

Our information draws on several onsite interviews with project participants, as well as information available from LLNL's Web site, project reports, and research papers. Lawrence Livermore National Laboratory (LLNL) is a research and development (R&D) facility, located in Livermore, California, and operated by the University of California under a contract with the U.S. Department of Energy. The Lab was established in 1952 to perform research on nuclear weapons and magnetic fusion energy. It is also responsible for the stewardship of the U.S. nuclear stockpile. Currently, it hosts six major programs: Defense and Nuclear Technology; Laser Programs; Non-proliferation, Arms Control, and International Security; Energy Programs; Environmental Programs; Biology and Biotechnology Research. Each of the specific programs is sharing support from four more general scientific and engineering directorates: Chemistry and Materials Science, Physics and Space Technology, Computation, Engineering

The Engineering Directorate, the focus of this case study, involves about 2,100 technical and administrative employees and is divided into two organizations, mechanical and electronics engineering. Both organizations share support from five activities, which cut across all areas of Engineering: Engineering Research, Computers, Recruiting, Continuing Education, and Safety and the Environment.

Formally, LLNL is part of the U.S. Department of Energy, and it focuses most of its research on defense related issues. The recent political changes that followed the fall of the Iron Curtain resulted in reduced financial budgets for defense issues, and a more stringent justification of research projects. Formerly extremely secluded institutions now see themselves competing with research facilities from all over the world, including the ex-nemesis Russia. At the same time opportunities for collaboration open up. With increased openness public awareness grows and this further increases the pressure to deliver high-quality results efficiently.

SETTING THE STAGE

Most of the work at LLNL is engineering related, i.e., the design and manufacturing of complex systems. Its outcome is usually based on tight collaboration among engineering specialists, not only within LLNL but also with partners of other research institutions, contractors, and suppliers. In some cases it even stretches beyond national and continental borders.

The exchange of information between the participants of an engineering process is difficult in general, especially when different steps such as design, procurement, payment, fabrication, testing, and evaluation are conducted in isolation from each other and communication is paper-based. In public institutions, such as LLNL, complex administrative processes often aggravate the problem. The resulting islands of information encounter difficulties when they need to interact and collaborate efficiently, and the overall process is slow, error-prone, and expensive.

At LLNL, two bottlenecks were considered especially problematic: the exchange of engineering and administrative documents and purchasing of parts for prototyping.

In 1995, the project team initiated by Cecil Jordan, superintendent of the Laser Engineering Division, sought to address both problems.

Inhibitors of Effective Collaboration

In engineering projects, interaction among specialists is especially important in the early steps when projects are defined and subsequently outlined in further detail, including the definition of the work breakdown structure.

Many engineering projects start out with a process called solicitation, defining the legal, technical, and environmental requirements necessary to accomplish project goals while developing a proposal leading to contract award. Participants exchange information about terms and conditions, contact iteration, specifications that contain technical information including engineering and blueprints. Collaboration in later stages evolves around the set up and changes in specifications, drawings, computer-aided design (CAD) models, tooling files, computer software, and other critical path items between the members of the project team.

Traditionally, teams meet in person for engineering reviews and exchange information across the table by marking up drawings, project schedules, and specifications. This process has proven successful as long as information is at hand because it allows immediate, interpersonal communication. However, interaction becomes cumbersome, time consuming, and costly as soon as process participants and information are dispersed. Nowadays common scenarios, such as the one where a system is designed in one country, manufactured in another, and delivered on a global basis are greatly inhibited when information and communication are primarily paper-based. Effective collaboration calls for the instant availability of documents, project schedules, and other items such as design libraries, tools, and models and testing data to all project members. The involvement of external development and contracting partners complicates the matter in the sense that external partners are usually supposed to access specific parts of the complete project information only.

At LLNL, engineers in need of items to build prototypes used to create requirement books that itemize deliverables, often illustrated with some hand sketches. Design engineers conceptualized the needs by generating drawings and related documents, e.g., using a CAD-CAM (Computer-Aided Design-Computer-Aided Manufacturing) workstation. The engineering documents were collected in paper form by the Engineering Record Center (ERC) which expended significant effort on their archival and release. Further iterations and reviewing processes between the communication partners required repeated and time-consuming printing and mailing of the documents. The use of paper-based specifications limited the amount of information that could be exchanged and increased the number of iterations necessary to reach complete understanding between experts in different locations.

In a complex engineering process close interaction between engineers is critical not only within the different phases, but it is also important that the output of each step is usable and valuable to the following phase. This ensures designs that are reliable, accessible, and easy to manufacture as well as to maintain later on.

Inhibitors for Fast and Efficient Procurement of Prototype-related Parts

After project conceptualization, design engineers decide whether to fabricate parts in-house or to acquire external vendor services, either to build a prototype or to manufacture the "final" system. LLNL's general policy is to outsource as many jobs as possible in order to keep internal overhead and operating costs down.

Besides weak collaboration functionality resulting from paper-based information exchange, the procurement of parts for prototyping was considered the other major inhibitor of an efficient engineering process.

LLNL's purchasing department used Electronic Data Interchange (EDI) systems for several years and considered them a valuable mechanism to process transactions in a cost and time efficient way. Restrictions inherent in the structure of EDI messages as well as the high setup and operations cost, however, call for large volume transactions of standardized, commodity goods. In the context of prototype-related items, it is very difficult to justify the high setup cost of EDI systems and their efficient use:

- Prototype-related parts are frequently non-standard with complex descriptions and specifications.
- Transaction volumes and order frequencies are low, demands are difficult to predict and often come on short notice.
- Procurement requires close interaction between the end user engineer and the vendor.

The purchasing process for prototype-related items at LLNL was considered as problematic for two major reasons: central procurement was involved in most of the purchasing acts, and again communication between engineering, procurement, and the vendors was mostly paper-based.

Prototype modules usually consist of many single items most of which required the processing of separate purchase order forms. With the procurement department being involved in purchasing of most of the items, the administrative overhead frequently exceeded their actual value. The additional "non-expert" layer of communication between the end user engineer and the vendor caused another problem (Figure 1). Vendor concerns about unclear requirement definitions and complaints about marginal quality and schedule impacts on LLNL's side were the frequent result.

The fact that the communication of engineering information within LLNL as well as between the Lab and external vendors was mostly based on paper also impacted the procurement process in terms of cost, time, and quality. To comply with regulatory requirements, LLNL often purchases items via a bid procedure, involving the following steps:

- Hand over request to the purchasing department for approval
- Set up Request for Quote (RFQ)
- Mail RFQ to prospective suppliers to solicit bids
- Select best bid
- Notify winning bid as well as the other bidders
- Send out comprehensive information and complete specification to the selected supplier
- Receive shipment and initiate payment

As a result of the complex process organization and paper-based communication, the procurement of prototype-related items typically took several weeks, in some cases even months. The average period for the purchasing of printed circuit boards, for example, was 20-25 days. In the case of an assembly fixture the process took 56 days with the steps from design release to contract award already accounting for 10-29 days. These figures do not include the time that LLNL usually took to initiate payment.

CASE DESCRIPTION

With Zephyr, the Internet- and Web-technology based system for the procurement of prototype-related items, LLNL pursued two main objectives: improve the quality of the engineering process by supporting the collaboration between engineers in design and manufacturing, and reduce time and cost of procurement.

To achieve these goals, the project team decided to move as many paper-based steps as possible, including payment, to electronic media. Today, Engineering's CAD systems are integrated with business systems that link procurement and finance, commercial vendors, and end users.

The project was initiated in an unusual way — unofficially and behind the scenes, by the engineering and computer departments, not by Procurement. In the beginning, there was no official funding and the system was developed by a couple of enthusiasts after dark. They initially attempted to introduce a common format and central location for the storage of engineering documents. The system promised easy transfer of documents between different CAD systems and thus improved collaboration between engineers. In a second step the team also considered business systems, as it became clear that the connection between procurement and designers was essential. By the time the prototype was ready and the benefits became more obvious the concept gradually gained acceptance from top management and eventually achieved four years of funding through the Department of Energy.

The Zephyr system is based on Internet technology to support communication around the creation of engineering data, procurement of prototype parts, and for fabrication. The issues of communication and cooperation that are covered by the project are common for large-scale R&D facilities. They are also becoming more important in business environments as organization-spanning cooperation increases. The following sections provide a more detailed look at the system.

The Zephyr system

Zephyr is a Web- and email-based system supporting the paperless, distributed workflow for the procurement of non-commodity items. It is the outcome of the initial idea to improve collaboration among engineers by making engineering documents centrally accessible. The system was first based on proprietary software that was developed by Lawrence Berkeley Laboratories. Internet-based technologies, however, proved to be very handy to establish a common communication

Figure 1: Information flow prior to Zephyr

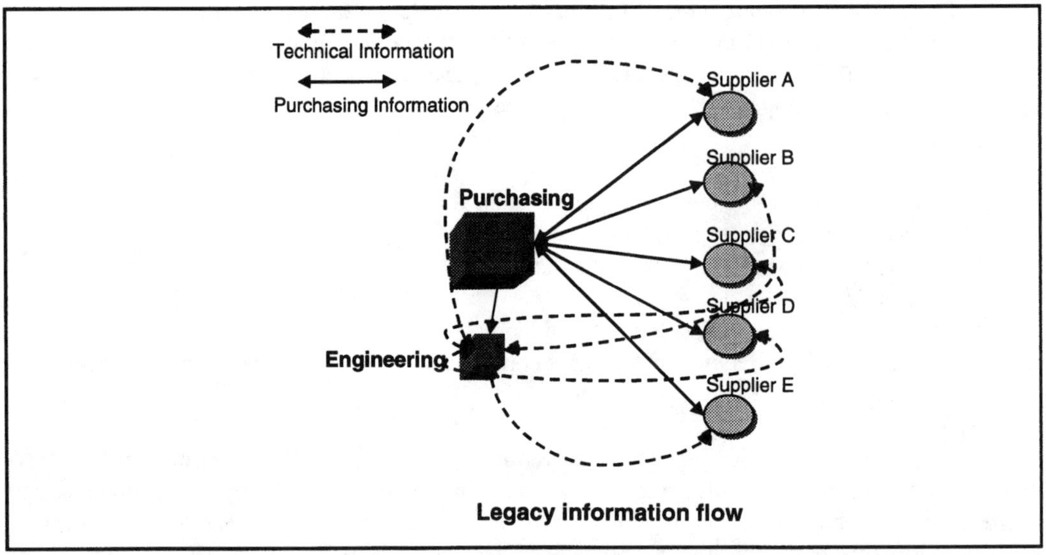

infrastructure and their availability gave the project major impetus. In short, the system stores complex data in a central database and directs the workflow around them via Web forms and e-mail.

Since March 1995, Zephyr has been piloted for the purchase of printed circuit boards. It is, however, scheduled for major extensions over the next years.

The workflow process supported by Zephyr is initiated after an engineering design package is approved for release followed by the decision that a prototype should be built. Subsequently, a user requests to purchase engineering items (Figure 2); in case of printed circuit boards, a couple of requests are issued per day.

To issue a requisition, fabrication data files describing the required design are transformed into

Figure 2: Zephyr's e-mail driven workflow process

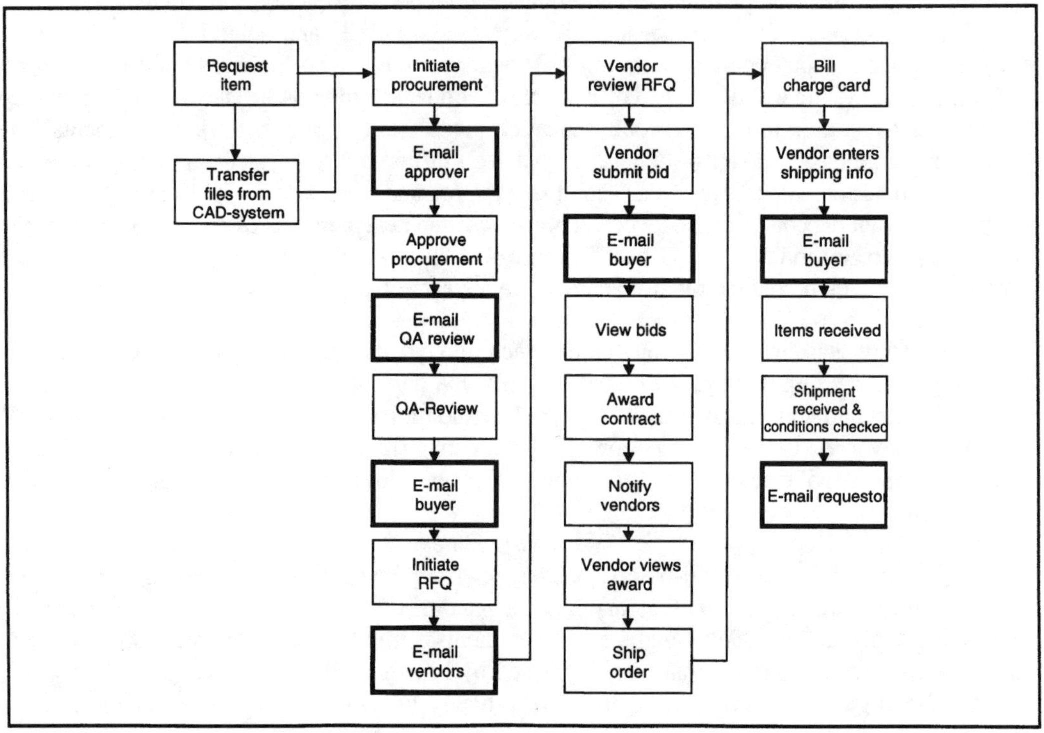

Figure 3: Information flow after the implementation of Zephyr

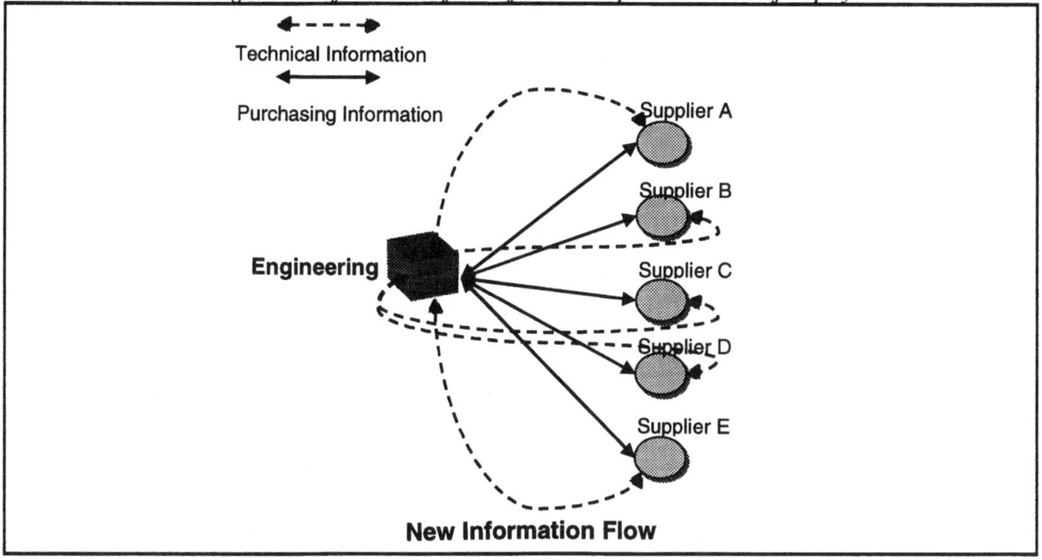

a commonly downloadable format and then transferred into the Zephyr database via the Engineering Document Transaction System (EDTS). Later on, the files can be accessed and downloaded via a Web browser, and printed. The system automatically assigns a job number for future reference and, depending on the size and other details of the job, e-mail is sent to the approval manager (within Engineering). While requisitions are not necessarily very detailed, they do contain a brief description of the required item and the desired delivery terms.

Subsequently, a "quote sheet" is generated from the fabrication data. This electronic form can be seen as a master document that following activities refer to and that is eventually sent out to the vendors. While it defines the job and contains enough information to allow vendors to bid on the offer, it does not enable them to actually build the prototype.

For procurement of items over $5,000 the procurement department has to provide its approval. While the designated Procurement Quality Approver is notified via e-mail, the actual file containing the details of the request remains on the server. After accessing the file via a Web browser, the approval manager reviews it, adds her approval/disapproval to the record, and resubmits the form. Items under $5,000, however, account for about 70% of the volume of custom manufactured items in Electronics Engineering. For these items Engineering can now initiate the bidding process without having to go through central procurement (Figure 3).

In the next step, a Request for Quote (RFQ) is generated and a group of selected vendors, typically 3 or 4, are notified via e-mail.

In line with the LLNL's internal requirements, all vendors of LLNL are pre-qualified. In order to participate in the Zephyr system, a supplier needs Internet and Web access and must be able to accept payments via corporate credit card (ProCard), which in LLNL's case is secured by the First National Bank of Chicago. While ProCard is LLNL's primary payment mechanism, the Zephyr system is generic enough to be easily adapted to other mechanisms such as blanket order agreements.

LLNL pre-qualifies all of its vendors using surveys to assess product quality and manufacturing capabilities. Vendors are ranked and categorized into groups of competencies according to their capabilities. The Lab subsequently keeps track of their performance using a performance database. Currently about 30 vendors have been pre-qualified for participation in the Zephyr system, 12 of which are actively using it. LLNL has been reducing the number of its suppliers over the past years. The current number of 3,000 vendors dropped from 6,000 a couple of years ago. Plans for the future indicate about 1,000 vendors.

The RFQ is handled by a Technical Release Representative (TRR), who selects the vendors, sets the deadlines for bids, and notifies the vendors via e-mail. Since not all - especially small - vendors are connected to e-mail all the time, he often uses the telephone additionally to ensure everybody knows about the invitation to bid. Informed by the "trigger e-mail," vendors access LLNL's web site where they find the specifications necessary to bid.

Security turned out to be a very important issue. Zephyr provides secure electronic forms for all process steps, e.g., for vendors to submit their bids. Bidders can only see the public quote sheet and their own bids, but not the bids from their competitors, and not even quote sheets for offers that do not apply to them. Only authorized staff inside LLNL can place quote sheets, or follow up on the shipping or receiving status. The design of secure areas inside the Netscape browser system was very important for the success of the project and required significant time and effort.

After reviewing the RFQ, vendors submit their bids. They do so by filling out the form they find on LLNL's Web site and uploading their bids. Again, notification about completed activities is communicated via e-mail. Sometimes the entire process takes no longer than 4 hours.

After the deadline, a representative at LLNL reviews the incoming bids, chooses the winning bid, and awards the contract. Again participants, i.e., successful as well as unsuccessful vendors, are notified via e-mail.

The successful vendor is given access to additional engineering data necessary to build the required parts. A ProCard number is assigned to the process which will be used subsequently as a reference number for payment. After completing the job the vendor notifies LLNL via e-mail that the item is now ready for shipping.

The shipment notice instantly triggers the payment process. Via the ProCard mechanism, payment gets credited to the vendor's bank in most cases within 48 hours and mostly using electronic funds transfer.

The vendor is asked as a courtesy to log back on to LLNL's Web site to transmit the shipping number and other tracking information. By capturing this information LLNL is able to monitor the shipment via the Internet.

The process is concluded upon receiving the shipment. After checking the condition of the items at the receiving dock, the end user is informed via e-mail that she can pick up her order. The loop of the triggered procurement process is closed.

To summarize, communication around the procurement for engineering prototypes is now entirely supported by the electronic Zephyr system. The end user community (Engineering) is basically responsible for the process and Procurement only becomes involved in cases of exceptions, such as for the approval of purchases over $5,000, to define the process infrastructure, e.g., by setting up supplier contracts, and to monitor supplier performance.

Pilot roll-out

The cost for the initial rollout of Zephyr amounted to $200,000. Resolving security issues and creating customizable restricted access areas, however, required significant additional effort over a period of about 12 months. About forty employees use the current pilot. Its success is based on several factors.

Procurement of commodities versus non-standard prototype items

While for the procurement of commodities and high-volume off-the-shelf items EDI solutions have been in place at LLNL for a long time, fabricated items for prototypes used to be ordered in traditional ways using paper documents and communication via telephone and fax. Low volumes, highly unpredictable demand, and the complexity of the items prevented the economic use of large-scale automated systems. However, the timely delivery of high-quality prototypes is a critical issue for the Laboratory.

The Internet and World Wide Web provided the basics to develop a system for the fast communication of complex engineering data and allowed for additional administrative changes that streamlined the entire process.

Printed circuit boards provided a good start to test the system because an industry-wide standard (International Graphics Exchange Standard) exists that helps communicate complex technical specifications in a clear way. Also, the project team assumed that vendors in this high-technology area would be versed enough with innovative technologies to reach adequate acceptance and participation.

Designed to handle standard situations

In order to keep development efforts within tight budget limits, the system is designed to handle standard situations, assuming that vendors deliver good quality on time. Exceptions have to be processed manually. As a result, the overhead costs are kept reasonably low.

Close and long-established relationships with vendors are a prerequisite for the arrangement to initiate payment upon shipment notice, clearly not an option for relationships that are created "ad hoc."

Signing up vendors and selecting pilot partners

A main success factor for Zephyr was to sign-up an acceptable number of suppliers. To avoid the time-intensive process of setting up new relationships, the project team targeted only vendors LLNL was already doing business with. The project team asked the end user engineers within the optics and mechanical fabrication departments to name their favorite vendors and then approached these vendors to find out about their general interest to participate in the system.

It turned out that small to medium-sized companies with up to several hundred employees were much easier to win than larger ones; they showed more flexibility in deciding about the changes that the participation in Zephyr would require. Several features of the system helped to keep the risk for the participants relatively low and made it altogether easy for the Zephyr team to sign up a reasonable number of participating vendors.

In order to participate in the system a vendor basically needs to establish access to the World Wide Web and e-mail and corporate credit card capabilities. Both can be obtained at relatively low cost: Internet-connections usually amount to less than several hundred dollars and are followed by access costs, which can be as low as $25 per month. Obtaining credit card capability costs about $500. In exchange, LLNL offers their vendors extremely fast payment as compared to the traditional weeks, and a sound basis for regular future business. Additionally, vendors are given the opportunity to access technical specifications in electronic form, which greatly facilitates their handling, and reduces errors.

Overall, about 30 suppliers have been signed up for participation in the system, most of which are small and medium enterprises. Twelve vendors are currently actively involved in Zephyr's pilot for the procurement of printed circuit boards.

Up-front negotiations

After the general sign up of system participants, some more technical details had to be agreed upon, such as the use of the system or the file format of the data to be transferred. Suppliers need to know the format of the files they are about to download from the Web in order to determine the requirements to transform them into their own internal systems, and they have to learn about the correct way of accessing and filling out the online forms.

These aspects are not complex, but it is important to consider them a priori in order to avoid subsequent problems, especially in cases where suppliers have little or no online experience.

Simple system architecture

Zephyr is based on a relatively simple IT architecture that makes it easy to handle and that limits access costs. It is also generic and flexible enough to be easily applied to different settings and processes. While all complex engineering data remains on LLNL's internal system, workflow processes are routed and steered via Web-based forms, supported by trigger e-mails.

Based on an open architecture, the system establishes a generic communication infrastructure system that supports the flow of information from the engineers' design boards to manufacturing, to procurement, and eventually onto the suppliers' screens, and that couples computer-aided design tools with electronic commerce applications. Thus, it supports the exchange of complex technical specifications and drawings between heterogeneous environments helping to improve collaboration between different engineering groups, as well as administrative tasks like internal approvals, bidding, and ultimately payment.

Customized system setup

The current setup of the Zephyr pilot is very much tailored to the needs of the Technical Release Representative (TRR) who plays a central role in the procurement of printed circuit boards. After receiving the requests from the end users (engineers), he performs several formal checks and then generates the quote sheet. This means, he initiates the procurement process, determines the complexity of the job, sets up the RFQ, chooses and contacts potential vendors, and finally makes the selection and awards the contract.

Among other factors, it was the setup of Zephyr that allowed LLNL to have the procurement of printed circuit boards administered by just one person instead of the 18 people that were involved in this job before. The flexibility of the system's infrastructure makes the customized approach easy, which ensures acceptance among end users and helps to apply the system to a variety of different settings.

Security Aspects

Security aspects turned out to be critical for Zephyr's success. While the project team was less worried about general break-ins, it was considered essential to ensure a variety of secure areas, granting different users access rights and data views according to their specific roles. The task was solved in part by using Netscape's Commerce Server and RSA secured forms, but also required significant development efforts by LLNL's staff. The team found it especially difficult to create a secluded Web server that allows individualized access to the stored data, but that does not allow external users to perform other functions such as anonymous login, ftp, or X-Windows. Simplifying the initial Zephyr pilot and making it compliant with LLNL's security requirements account for a major part of the overall project cost.

To determine the adequate level of security for different areas of the system in terms of feasibility and bottom-line value, LLNL performed cost-risk analyses. Time dependency of the data was chosen to be the major indicator of security requirements — the longer a piece of information stays valid, the more strongly it needs to be protected. To test the system, LLNL used internal computer experts who performed a variety of possible break-in attempts.

Administrative Changes

The full benefits of the new system result as much from the implementation of the Zephyr pilot, i.e., the use of information technology, as from administrative changes of the underlying procurement process.

In the new setting, the end user of the requested parts are empowered to do most of the transactions within their own domain without having to use purchase orders and to go through the procurement department. The latter is now in the position where it can concentrate more on strategic issues, such as establishing and managing vendor relationships.

Although the benefits are obvious today, the project which was set up by Engineering and Computational Services initially encountered resistance from Procurement, which felt threatened by circumvention. A re-organization in central Procurement actually helped jumpstart the project. The new procurement group shared the vision with Engineering about the benefits of streamlining the process and relying on outsourcing as much as possible.

Benefits

The single fact that a job which was previously done by 18 people is now handled by one person makes it obvious that the new system brings enormous benefits in terms of process streamlining. Not all of the benefits can be quantified, though.

Time

One of the major project goals was to reduce process lead-time. This objective was achieved to an impressive extent, with time savings ranging from 60% to 90%.

- The procurement of printed circuit boards, which traditionally took 20-25 days, can now be done in less than 2 days.
- In a series of pilot procurements of non-standard optical components, the bid package, RFQ, and award process cycle took less than 5 days using the combined Zephyr and ProCard process compared to much longer cycle times associated with paper-based processes.
- In the case of an assembly fixture the procurement cycle time dropped from the typical 56 days to 5.5 days.

The time savings are not only resulting from the substitution of paper-based communication by the electronic Zephyr system, but to a large extent also from administrative changes. Regarding printed circuit boards, those changes alone reduced the purchasing process from weeks to days. The use of Zephyr then further squeezed lead-times into what is sometimes just a matter of hours. Additionally, LLNL receives the incoming bids now in a standardized format. This reduces potential errors and makes bids easier to compare.

Fast delivery of research results and prototypes is essential in an environment where results are often not measured in hard dollar figures. It enables the Lab to acquire funding for additional projects and helps to improve the overall research quality.

Cost savings and improved product quality

As important as the time reductions, significant cost savings also occurred. In some cases these are directly linked, e.g., when a system allows reducing the level of staff necessary to perform a given task. In the case of Zephyr, administrative and IT related changes together allowed the reassignment of 17 people involved in the procurement process of printed circuit boards. The task is now accomplished by just one person.

Initially, the internal financial department raised some objections because payment upon receipt was preferred rather than on announced shipment. The system development team, however, could explain convincingly that the Lab has been maintaining relationships with the selected vendors over a long period of time, and that the vendors were very aware of the fact that if they did not meet the expectations they risked losing future business.

The fact that payment is now initiated much faster than in the past actually allowed LLNL's procurement officer to renegotiate contract conditions, resulting in "same as cash" discounts worth $120,000 in just two months.

In the past, some vendors, especially smaller ones, were not able to do business with LLNL at all because of its slow payment cycle that could easily take more than 180 days. With the new system LLNL can in principle choose from a larger number of vendors. This increases competition and opens new opportunities for cost savings and improved product quality.

Additional cost (as well as time) savings are gained from the improved communication between the end user engineer and the suppliers. Misunderstandings and the cost related with it are now avoided because the "experts" communicate directly with each other and the former additional administrative information filter is circumvented. Moreover, IT-based communication is often more precise and the data transmitted is more detailed than in the case of paper and fax. Also, by the use of forms, the Lab as well as the vendors are forced to submit all the information necessary to complete the job.

As an outcome of the standardized bidding process, bids are now easier to handle and to compare. With the same effort, LLNL could now obtain more bids for one offer. Cost savings might then result from the increased competition and the wider range of the bids. This variety is supported by the usage of the Web, because it is now relatively easy for prospective vendors to submit bids, independently from their geographic location. LLNL has not made use of this feature yet, however.

Improved communication and collaboration

The elimination of paper based communication between LLNL's engineers and the suppliers and the central storage of documents that are now accessible to everybody involved in a design, also helped improve collaboration. It ultimately enabled higher-quality results and accelerated the decision process. Truly concurrent engineering is greatly facilitated, involving engineers and other experts within the Lab as well as external partners. This includes suppliers who now have the potential to play a much more active role in the projects. Eventually, the design products will more closely comply with LLNL's general request of designing prototypes that are efficient to manufacture and easy to maintain.

SUMMARY AND FUTURE CHALLENGES

Zephyr is an integrated computer system that enables the seamless flow of information from the design boards of the engineers to manufacturing, procurement and eventually onto the screens of the suppliers. By coupling computer-aided design and manufacturing tools with electronic commerce applications, the system allows the exchange of complex technical specifications and drawings, supporting collaboration between different engineering groups, as well as the processing of administrative tasks, e.g., internal approvals, bidding, and payment. While complex data stays on an internal LLNL database server, workflow processes around them are routed and steered through Web-based forms and electronic mail. The system enables rapid agreement on design changes among geographically dispersed project participants. It speeds up internal approval as well as procurement processes through external links with parts suppliers.

The case shows how organizational redesign and the use of information technology together can lead to significant benefits. From the start, the project team saw IT as a means to overcome inefficient organizational structures and procedures. Ultimately, it was the combination of streamlining responsibilities and procedures in the procurement process with using the Internet and e-mail that made LLNL more efficient and competitive.

The fact that the Zephyr system is largely based on open standards technology, as well as its general architecture, make it a very generic concept, which is easily adaptable to other settings, inside and outside of procurement and engineering tasks.

The case also shows a very pragmatic approach—procurement via Zephyr is not a fully automated process: it does in fact require significant human intervention. Although there are plans to take some further steps in that direction, e.g., automating the supplier selection process, the system deliberately does not cover all possible situations. Rather, it is designed to handle standard situations. Although exceptions, such as the situation when vendors fail to fulfill orders in a correct or timely manner, have to be detected and taken care of "manually," the fact that the whole set up is based on long-term and ongoing relationships prevents these situations from getting out of hand. As a result, the costs and efforts to develop the system could be kept reasonably low; the outcome is a highly flexible concept that has been rewarded with high acceptance from users inside and outside the Lab, and has gained great attention from top management.

Zephyr has been implemented as a prototype for the procurement of printed circuit boards, mainly demonstrating the feasibility of the underlying concept. As such, the system covered only a small part of Engineering and has not been integrated with other Lab wide information systems. It did show significant benefits in terms of improved collaboration, shorter engineering cycles, reduced procurement time, and lower overall costs. In late 1997, LLNL's top management awarded the project team "for developing a revolutionary new approach to engineering procurement with shows promise for emulation at the Laboratory and beyond." The recognition is a thrilling experience and late satisfaction for Cecil and his team, who set out to develop their vision in secret and after hours.

As a result of its initial successes, the system will soon see major extensions in terms of functionality, scope, and reach. Recently, LLNL has been approached by a large Midwestern bank concerning its interest in licensing the Zephyr system. At LLNL itself, the 1998's plans for Zephyr continue to focus on collaboration, both national and international, aiming at expediting technology development. Eventually, Zephyr will become fully integrated into LLNL's institutional business system. Any authorized vendor will then be able to access to the system Lab-wide.

This decision was driven by the concern to expedite engineering procurement related to the National Ignition Facility (NIF). This $1.2 billion facility will provide an alternative to traditional testing of nuclear weapons banned in 1996. NIF will use laser technology and computer simulation as research methods, and additionally conduct other basic research, e.g., on medical uses of laser technology. Operations are scheduled to begin in 2003. Construction and manufacturing will account for over 75% of the project's cost and involve alliances between the public and private sectors, including not only LLNL but also several other national laboratories and high-tech vendors.

The Zephyr project team thus faces the challenge of not only having to enlarge the system to handle transactions in excess of $200 million, but also to show its viability in a large-scale "real life" setting.

ACKNOWLEDGMENTS

To collect data for this case study, we conducted several interviews with a number of people from two of LLNL's central Directorates, Electronics Engineering and Computation. Together they are responsible for the initiation, implementation, and, currently, the use of the system:

- Rudy Cavitt, Leader of the Laser Programs Design and Drafting Group (part of the Laser Engineering Division)
- Ray Gonfiotti, providing field support for the Electronic Manufacturing Group
- Cecil W. Jordan, Superintendent of the Laser Engineering Division
- Kenneth V. Luu, Leader of the Electronic Manufacturing Group (part of the Electronics Engineering Technologies Division)
- Fred E. Warren, Project Leader in the Computation Directorate

The interviews were conducted by Peter Buxmann and Judith Gebauer, September 16, 1997, and October 3, 1997. We want to thank all participants for their kindness, time, and expertise.

BIOGRAPHICAL SKETCHES

Judith Gebauer joined the Fisher Center for Management and Information Technology, U.C. Berkeley, as a research fellow in August 1996. Her research interests include the management of emerging technologies and the impact on organizations, innovative forms of organizations, the changing role of intermediaries, and electronic markets. Dr. Gebauer plays a leading role in the Fisher Center research project on electronic procurement and inter-organizational electronic commerce. Dr. Gebauer holds a Ph.D. (1996) and a master in Economics (1991), both from the University of Freiburg in Germany. More information on her work can be found at http://haas.berkeley.edu/~gebauer. Dr. Gebauer can be reached at gebauer@haas.berkeley.edu.

Heike Schad is a PhD student at the University of St. Gallen. She works as an associate team member at the Institute for Information Management and is part of the Competence Center Electronic Markets (CCEM). Ms. Schad holds a master in business administration and computer science (Dipl.-Wirtschaftsinformatiker (FH)) from the Fachhochschule für Technik und Wirtschaft in Reutlingen, Germany as well as a master in information science (Dipl.-Informationswissenschaftlerin) from the University of Konstanz, Germany. During her PhD studies Ms. Schad was affiliated with the Fisher Center at the Haas School of Business, University of California, Berkeley. Her research interests are in the field of Electronic Commerce and its impact on organizational processes and vice versa with a focus on: Inter-organizational business process redesign (ioBPR); business network redesign (BNR); Inter-organizational systems (IOS); virtual organizations and business network; and electronic commerce. E-Mail: Heike.Schad@unisg.ch

Building a Paperless Service: Making the Internship Connection

Theresa M. Vitolo
Pennsylvania State University, USA

Aaron J. Sparks
Deloitte & Touche, USA

EXECUTIVE SUMMARY

Central Ridge University[1] (CRU) is a large, research-oriented university composed of 10 major colleges, connecting over 15 branch campuses, and offering various continuing, distance, outreach and executive programs. One of the major colleges of CRU is The College of Business Administration (CBA). CBA maintains an optional internship program for its undergraduate and MBA populations. CBA strongly recommends its students to experience the corporate world by doing an internship before their graduation. Also, CBA students seek the internship experience to complement their academic courses.

The Internship Office of CBA is charged with the responsibility of facilitating students' contacts with corporate clients. In order to do so, the office must gather, record, and disseminate information to students and to corporations. In addition, the information must be kept private, accurate, and comprehensive. Typical information processes existing in the Internship Office relied heavily on paper entries, paper duplication, and mailings. However, in the Internship Office, the use of technology was lagging. An end-user within the Internship Office was enlisted to develop an information technology solution to the paper-intensive tasks of the office.

Issues of the case study include (1) the organizational behavior issues to overcome when implementing information technology even in an enthusiastic and sympathetic organization, (2) the difficulties in defining a system by an end-user even a sophisticated one, and (3) the obstacles of implementing a satisfactory system under tight time and security constraints even with the cooperation of a systems department.

BACKGROUND

Central Ridge University (CRU) is a large, research-oriented, public university composed of 10 major colleges, connecting over 15 branch campuses, and offering various continuing, distance, outreach and executive programs. Through this massive educational network, approximately 100,000 individuals each year study in a program affiliated with CRU. In order to keep pace with the demands of its broad-based and popular programs, CRU is an aggressive leader in deploying information technology in order to realize its mission to provide high-quality education to its students.

About five years ago, CRU instituted email accounts for all of its students, staff, faculty and administrators — regardless of their discipline, interest, or desire. CRU wanted its primary participants to have an electronic, always available, always vigilant access to one another. From the president of the university to the night watchman to in-coming freshmen, CRU issued email accounts. To support the effort, introductory and training sessions were provided to familiarize everyone; electronic kiosks were established for quick communications; state-of-the-art, high-capacity labs and networks were installed to handle the increased resource demand; information centers were staffed to be readily available for end-users' questions and university-wide licensed, communications software was offered to all university individuals and units. What resulted was a university-wide temperament for doing work, academics, communications, and appointments electronically without constraints.

One of the major colleges within CRU is The College of Business Administration (CBA). CBA accounts for about 10% of the student population of CRU, approximately 10,000 students. Its offerings span baccalaureate, masters, MBA, doctoral and executive programs through resident and distance education. Several of its departments are nationally and internationally recognized, often being forerunners in their fields.

A significant feature of CBA is the strong ties it fosters with corporations. Through a variety of programs, CBA cultivates relationships with corporations. Likewise, corporations eagerly strive to align themselves with CBA. The Internship Program realizes one strong symbiotic association between CBA and corporations. Initiated in 1978, this program helps CBA students find four- to eight-month work assignments at corporations prior to graduation. Students receive the benefit of practicing the skills from the classroom in a "real" work environment. Companies receive the benefit of assessing a student's skills without a long-term employment commitment. In fact, during downturns in the economy, companies use their internship population as pre-screened employees and students can usually rely on their internship site for a job offering.

The Internship Office manages the Internship Program of CBA. The office has counter-parts in the other colleges of CRU, but acts independently of these other internship offices in all aspects of its functioning. The Internship Office is one part of a group of services handled by the Corporate Support Alliance of CBA, an administrative unit reporting directly to the Office of the Dean. The organizational structure of CBA is presented in Figure 1; the Internship Program is presented in Figure 2.

The organizational linkages within the Corporate Support Alliance enable direct interactions between students, faculty, the business school, and the university with corporations — at the executive, division, or department level. Leadership of the Internship Program, however, is the

Figure 1: Organization Chart for CBA and Primary Administrative Units

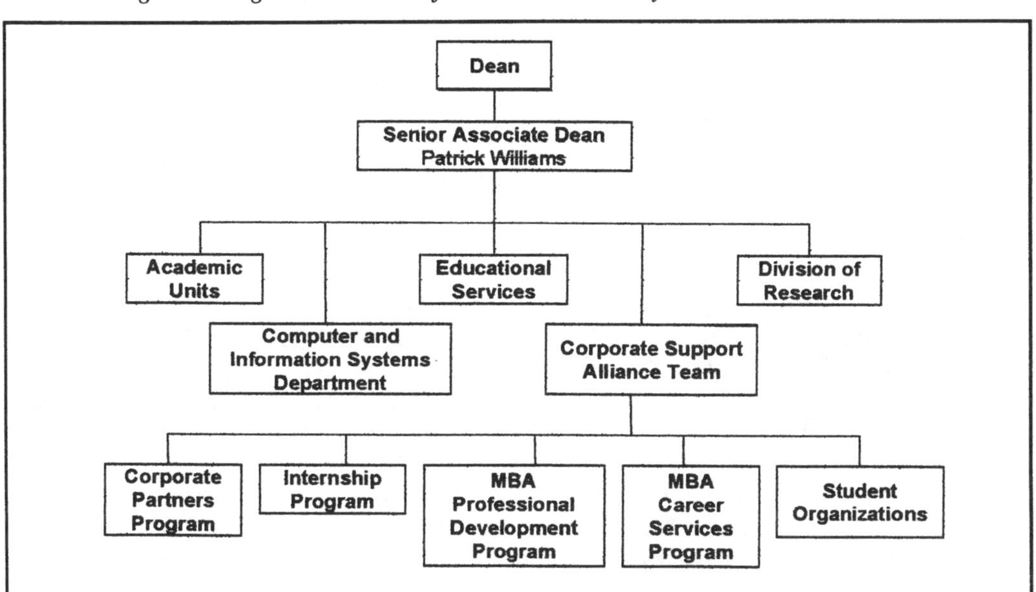

Figure 2: Organization Chart for the Internship Program

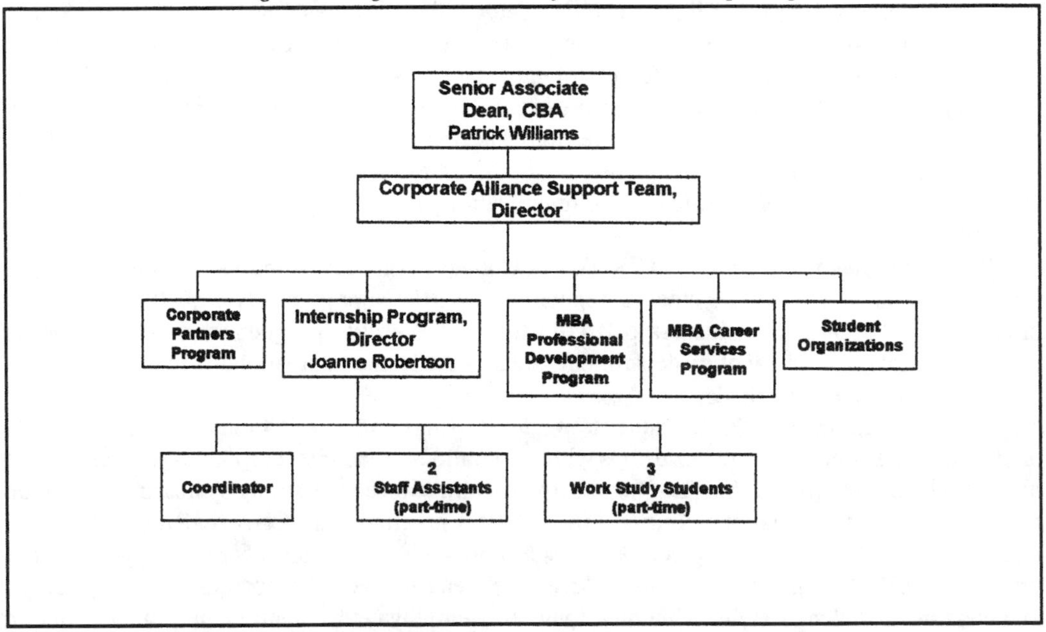

responsibility of its director, Joanne Robertson.

Specifically, the Internship Program provides services for two primary clients — students and companies. The first clients, the students, must be pursuing a major or a minor offered by CBA and must be enrolled at the main campus location. The location constraint exists to ease the interviewing process. The program does not place any grade point average criteria on students seeking its services.

The second clients, the companies, differ by size, industry type, and geographic location. The Internship Program has partnered with firms listed as Fortune 500 as well as with small firms local to the university. Global corporations with international hiring needs also connect with the Internship Program. The participating companies gain the advantages (1) of providing a source for temporary employees as well as for future, permanent employees, (2) of training and orienting employees through a low-cost, but high-quality avenue, and (3) of increasing the employee retention rate by enthusing their permanent workforce with the energy of the new hires.

SETTING THE STAGE

In the 90s, the Internship Program experienced tremendous growth in its student applicants and in its corporate affiliates. In the 1992/93 academic year, 633 undergraduate students had applied; 303 were placed. By the 1996/97 academic year, the number had nearly doubled to 1,245 students applying, 596 being placed. Similarly, the companies seeking interns also grew. In the 1996/97 academic year, the program saw a 23% increase in the number of actively recruiting companies to a current level of 235.

Like its parent organization, CBA usually advocates information technology options. Also, CBA students — indoctrinated with the electronic dogma of their university — seek electronic access for their tasks that included communicating with the Internship Office. The Internship Office, however, is a paper-driven office. By 1996, its entire annual postal budget of $4,000 was used to cover its mailings. While this situation is not unthinkable — given the amount of postal traffic required by the processes, it is unthinkable — given the electronic expectations of the students, of the companies, and of the university.

Even as an administrative unit, the Internship Office lacks information technology. Connectivity to the Internet and to CBA-affiliated intranets is non-existent. The office keeps all information about its clients manually, not utilizing database technology. Email access for the director of the office is through a mainframe mail utility, EMC2. The office staff had no direct email access. For the most part, the office computers are word processors. Capability with computer-based office automation is not a requirement of the staff positions.

This backwardness on the part of the Internship Program is not lost on the director, and these points are addressed in her strategic plan (submitted in mid-1996) for the office. Improvements in the technology base, increased space allocations, additional staff appointments, training opportunities, and more faculty involvement are the major items desired by the director. The director position needs to be updated, too. Rather than being a 10-month position, the position of director should be extended to a 12-month appointment. In the early 1980s ten months was sufficient time to handle the flow of traffic for internships, but with the increased volume of clients — students and corporations — and with maintaining paper-based processes, the Internship Program is becoming overwhelmed. Currently, the duration of the director's appointment matches the duration during which the Internship Program is open to students.

Changes would not be easy, however for a variety of reasons. Due to the organizational connection of the Internship Program to the Office of the Dean, any improvements to the Internship Office require direct consultation with and approval by the Office of the Dean. As a high-level administrative unit, the Computer and Information Systems (CIS) Department of CBA handles all technology improvements. (Its organizational layout is shown in Figure 3.) Due to the high profile of the Internship Program with corporate clients, changes in the processes of the Internship Program need to be handled judiciously. Upgrading the Internship Program would require administrators, directors, and top-level systems staff coordination and approval.

Although Robertson had directorship power over the Internship Office, the inclusion of information technology brought the CIS Department into the scenario. More importantly, since the

Figure 3: Organization Chart for the Computer and Information Systems Department

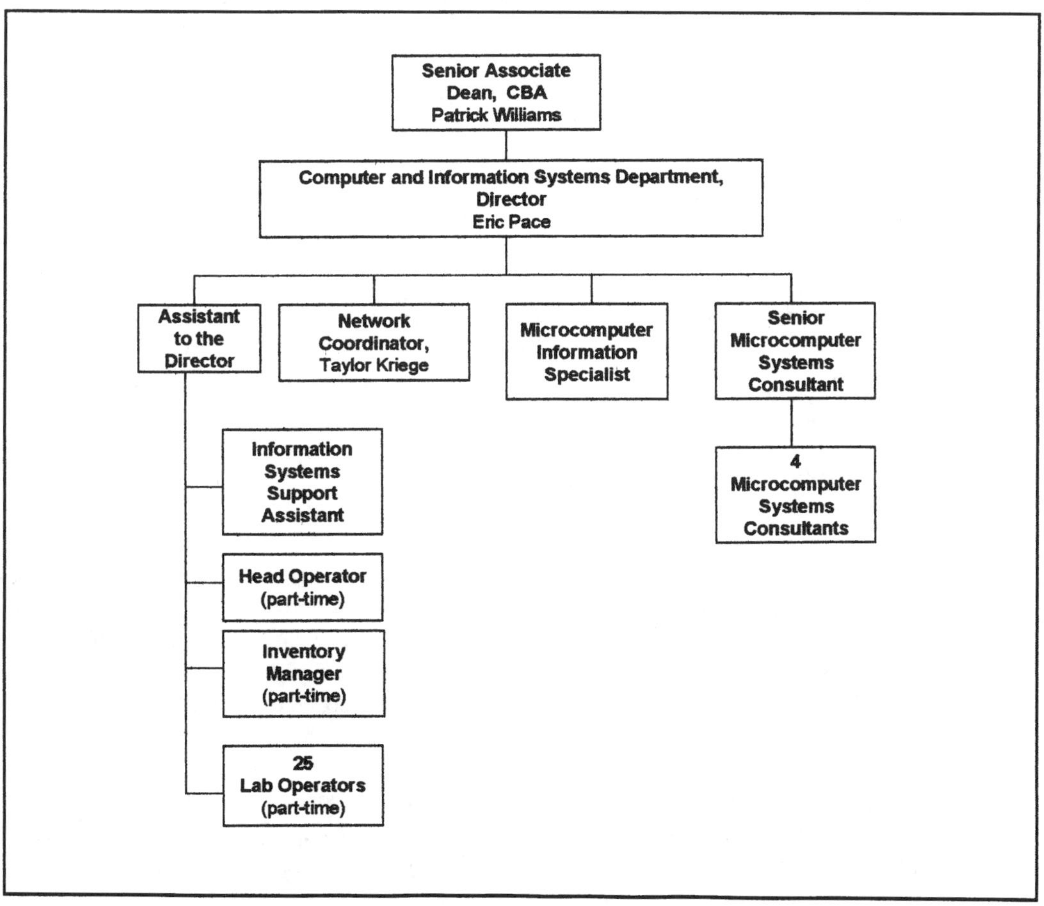

sought information technology required the network, Taylor Kriege entered the picture. Kriege's involvement is the linchpin in any technology solution requiring network services.

Kriege handles all new development of network-based processes for CBA— such as developing web-based services for the Dean's Office of CBA, initiating distance education links for the college, and establishing on-line course-links for high-profile MBA classes. His services are indispensable to CBA. Knowing this, he becomes a powerful figure in any network-based development project. Pace is not oblivious to the subversion power wielded by Kriege. Unfortunately, CRU does not have a huge technology-based population from which to draw employees nor does CBA have a large budget to coax high-demand professionals to its employ. In many ways, Kriege holds CBA hostage.

While Kriege's power is obvious, his leadership is completely lacking. Technical staff members who must interact with him in order to perform their jobs have a higher turnover rate than other staff members do. No one is directly under Kriege because he will not share his domain and because he repeatedly has alienated individuals who have temporarily been assigned to him.

Much has been written about the cultures and subcultures of organizations. Technology-based units often have a discernible subculture (Gregory, 1983; Krackhardt and Kildruff, 1990; Jermier et al., 1991). The subculture may function productively for the good of the organization or it may function maladaptively to the detriment of the organization. Kriege's subculture is whimsical — and as such, is a random element in any development project. Since CIS has no parallel resource for Kriege's competency, he goes unchallenged. For the most part, this arrangement is fine because his competency usually is highly in demand by units outside the technology group, dependent upon Kriege.

Kriege maintains his power over the organization by keeping himself as a limited, necessary resource. He completes projects of a high profile — such as those for the Dean's Office — in record time. If a project requires several technology competencies, his effort may be reduced. In these situations, the blame for the slowness in the project's implementation is spread across several parties. Directors and individuals outside of the technology group — who do not understand the interactions and needs of a technology-based project — will not be able to discern who is not contributing, who is causing the project to stall. Often individuals outside of the technology-based subculture see systems as being highly idiosyncratic. This perspective enhances Kriege's position with the necessary camouflage to be discriminating in his project priorities. Kriege also maintains his control from erosion by end-user development because of his responsibility over network-based processes. If the development effort requires the network, then the development effort will come head-to-head with Kriege.

Kriege's manipulations are not lost on Pace, but neither are his strengths. By judiciously selecting to be a master and a resource on high-profile projects, Kriege has protected his power base. While Pace and Kriege may disagree, the fact still stands that when Kriege performs well, Pace also shares in the success.

CASE DESCRIPTION

The Processes and Their Problems.

By the fall of 1996, the Internship Program had begun to concretely analyze its technology deficits. In order to gain insight into the specific changes, the director allowed a team of MBA students to perform a process improvement study on the program as part of a class requirement of theirs. The recommendations by the MBA-team paralleled what the director had previously outlined in her strategic plan, reinforcing her stand with CBA: the office needed to revamp its processes and to take advantage of internet-accessibility.

The processes of the Internship Office are shown in Figure 4.

Process 1: Student Registering. Each student is required to complete a personal data form, providing contact information, academic records, and work experience. The form must be updated each semester. As part of the Student Registering Process, the student is instructed

Figure 4: Level-O Data Flow Diagram of Internship Program Processes

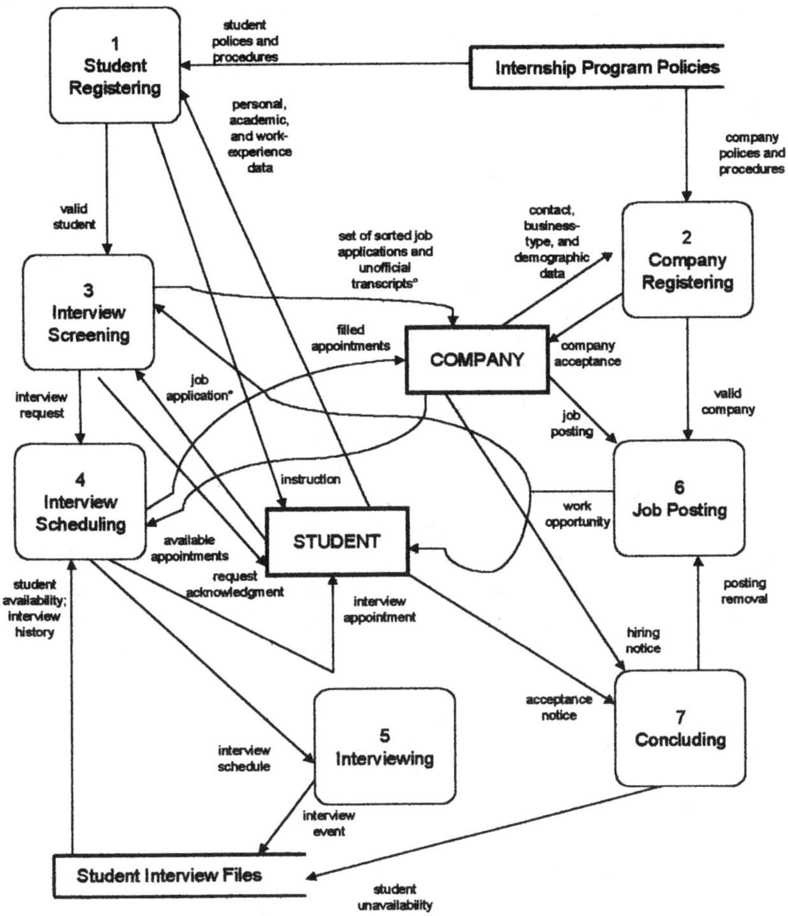

job applications*—two-page document, each student averages approximately 7 applications a week.

set of sorted applications*—a job posting will generate between 200 to 800 applications, each which must be accompanied by the student's unofficial transcript.

about the policies and procedures of the Internship Office. This process identifies a valid student who may request interviews with companies.

Process 2: Company Registering. Each company must formally request to seek interns through the Internship Office. The company provides contact information, business-type identification, and demographic data. A "company" may be the corporate unit, a division of a corporation, a government unit, or even an academic research unit. The companies and their positions are screened to insure educational, well-managed, and beneficial experiences for the interns. Upon completion of this process, the company is notified of their acceptance into the Internship Program.

Process 3: Interview Screening. A student submits a completed job application for a position — physically posted on the bulletin boards of the Internship Office — to a drop-off box. The applications are collected every Friday. Upon collection, the applications are sorted by position, requests are checked for validity, unofficial transcripts are included, company sets are mailed, and students are notified of the mailing.

Process 4: Interview Scheduling. Within the next two to three weeks, a recruiter for a company would receive the set, review the applicants, select a set of candidates to interview, and identify a set of dates to have campus-based interviews. Upon receiving notice of the company's intentions — notified by fax or phone, the coordinator makes a phone contact with all students in the set and informs them of their status. If a student has been selected for an interview, the student must physically come to the Internship Office and select an appointment time.

Process 5: Interviewing. The Internship Office manages the schedule of on-campus interviews. Companies and students meet in rooms arranged for by the Office.

Process 6: Job Posting. For each position, a one-page job description must be submitted and accepted by the Internship Office, at least two weeks prior to its posting. The acceptable positions are catalogued and posted.

Process 7: Concluding. Once a company selects an individual for a position, the Internship Office is notified and the description is physically pulled. If the selected individual is registered with the Internship Program and accepts, his availability status is changed, future interview dates are cancelled, and specifics about the offer (such as salary and work site) are collected.

Making the internship connection between students and companies requires several documents to be received, catalogued, verified, acknowledged, duplicated, and mailed. Only in the instance of notifying the coordinator with the list of selected interviewees (Process 4) is faxing an option. Most of the acknowledgements of events made to students and to the companies are handled by phone.

The students, the companies, and the Internship Office, itself, are inconvenienced by the time and effort constraints implicit in the processes. The Internship Office is located on the edge of campus and is not on the route of the average business student's travels. A student physically came to the office (1) to register with the program, (2) to check the postings, (3) to apply for positions, (4) to schedule an interview appointment, and (5) to be interviewed. A company physically contacted (through mail or fax) the office (1) to register with the program, (2) to submit job postings, (3) to supply an interview slate, (4) to supply interview dates, and (5) to interview. For both the company and the student, only the interview needed to be a physical contact; all other events could have electronic substitutions.

These typical, clerical procedures were not effective as the Internship Program grew. The increase in the office traffic, alone, made the office difficult to access, let alone, to work there. Staff members felt increasingly pressed for time. Some students stopped using the service of the Internship Program because of its time-consuming, tedious procedures. Companies complained about the time pressures placed on them to provide early notice of a position and then also to provide quick turn-around on their hiring decisions.

Proposed Solutions

In early 1997, Joanne Robertson, the director of the Internship Program, had her (part-time) staff investigate pre-packaged software programs that would incorporate database and web technology together. Her goal was to provide the office with a paperless system that required minimal to no amount of technical expertise. The selected package was a product by FirstPlace Academic Software, Inc. (Throughout the remainder of the case, the product will be referred to as FirstPlace.)

Knowing that any technical changes in the office would require the help and support of the CIS Department, Robertson requested the network coordinator, Taylor Kriege, to evaluate the feasibility of the package. FirstPlace Academic Software assured Kriege that the software would be compatible with the Banyan network used by CBA. When asked directly by Kriege, "Is FirstPlace able to work with a Banyan networked system," the reply was consistently, "Yes." With these assurances, the CIS Department gave their administrative approval for the purchase of FirstPlace. In the summer of 1997, the Internship Office used its funds to purchase FirstPlace, sent its staff for training, and scheduled work orders for the installation. Unfortunately, FirstPlace and the Banyan network were not immediately compatible and could not be made compatible — a fact that did not become verified until late 1997.

In the fall of 1997, the Internship Program hired a technology-savvy, work-study student, Jacob Billings, to develop a custom-configured, interim solution until the difficulties with the purchased software could be resolved. Billings was competent with systems, but his major, accounting, was not a systems-related one. Unfortunately, at this time, the Internship Program has entered another fully packed, 10 months of operation. No solution could be speedily implemented due to time constraints on the Internship Program, its staff, and the CIS Department staff.

Billings proposed a system with the limitations and capabilities of the Banyan network, the Internship Program, and the CBA community in mind. Data was to be collected from students and from companies via web-based forms. The forms would reside on a secure-socket web site. (A secure socket is an encrypted portion of the Internet.) Pages and data transferred from a user's terminal to a web server were to be encrypted. The web server would use a Perl script to transform the data into an email message. The Perl script would call the encryption program PGP (Pretty Good Privacy) (Zimmermann, 1995), would encrypt the email, would send the message to a staff member and would unencrypt the email. The email message would be handled by macros within a Microsoft Access (Access) database to convert the email text into password-protected, table data.

Job postings could be collected from companies via paper or electronic form. The information would be converted into an HTML document and electronically posted to the web. Students would review these postings and would submit an electronic form to request an interview. The application data and unofficial transcripts for the set of all students requesting interviews for a position would be sent from the file server to the company's fax using Microsoft Exchange.

In fall 1997, the Internship Office believed it had a choice between two alternatives: (1) wait until the pre-packaged software became useful and continue with the manual system or (2) wait until the pre-packaged software became useful and in the interim implement the custom-built system. In order to make the best decision, Billings analyzed both in terms of the option's operational and technical feasibility. (See Whitten, Bentley and Barlow, 1994, pp. 811-833 for an explanation of how to perform these feasibility assessments.) Schedule feasibility was less of a concern at this time; another 10-month interviewing cycle had begun, making all other activities conducted by the Internship Office of secondary importance — including process renovations.

To classify the development situation at the Internship Office, he also examined it with respect to James Wetherbe's (1988) PIECES framework. He conducted these analyses to discern if the FirstPlace package should have ever been pursued or if a custom system would have been better. The analysis was of interest to him personally because (1) he wanted to understand project management methods from an IS perspective rather than from a pure accounting perspective and (2) he wanted to understand how the custom system fared with respect to the packaged option.

Also, economic feasibility was not relevant because of the sunk costs of the purchase and training. If the CIS Department could identify how to interface FirstPlace with the security protocols of the Banyan network, then the investment in FirstPlace stood a chance of continuing. If FirstPlace could not interface with the security protocols of the Banyan network, then a different alternative — either manual processing or another automated alternative — would be necessary. However, when Billings did his feasibility comparisons, he was seeking to identify if FirstPlace should ever have been a target system for implementation. At the time he started his comparisons, the technical difficulties with FirstPlace were hoping to be resolved. Ultimately, security and ease-of-use would be the major issues in the decision.

Operational Feasibility Comparisons

The performance of the new system is a key consideration for the Internship Office, but both options give the office better throughput and response times to its processes. The office would not require firms or students to use an Internet connection, but would offer the choice. With this strategy in mind, the Internship Office could enhance their operations, could respond to the wishes of their clients, and could continue to place students in meaningful positions.

In terms of enabling the office to use the collected data in a timely and appropriately formatted mode, the custom system has an advantage. At the end of each day, macros of the Access application would convert the collected data into an Access form. Then, the Internship Office could use the query and report generating capabilities of Access to probe and to display the data. FirstPlace would gather the data and make it available to end-users immediately through a Paradox database. Unfortunately, the query interface and report generator of FirstPlace are not able to compile and to display the data as needed by the Internship Office. While the custom system requires a day to structure the data into usable form, more than a single day of staff time would have been consumed in order to structure the data available through the FirstPlace version.

In terms of how either system would impact the economics, efficiency, and service of the Internship Office, both are acceptable choices. Both offer to reduce the paper costs, to reduce the postal expense, and to save the costs of the staff's time for sorting and mailing applications to companies. Both systems offer increases in efficiency in other ways, too. Both offer a scheduling feature, effectively eliminating the need for students to physically travel to the Internship Office for interview scheduling. Both collect student and company data online through Internet interfaces. This collected data is available for companies as well as the Internship Office to review online.

The major difference between the two systems is the security control offered by the option. Due to privacy issues, student data must be safeguarded with the utmost care. The custom system transmits data in an encrypted form during the entire transmission. Forms and data would be sent from the server to the browser using secure socket protocols. Data sent back to the server would use a secure socket protocol, also. FirstPlace would transmit the data as plain text. Forms and data transmitted from the server to the browser are unencrypted. Furthermore, FirstPlace did not have the capability to use secure socket web pages or any other type of encryption.

Finally, with respect to flexibility, the custom system would be better. The Internship Program needs a process solution to adapt as the program changes and grows. A software package would offer changes through its updates and new versions, but these changes may not produce features desired by the office. FirstPlace is limited in its server option and in its Web-site software. Alternatively the custom system offers greater potential for flexibility because it is customized. An individual as technically proficient as the work-study student, Jacob Billings, could add any changes. Also, making any changes to the custom system would be easier to make as compared to trying to revamp a purchased package to fit a unit's data needs.

Billings did not apply business re-engineering methods to the processes of the Internship Office. In the field of business re-engineering the conventional wisdom is to inspect current processes for their efficacy rather than simply automating — applying information technology — to them. In the case of the Internship Office, the flow of data and information was prescribed in order to protect the privacy of the students and to control access to the companies.

In any software develop initiative, management support is required for its success. Robertson was in favor of any system to improve her office. Eric Pace, director of the Computer and Information Systems Department, who had final responsibility for the security of any CBA-related system, could not accept the security lapse inherent in the FirstPlace option. With his insistence and support, the custom system took sway over the pre-packaged option. With Pace and Robertson favoring the custom system, Patrick Williams, senior associate dean of CBA, provided his administrative backing to its development.

Equally important as management support is end-user support. Initially, the staff members of the Internship Program were concerned that a custom system would be inherently more difficult to use than a packaged system. Also, they felt better prepared to use the packaged system due to their training over the summer. The staff members were particularly concerned about their ability to formulate queries.

Although they were hesitant about the ease of use of the custom-built system, pieces of the custom system became part of the Internship Office by late fall 1997. As the staff members were freed from some of the day-to-day processing requirements of the paper-driven system, they were able to focus on serving the clients, on evaluating the strategic role of the office, and on learning the manipulations of the custom system. With Billings present in the Internship Office on a daily basis, the staff had immediate answers to any of their questions. As Billings added pieces to the custom system, he would walk the rest of the staff through addition and handled their concerns as they arose. More so, as the staff began using the custom system, the information-retrieval potential of the system became more clear to them. Since the staff knew the data and processes intimately, their questions were regular, parametric queries. Then Billings easily added any new parametric queries to the custom system. Summary reports and queries were also well articulated due to the clear boundaries of the purpose of the Internship Office. Formal training on the custom system was never scheduled; the staff learned the system by using it, by its straightforward design, and by the presence of Billings as its technical liaison. Usage of the system soon swayed the staff in favor of it and eased any concerns they had about it.

Technical Feasibility Comparisons

To assess the technical potential of the two options, Jacob Billings compared the practicality of the solution, the availability of the solution's components, and the availability of relevant technical expertise. He compared the systems early in fall 1997 in order to promote his custom system while the problems with FirstPlace were being examined. Practicality is an important aspect of any undertaking. Both options provide the Internship Office with mainstream, reasonable approaches. Both options need an additional component to be procured, but in neither case, is the lack of the component a barrier to its implementation. FirstPlace requires the specific Web-site software to be downloaded from a Web site; the custom system requires a secure socket Web site and the PGP software.

In terms of technical expertise, both systems have their disadvantages. The network coordinator, Kriege, could easily install FirstPlace, but the Internship Office staff did need to go through the training program provided by FirstPlace Academic Software — which they had done in the summer of 1997. The implementation tasks of the custom system are split between Kriege and Billings. Kriege handles the secure socket issues. Billings handles the specific tasks required by the Internship Office to interact with the web-interface such as building the database, training the staff, writing the data conversion scripts and testing the system. While the high-level skills of Billings would be necessary for the development of the custom system, a competent staff member could handle its daily operation and maintenance.

In terms of person-hours required — and the associated costs — for developing the two systems, the FirstPlace option required approximately 8 hours for installing and testing and 80 hours for the training of the two staff members for a week off-site, totaling 88 hours. Due to the expense of the individuals involved and the travel expense of the training, the packaged-software option costs about $10,000. The custom system required approximately 50 hours for database structuring, 5 hours for installing the secure socket, 20 hours for creating and testing the scripts, 20 hours for the web-site construction, and 10 hours for the Perl script addition, totaling 105 hours. Training on the custom system was not evaluated because the staff never had to suspend their normal work responsibilities for training sessions. The training was "on the job". Since the major portion of the work was handled by the undergraduate assistant, the expense of the custom system is about $2,500.

Organizational Interactions.

Through most of fall 1997, the custom system was projected as the interim solution while the suitability of using FirstPlace with a Banyan network was investigated. When it became apparent that FirstPlace was not compatible with the network and could not support secure processing, the custom system became the target to implement. Robertson made this decision once Kriege could not obtain any technical confirmation to indicate that FirstPlace would work — ever — with a Banyan network.

However, this change also caused the Dean's Office and the CIS Department to become more directly involved in the process. Coordination meetings routinely involved Williams, the Senior associate dean; Pace, the director of the CIS Department; Robertson, the director of the Internship Program, and Billings, the undergraduate assistant, but not Kriege, the Network Coordinator. In other circumstances only Pace, Robertson, and Billings would have met. Williams entered the meetings as a high-level connection between two units under his authority, the Corporate Support Alliance Team and the Computer and Information Systems Department. What had begun — as simply an effort to upgrade a service unit of CBA — became a focus of the administration.

Since the Internship Program is under the direction of Robertson, she had the responsibility for coordinating all of the many tasks required for the system's construction. However, she had no authority—and was given no authority by Williams—over the individuals outside of her unit, within the CIS Department, who would handle the network aspects of the custom system. With this difficulty, the timeline for the custom system was solely dependent upon the auspices of the CIS Department since all tasks directly under Robertson's control had been completed.

The CIS Department did not see its tasks on the custom system as a priority, but placed the work on its schedule. Ironically, the only task to be placed on the schedule was the Perl script since the secure socket web site had been completed in November 1997. Without the Perl script, the custom system was stalled. For the want of this simple piece of code, the final stages of the custom system could not be realized. The Perl script allowed the custom system to interface with the network, Kriege's domain.

Billings could have easily developed the script, but was not permitted. CBA departmental protocols firmly classify systems-related tasks as being either inside or outside the domain of the CIS Department. Once a task is classified, the culture of CBA dictates that progress on a task is under the authority of the CIS Department director, Pace. However due to the established delegation of authority within CIS, Pace left the scheduling and work decisions of network-related tasks to Kriege's discretion.

Pace's laisser faire attitude towards Kriege's domain replicated the culture across and within the service units of CBA. Pace might get annoyed with Kriege, but he would not interfere. Just as Pace would not directly interfere with Kriege's domain, Williams would not directly interfere with Pace's domain. Robertson and Billings could only petition Kriege for a change in the work schedule, but could not induce the effort. Williams expressed displeasure at the lack of movement on the project, but unless he was willing to revamp the lines of control and authority throughout CBA, the system would be stalled. This situation was made evident at a meeting between Williams, Robertson, Pace, and Billings (note, Kriege is not present):

> *Williams (pointing to Robertson and Pace): I want you two to stop complaining about each other and get the job done.*

With this, Williams is essentially allowing the normal power struggles to continue. Alternatively, he could have allowed Robertson to out-source the work pending from the CIS Department. However, out-sourcing the work would have been a change to the CBA organizational behavior, too, and not ever really considered.

System Status

By the end of fall semester and Billings's graduation, the custom system was not fully completed. In fact, by June 1998, the system still had not been completed.

CURRENT CHALLENGES / PROBLEMS FACING THE ORGANIZATION

When the Internship Program began to upgrade its processes, technology seemed to be the greatest hurdle. Twelve-months later, the greatest hurdle was the organizational behavior of CBA. Technology can be bought, learned, or "worked-around." If the flow of control within an organization remains strictly along lines of authority, then the success of a project may not lie in the technical abilities of its participants.

The main challenge expressed by this case is the one faced by CBA: How to deploy interdepartmental systems without involving the highest-level of management that links the departments while maintaining standard lines of authority. Many organizations, like CBA, rely on a systems division to realize information technology within the corporation. These organizations rely on the director of the systems division to be aware of the importance of a project, to be sensitive to changes in the importance, and to communicate these changes to the staff. Once Williams became a participant in the coordination meetings, the importance of the project had increased by default. Also, the presence of Williams in the meetings implied the project now spanned departmental lines — the project would effect all units under Williams's authority. Yet, Pace did not effectively communicate this change to his staff, particularly Kriege. And yet, Robertson was not empowered by Williams with his authority to influence the staff. Once, hierarchical lines of authority are reinforced, the success of an interdepartmental operation will be under the good will and auspices of the systems group.

ENDNOTE

[1] "Central Ridge University" is a pseudonym as are all names and labels in the case. In all other respects, the case is an accurate depiction of the situation. The data comes from authoritative, but private, sources.

REFERENCES

Gregory, K. (1983). "Native-view paradigms: Multiple cultures and culture conflicts in organizations," *Administrative Science Quarterly,* 28, 359-376.

Jermier, J.; Slocum, J.; Fry, L. and Gaines, J. (1991). "Organizational subcultures in a soft bureaucracy: Resistance beyond myth and facade of an official culture," *Organizational Science*, 2, 170-194.

Krackhardt, D. and Kilduff, M. (1990). "Friendship patterns and culture: The control of organizational diversity," *American Anthropologist,* 92, 142-154.

Pheysey, D. C. (1993). *Organizational cultures: Types and transfromations*. London, New York: Routledge.

Wetherbe, J. (1988). *Systems analysis and design: Traditional, structured, and advanced concepts and techniques.* (3rd ed.) . St. Paul, MN: West.

Whitten, J. L.; Bentley, L. D. and Barlow V. M. (1994). *Systems analysis and design methods.* (3rd ed.). Burr Ridge, IL: Irwin.

Zimmermann, P. R. (1995). *The official PGP user's guide.* Boston: The MIT Press.

BIOGRAPHICAL SKETCHES

Theresa M. Vitolo teaches management information systems at Penn State, University Park. She earned her doctorate from the Interdisciplinary Department of Information Science at the University of Pittsburgh. Prior to her teaching career, she had worked as a systems analyst / programmer on a variety of systems development projects. Her interests include intelligent interface design and other issues in the field of human-computer interactions.

Aaron J. Sparks works for Deloitte & Touche as a consultant. Prior to working in industry he was earning his academic degree in accounting. He has been extensively involved in a variety of systems development efforts from the end-user development perspective. Currently his industrial projects have centered on SAP-related assignments.

Problems, Their Causes and Effects in the Use of Information Systems: A Case of a Scientific Library

Katariina Jalonen, Mika Kirveennummi and Vesa Torvinen
University of Turku, Finland

EXECUTIVE SUMMARY

Information technology (IT) has radically improved many aspects of organisational activities. Computer-based information systems (CBIS) are constantly developed more effective and efficient. Development of a new CBIS is justified by higher quality of work, more efficient work processes, and more flexible work practices. However, introduction of the new IS can produce a variety of problems.

This paper describes the problems observed in the use of a library CBIS in a Finnish scientific library. The results of the study illustrate well the environment which should be understood by the designers of computer-based information systems. One important prerequisite of good (re)design is a wide understanding of the problems that may prevent effective use of a CBIS. In this study we introduce a classification based on the problems which were found in the case. The classification describes the causes and effects of the observed problems.

BACKGROUND

The study was carried out in a scientific library which hierarchically is a special unit of the University. The library is administrated by a chief librarian and governing board, consisting of the chief librarian, one member appointed by the rector of the University, and six members appointed by the Council of the University.

The library is organised into 11 departments. Four departments deal with administrative and acquisition related tasks. One of the departments is the central library which is in charge of the general collections. The remaining six units are faculty libraries (mathematical and natural sciences, humanities, education, medicine, law, and social sciences), which are responsible for specialised collections.

In general, the activity of a library is characterised by two different work processes. The goals of these processes are different and, in some sense, also conflicting. On the one hand, the library increases and maintains its collections. On the other hand, the library serves the customers by lending volumes from the collections. From the first point of view, the situation would be ideal when the collections are complete and all the volumes are in the shelves. For customer services, it is important that the volumes are borrowed by the customers. In an ideal situation, all volumes would be checked out. For the organisation as a whole, the most important thing is to achieve both goals in an optimal way.

The main stakeholders of the library's activity are different types of customers, publishers, brokers, bookstores, other national and international scientific libraries, Customs, and university administrative bodies. Typical customers are university researchers, teachers and students. The library's work practices are partly defined by the nationwide community of scientific libraries in Finland. Publishers, brokers, bookstores and Customs are related to the acquisition process of the library.

The library has 118 employees (see Table 1). Half of the employees hold a university degree which suggests that the organisation has a strong expert flavour. The organisation is female-dominated, and consequently the wages are low.

There are five main professional groups in the library. The largest group consists of assistant librarians. They are employed in almost all the departments, they are a little younger than the other employees, and about 40% of them are doing either their professional training period, civil alternative service, or government supported work period because they had been unemployed. Assistant librarians work typically in customer service or maintenance of the collections, they have low income, and do not have any education in library science or related disciplines.

Librarians are the second largest group of employees. A typical librarian is female, has a permanent position, and holds a university degree in library science. As a group, librarians are not specialised in any specific type of tasks, even though administrative tasks were often included in their jobs.

Library secretaries form 13% of the staff. Typically, library secretaries have a temporal position and a library science related degree. Library secretaries work in customer service, and do cataloguing, acquisition, or maintenance related tasks.

The library has 13 library clerks. A typical library clerk is female, somewhat older than the other employees, has a permanent position and a library science related university degree. They do mostly cataloguing related tasks, but also maintenance and customer services.

Of the minor professional groups the librarians who are specialised in information service, are the most important. This library employs only two information specialists. They are highly skilled in the use of different types of information systems, they train the customers, and they are better paid than the other employees.

From the perspective of librarianship some alarming, rather global changes have taken place in the library industry. Firstly, the increasing use of IT in publishing is a challenge for the libraries. For example, electronic publications have already radically changed the traditional domain of the library, causing uncertainty among employees as they have to modify their work practices. Secondly, the development of IT has made it possible for the customers to take over traditional tasks of librarians. The tendency to increase self services is seen as a threat for the proficiency of librarians. The employees are frightened of the changes because their professional status is weakening. The pessimistic view of the future of the profession is embodied in low-educated assistant librarians. Their number has grown radically which reflects the weakening position of librarians. Furthermore, the number of the top professionals in librarianship, i.e. the information specialists, is next to nothing.

Table 1: Employees, collections and CBIS

Staff	118	Collections, total	2,100,000
Professional groups		• annual accumulation	50,000
• information specialists	2%		
• librarians	26%	Central CBIS	
• assistant librarians	35%	• employees using the system	84%
• library secretaries	13%	• working hours used on the central system	55%
• librarian clerks	11%		
• others	13%	• working hours used on other systems	16%
Education			
• university degree	51%		
• lower education	49%		
Gender		Collections (1997)	
• male	32%	• in CBIS database	32%
• female	68%	• manually updated	68%

The use of IT in the case organisation was the primary interest of the study. A central CBIS supports the core activities. Altogether 84% of the employees use the system, and the average use of the system was 55% of working hours (see Table 1). Furthermore, the organisation applies around ten other important computer applications. The employees who use the central system in their work also use these applications around 16% of working hours. In other words, 84% of the employees spend 71% of their working hours using some CBIS, which suggests that the organisation is highly dependent on information technology.

The importance of the CBISs is more obvious if we look at the pool of information which is stored, updated, sorted and searched by the employees and the customers. The total amount of volumes in the library is huge, around 2,100,000 copies. In 1997, there were about 673,000 volumes in the CBIS, and the rest of the collections filed in manual card catalogues. However, not only is the pool of information about the existing collections huge, but also the information concerning orders and acquisition of new items is significant. The collections increase by about 50,000 copies per year. Furthermore, the information related to the customer service process, i.e., ordering, borrowing and returning publications, is an important part of information flow from the perspective of information technology.

SETTING THE STAGE

The library system has been developed by an international software house. It is used in every scientific library in Finland with tens of thousands of users in the country. Further, different releases of the system are applied at least in several European countries and in the USA.

The introduction of the IS has changed the nature of the activities in scientific libraries in Finland. The system has facilitated totally new forms of cooperation and communication between the libraries. It has integrated the departments within each library. In addition, the system has allowed more effective utilisation of collections by the customers.

Although the system has radically improved the quality of library activities, the users have encountered a number of problems. However, no complete understanding exists about the quantity or quality of the problems related to the system. Furthermore, the organisational structure needs apparently to be redesigned, which would set new requirements for the system.

CASE DESCRIPTION

The organisation invited a group of IS researchers to study the deployment of the CBIS in the library's work processes. The research group spent nine months in the organisation in order to gather information about the problems perceived and experienced by the users. Data were gathered in several ways: by using the system, by examining the documents produced by the organisation, by observing, by performing two questionnaires, and by interviewing. One of the questionnaires was filled in by the employees (all employees, n=118) of the library. Customers' viewpoints were studied by the other questionnaire (n=200). Further, 15 employees representing different phases in the major work processes were interviewed.

Parallel to the research, an organisational development project was started. The goal of the development project was to enhance the effectiveness and efficiency of work processes. The development project modelled the work processes in order to document the existing work practices, to compare the differences between the organisational units and to analyse the problems in the existing work processes. The results of this project will be used as the starting point in a future project, which will concentrate on the reorganisation of work processes.

The activities of the library were analysed both as a whole and as separate work processes. Two main objectives were found when the activity was observed. On the one hand, the goal of the library is to accumulate and store the collections, which was manifested as comprehensive and well-documented collections. On the other hand, the goal is to utilise the collections which takes the form of fast and flexible circulation of copies to the customers. These goals of the library are in a way contradictory. To oversimplify, the goal is achieved when either none of the copies is checked out or all the copies are checked out.

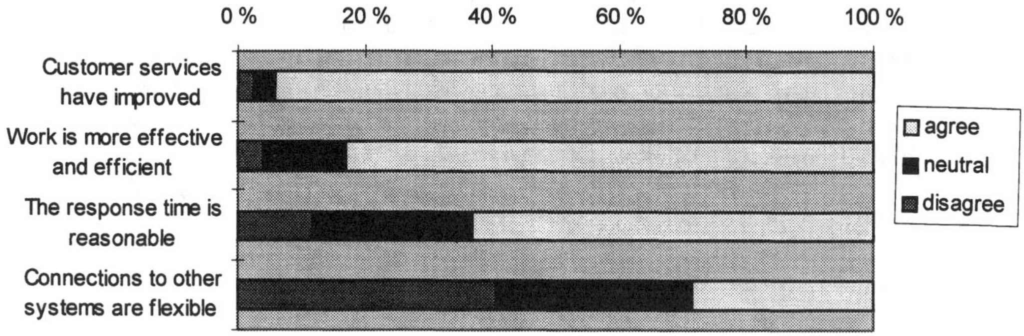

Figure 1: Some opinions of the employees about the influence and characteristics of the system (n=99).

These two goals were manifested in the main work processes of the library: the acquisition and service processes. The researchers modelled the processes by using generalised process diagrams. The diagrams included information about the sub-goals of work phases, and the ways to achieve the goals in different organisational units. The tools (e.g., information technology) which were available to support the performance of work tasks were analysed, and the problems in their use were gathered for analysis.

PROBLEMS FACING THE ORGANISATION

The CBIS had an important role in the library's activities. The benefits of the system were visible in the every day work of the employees. Most users thought that the system had improved the customer service. Further, the users considered that the system had improved the effectiveness and efficiency of work. The response time of the system was reasonable for most of the users, but the connections to other systems were not flexible enough in some tasks. (See Figure 1.)

The employees were able to tell quite accurately about the problems in the use of the system. However, evaluation of the causes and effects of the problems seemed to be more difficult. Because the knowledge about the problems was fragmented, the users had difficulties in seeing the overall pattern of problems. They had to tolerate these problems in their daily routines and find alternative ways to perform their tasks in order to avoid the problems.

The most obvious problems were of a technical nature or indicated weaknesses in the user interface. However, problems also derived from the way of applying the system, lack of user skills, and even from the organisation of work processes. Because nobody had a clear view of the possible causes and effects of the problems and the overall pattern of them, the organisation tried to compensate for the problems by improving the employees' skills in the use of the system.

Next, the daily problems facing the users as well as the causes of the problems are discussed in detail. The problems can be classified into five main groups: technical problems, problems in the user interface, problems in the usage of the system, problems related to skills and knowledge, and problems related to division of labour and work processes. Some instances of the problems and their causes belong to several main groups.

Technical problems

The problems classified as technical problems were derived from both in hardware and software. Hardware problems were related to efficiency and telecommunications. Efficiency of the hardware became apparent in continuous starting and closing different software applications during the work. Also, breakdowns in telecommunications caused serious harm to the work. Because many employees worked up to 90% of their working hours with the system, the breakdowns in telecommunications constituted a serious problem.

Obvious errors in the software caused serious problems in the work. Most of the problems were accumulated to the service process. Software errors covered, for example, unexplained response of software while pushing a certain button or while working with a certain group of customers, the malfunction of a control parameter or other illogical behaviour of the software. A part of the software did not work at all. From the users' point of view, errors in the software were especially difficult because they appeared always unexpectedly and were related to certain work situations. Some errors were found only by chance. Thus, nobody in the organisation knew the harm the error had already caused or its final effects. Because the organisation could not correct the errors in the software package, the users demanded a great deal of creativeness in order to cope with software problems. It was only possible to report the errors. Consequently, the only way to handle the problems was to tolerate them or to get around as well as possible.

> "...and then there are these bugs that some programs don't function as they should. Just like this one. If a book has not been returned and you can see that two reminders have been sent. In that situation you could send a bill. But you can't. You have to contact the customer first because you can't be sure that the reminders really have been printed out by the system... it's quite unfair that you get a bill, if you haven't got even a reminder."

There were a number of serious shortcomings in the software which actually cannot be called errors. For example, some data fields and pieces of software which would have been of value for the activity were totally missing. The integration between different parts of the software was insufficient. Also, some parts of the software simply did not fit into the work practices of the organisation, because the software was originally developed to support a different organisational culture.

In all, the organisation could not influence the features of the CBIS which caused the technical problems. This was very frustrating for the employees, because even if they were aware of the problems, it was practically impossible for them to solve the problems. One employee described the process of correcting the software errors as follows:

> "Yes, those [development suggestions] have usually gone through these teams and then they are treated on the European level and then only after that they have been sent to the USA as suggestions. Generally they have been ranked so that the developers can decide the order of implementing them and....and.... it is not a very quick process... to get further that way. One could be about these different releases of the library system that only some of the bugs have been managed to get rid of. And it has happened that when one bug has been removed it has created others somewhere else. Always, when you start working with an updated version you have to be very sharp, because you never know what has changed."

The problems related to the technical aspects of the CBIS actually were manifested in two ways: the system did not behave as expected or some information "disappeared." In addition, the technical shortcomings caused unnecessary work, and some tasks were put off or totally given up. In some situations, the necessary information was hard to find which hindered the customer service.

User Interface

The weaknesses in the user interface caused various problems. Generally, the user interface was found very uninformative. This was not a surprise, because the interface was character-based and the system was used by line commands. The commands usually consisted of only one or two characters. Further, the commands referred to English words and thus were not so easy to remember for Finnish workers, which decreased the usability of the system.

To compensate for the weaknesses of the user interface, a number of user manuals had been written in the organisation. However, the manuals were of poor quality. Only 7% of the employees reported that they use the manuals to cope with problem situations. The weaknesses of the user interface, the inconvenient way of using the system, and the poor quality of the user manuals together resulted in that it was difficult to learn independently to use the system.

> "Well, this really is a very complicated system, although I've got used to it and don't have any good point of comparison. I have always "floundered" with this record format. Earlier, I worked in another library and there was also the same system. It was a long time ago. But it is a fact that

you have to browse through many screens. Of course I'm not the right person to say something is complicated, but this system really is constructed quite..."

As mentioned earlier, the actual operative use of the system, for example, remembering a command was difficult. Furthermore, certain tasks were quite troublesome to perform. For example, editing an existing data field was so difficult that the users preferred deleting the old data and rewriting the whole thing. In some cases, the users experienced the manual system even better than the computerised.

"...[correcting mistakes] could be simpler. I'll give you an example here. If I want to add something, for example, to line 6, first I take the line up and then I move the cursor with the space key [to the right position]. And if I want to add 'm' here I have to write 'i' [insert] under the next letter and then 'm'. Now it's there and then I have to save this record."

The shortcomings and weaknesses of the user interface had various consequences. Some data in the database were insufficient or faulty, and sometimes very difficult to search for. Also, finding the right command was difficult and sometimes the work practices turned complicated. Difficulties in interpreting the meaning and functionality of data fields reduced the quality of the interface. Furthermore, the employees had to bear in mind many details related to the system. The user interface caused extra work and delayed some tasks. Because the user interface to be used by the customers was also very primitive, the shortcomings caused extra work to the stressed staff. In this way, the low quality of the interface weakened also the standard of customer service.

Usafe of the CBIS

Difficulties in defining the value of certain software parameters caused unexpected behaviour of the CBIS. In some service-related tasks, the users suddenly observed loss of information which hindered the coordination of the activity. For example, information of claims was lost when a customer returned a loan, irrespective of whether the returned item was the cause for the claim or not. The consequence was that the librarians had no possibilities to control that the customer paid the notification fee. Therefore, the problem caused some loss of money.

Timeliness of the information in databases was not adequate. Firstly, some tasks were performed manually, because it was found better and more practical. Thus, all the information in the library database was not up-to-date. Secondly, there was no program for removing large amounts of data. Further, it was difficult to provide information for the customers because of missing data fields. To conclude, many things, such as manual activities, human mistakes, and misinterpretations of the data caused false notions of the true situation. Also, it was not clear who was responsible for the contents of information, which did not help in avoiding problems in the correctness of the information.

All necessary data were not entered into the database because there was shortage of resources in the library. Consequently the quality of the customer service was not the best possible, because customers could not rely on the correctness of the data. Further, manual performance of certain activities in some units multiplied the effect. Problems in defining certain data items weakened the customer service by delaying the circulation rate of the material. The possibilities of the customers to utilise the CBIS effectively were impaired by superficiality of customer manuals and by lack of customer training. Sometimes, customers were not able to use the CBIS at all, because the system had without any reason cancelled the rights to use the library. Also, the contents of the database often failed to satisfy the needs of the customers.

Finding a piece of information sometimes required browsing through several screens. Some tasks required so much browsing that work practices were unnecessarily complex. Editing was so laborious that simple modifying operations had to be performed by deleting and rewriting. In some units, the work was organised so that the material had to be transferred back and forth between different units.

Many work tasks were to be performed manually because the CBIS was found difficult to use.

Further, 'working around' was performed in computerised tasks, resulting in unnecessary operations. The reasons for working around were related both to the functionality of and deficiencies in the software. In some cases, the functions of the CBIS had to be applied to fit the work tasks. Sometimes, tasks had to be delayed in order to be able to complete them. Awkward editing caused unnecessary deleting and rewriting. Also, the value of certain data item had to be changed for a moment, so that the task could be completed. Performance problems in hardware forced the workers to close and re-start applications, if they wanted to use several programs. One employee described working around as follows:

> "I've invented this kind of a trick, when I want to print the whole reference. In the system there is the 'print' command which prints a couple of lines, but I'd like to print all of them, the whole [index] card. I've put here the 'print-to-file' command, and now all that I take with 'print screen' will be printed to a file. And in the library system I use ... then I will use Windows and ... when I get the reference to the card screen I will take the 'print screen', and it goes to the file. Then you have to edit it. All the stuff at the screen goes there. That is why it is difficult. "

A lot of problems were found in the usage of the library system because information was hard to find. First, the commands were complicated which made searching for information difficult. Second, searches were difficult to perform, because the search command could often be applied only to impractical data fields. Sometimes it happened that for an unknown reason the restriction of the search space had changed without the user noticing it. Therefore, only part of the searched information was found and the user never realised the mistake. Manuals would often have been needed when using more complicated commands. The manuals, however, only provided information about rather simple operations. Furthermore, customers who used the CBIS did not often know the variety of data the system included.

The characteristics of the library system caused that some tasks were delayed or not performed at all. In consequence, the work process was performed slower or the customer service was impaired. In some cases people simply avoided using the CBIS, because they had had problems in using it.

Problems in the usage of the CBIS caused unnecessary work tasks. For example, when ordering for material, customers typed inadequate or faulty commands, and the system did not work as expected. However, the customer learned about the error only when coming to collect the requested material. Because of the complexity of the CBIS, work processes were often divided to very small parts, which increased the number of coordination tasks, i.e. starting, stopping, and transferring.

To conclude, the usability and flexibility of the CBIS were not at a very high level. The fact is indicated by deficiencies in the database, problems in finding material, and the need to remember a lot of details. Many tasks included working around, and the same tasks had to be performed several times. Also, delayed tasks and impaired customer service were, at least in some cases, consequences of the properties of the CBIS. In the worst cases, the problems were so difficult that the CBIS was not used at all.

Skills and knowledge

Poor technical quality of the CBIS set additional requirements on users' skills. Some of the problems in the usage of the CBIS can be explained by shortages in skills and knowledge. This put strain on the more skillful employees and caused dissatisfaction for the customers who could not use the system by themselves. Further, the number of temporary employees was high in the library, which increased problems related to skills and knowledge. In some cases the employees did not have a shared view on the usage, e.g., in which form data should be entered into the CBIS. The lack of shared view resulted in extra tasks and made the internal atmosphere worse.

The users had to learn quite complicated commands of the CBIS by heart. On the other hand, the users also had to remember a number of fields, their meanings, and the order in which to fill them in. All these issues were an excessive requirement on the memory of the users. Furthermore, the issues were often not related to the contents of work, but only to the use of the CBIS.

Both customers and employees had considerable difficulties in finding the right commands for different purposes. These difficulties caused the need for training of both groups. For example, a librarian who led customer training sessions evaluated that the customers had considerable difficulties in finding all relevant references from the database. She thought that the major reason for this was that the customers could not use the right commands for the right problem:

> "Yep, it's the 'cutting'. All of the customers don't get it. That she didn't know that the [search] term should be cut. They just do those quick and dirty searches. And they find something and accept it, but... How would it be... I think that it is at least every third reference they don't find. It depends on the topic. In some cases, you don't find anything, no matter how many searches you do... but it requires teaching and participating. Always when you see somebody trying to do searches, you have to stand next to her, ... and I always keep asking them if they have tried to search by cutting the term."

So, problems in skills and knowledge were realised in many different ways. Having a good command of the CBIS use requires a lot of rote learning. Information and functions were hard to find. Tasks which the customers basically could have performed themselves burdened the staff. In practice, customers did not find all possible references without assistance, or could not use the commands at all. These issues caused extra need for training. On the other hand, customer service was impaired because of shortage of resources. As its worst, problems in skills and knowledge led to job dissatisfaction and in some cases it was decided not to use the CBIS at all.

Work processes and division of labour

Overlapping tasks derived from several sources. Inadequate integration of the CBISs used in the same work process resulted in typing of the same data several times. Shortcomings in the functionality of the system forced the users to write many lists and statistics manually. Also, the structure of the records required that the same data had to be typed several times in different fields. Old manual work practices were still used in many units, either along with the computerised system or alone. This was, at least partly, due to negative attitudes towards information technology. Fragmented work processes caused repeated copying and deleting of data. Also, dissatisfaction with colleagues' work, e.g. in cataloguing, was found to be one reason of repeating tasks.

In some cases, the work practices led to problems in the usage of the CBIS, for example, varying cataloguing practices both inside the organisation and in other cooperating libraries. These problems were rather interesting, because their sources may have hidden in other phases of the work process than the one in which the effect was perceived. Also, obvious failures in work tasks naturally led to, e.g., erroneous data in the CBIS.

Several CBISs were linked to the library system. Considerable problems in the usage of the system were discovered because of the poor integration of the parallel CBISs. Repetition of tasks was observed in many different forms:

> "Yes, there is a problem ...if you think the material which comes from the acquisitions you can not utilise it well ... ... the record which comes from there ... because the form of the record is so different and you should fix it. And then in 'LINDA' [national scientific library database] there can already be a complete record, so it is useless to do here [library system], because first we create a record to 'LINDA' and then we copy it here"

One serious problem originating from organisation of work was related to communication between employees. When the work processes were studied it was found that highly specialised division of labour resulted in barriers in information flows, and fragmentation of skills and knowledge to certain employees. At the same time, other employees who were responsible for other phases of the work process, would have benefited of the skill or piece of knowledge in question. Broader tasks would clearly have helped in reducing these problems in the usage of the CBIS.

DISCUSSION

It is often thought that IS designers suggest only technical solutions to organisational problems. Our study, however, indicates that the users also tend to suggest technical improvements, whether the causes of problems are technical or not. The large number of problems that were found in the case organisation indicates the importance of accurate analysis of use related problems. Means for enhancing the utilisation of CBIS have to be discovered in order to perform the work effectively and to achieve the set goals. Improvements in design require that both the designers and the users take an open-minded view on the various causes of problems.

The classification of use related problems does not offer any solution for the organisation. One way to analyse the problems and their effects on activities is to place them in work process diagrams. Figure 2 illustrates six problems with their relationships that were found in the case library's processes. Problems a, b, c, d, e and f were perceived by four employees. In fact, the problems were related to each other, although the employees did not completely realise it. The real cause of the problem chain hides in the ordering phase of the acquisition process (Problem a) and is reflected in other parts of the process and even in the service process. In this case, poor integration between the acquisition system and the library system was the ultimate cause. The solution would be either better integration of the systems or introduction of an acquisition module to the library system. However, all the problems were not technical which may require also organisational changes.

Even if the problems could not be explained in relation to each other, the process oriented analysis allows identification of the most problematic work phases. Further, there are usually several computer applications in a real work situation which have to be evaluated together. In the library case there were three main applications in the acquisition and service processes. The analysis revealed two most problematic work phases: cataloguing in the acquisition process and searching in the service process.

The analysis becomes even more complex than described above, because the organisational aspects have to be considered besides the technical issues. In the library, the need for organisational redesign was indicated by two obvious signals: the shortage of resources in customer service and the employees' dissatisfaction on work conditions. The questionnaire revealed a clear imbalance in the allocation of the resources between the work processes (Figure 3). Each employee evaluated the proportion of working hours by a list of work tasks which included about 100 different tasks. The tasks related to acquisition took almost half of the working hours in the organisation while the tasks related to customer service took only one third of working hours. This can be explained by the existence of important collections which needed a lot of acquisition related tasks and which had a value of their own, even if they were not used by customers. Another interpretation would suggest the need for organisational redesign. The relatively high number of working hours spent to administrative tasks reflects the high number of organisational units with their own administration, as well as the bureaucratic organisational culture.

There was a clear need for change indicated by the employees (Figure 4). More than half of the employees agreed that the number of employees in the organisation was insufficient in proportion to the work to be done. In other words, the employees were under stress and burdened with too much

Figure 2: The relationships of the problems

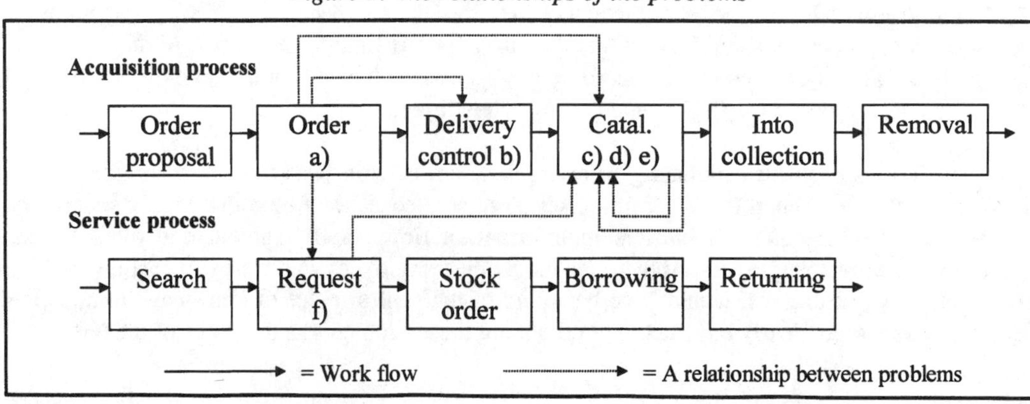

Figure 3: Distribution of working hours (%)

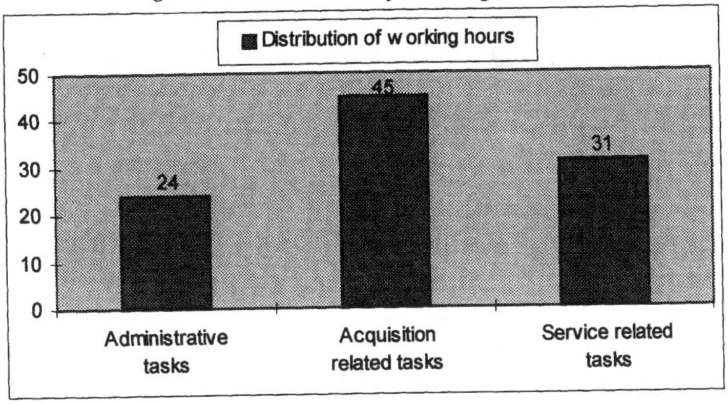

work. Furthermore, only 20% of the employees agreed that the work tasks were well organised. There was something wrong in the organisational climate, too. Less than 20% of the employees agreed that the organisational climate was open, encouraged the employees to present new ideas, and that information was shared openly in the organisation.

The multifaceted nature of the observed problems indicates that CBIS should not be analysed as a separate entity, but as an inseparable part of the work. It is hard to imagine that the problems found in this study would be unique or that the number of problems would be somehow unusual. On the contrary, we suggest that problems in the use of information technology are underestimated both in research and practice. Therefore, a deeper and more holistic understanding of problems should be pursued.

There seem to be five development areas of which the organisation would benefit the most. These development ideas are based on the case study and some later analysis of possible organisational changes. The ideas are not necessarily in line with the views of the organisation, but reflect the results of the researchers' analysis. The ideas are presented here as the areas which probably have the highest potential to benefit the organisation. However, there is still an analytical gap between the problems reported in the previous sections and the ideas for changes presented here.

First, the user interface should be replaced by a graphical one. This investment would be of benefit for both the employees and the customers. The graphical interface would simplify the work of employees who reported difficulties in, e.g., editing the data fields, memorising the right commands, and browsing through several screens. On the one hand, customers and temporary employees would learn the graphical interface more quickly which would reduce the need of user training and manuals, as well as the number of interruptions caused by the novice users asking for help in problematic situations. On the other hand, the lower skill requirements would offer new solutions for the organisation of work tasks. Some tasks were performed exclusively by certain employees because the system was difficult to use. Graphical user interface could diversify the work

Figure 4: The employees' opinions about the number of employees, organisation of work tasks, and the organisational climate (n=118).

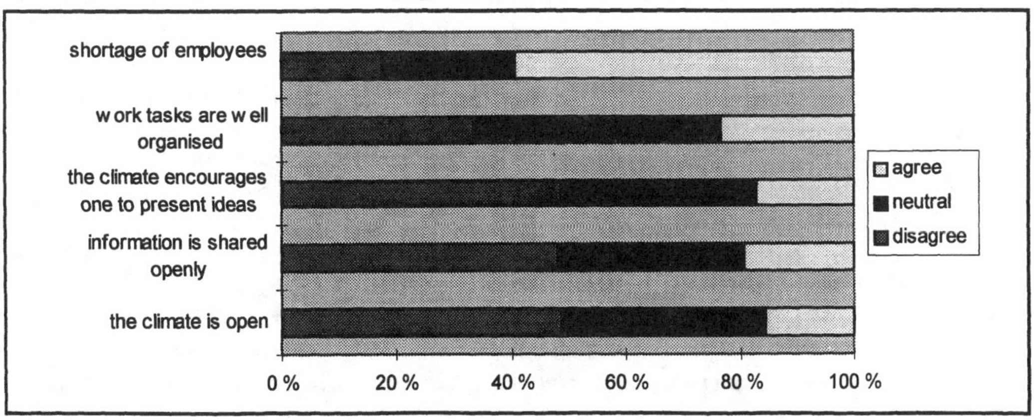

of employees because specific tasks would not be so difficult to perform.

Secondly, better integration of various CBISs would offer a possibility to transfer and utilise existing data, for example in cataloguing tasks, more effectively. This would require enhanced integration of the acquisition system and the library system in such a way that the quantity and quality of the data transferred between the systems would be easily utilised and edited. The effect of the same problem is seen in a wider scale when working with the national scientific library database. Better integration would simplify the work practices and diminish the number of certain operations.

Thirdly, both national and international co-operation provides significant opportunities in reducing the cataloguing work carried out by a single library. One possibility would be shared catalogues of scientific libraries in Finland. In this way the cataloguing data would be entered only once into the CBIS. Of course, for effectivity reasons the data should be replicated over several physical locations. International cooperation which already is under examination could provide most of the cataloguing data in electronic form directly from the publishing companies.

Fourthly, the reorganisation of work and tasks offers several development possibilities. Work processes could be performed more effectively and efficiently, if some subsequent tasks were done by one person instead of splitting the tasks between several employees. Furthermore, organising the work processes in a more sensible way would diminish unnecessary transferring of material and reduce the danger of losing information in additional coordinating tasks. Also, totally new forms of organising and conceptualising the activities, like new types of services, could contain potential for change.

Lastly, many of the manual catalogues could be computerised. For example, lists of terms which are used in describing the nature of a publication and in the search of them could be updated centrally. Further, integration of these lists into the library system would probably reduce the work load and the amount of mistakes made in cataloguing tasks.

FURTHER READING

Avgerou, C. (1995). Evaluating Information Systems by Consultation and Negotiation. International *Journal of Information Management*, 15(6), 427-436.

Kortteinen, B., Nurminen, M.I., Reijonen, P. & Torvinen, V. (1996). *Improving IS deployment through evaluation: Application of the ONION model. 3rd European Conference on the evaluation of IT*, University of Bath, UK, 175-181.

Landauer, T. K. (1995). *The Trouble with Computers.* The MIT Press, Cambridge, Massachusetts.

Lyytinen, K. & Hirschheim, R. (1987). Information systems failures - a survey and classification of the empirical literature. *Oxford Surveys in Information Technology*, 4, 257-309.

Symons, V. J. (1991). A review of information systems evaluation: content, context and process. *European Journal of Information Systems,* 1(3), 205-212.

BIOGRAPHICAL SKETCHES

Katariina Jalonen is research assistant at the University of Turku. She received her Master's degree from the University of Turku and is presently preparing her doctoral thesis. Her research interests include philosophical foundations of information systems and evaluation of information systems in organisations. E-mail: katariina.jalonen@cs.utu.fi

Mika Kirveennummi is assistant professor at the University of Turku. He received a master's degree from the University of Turku and is presently preparing his doctoral thesis at Turku Centre for Computer Science. His research interests include participative approaches to information systems development, evolutionary information systems development, and evaluation of information systems in organisations. He has published articles in several international conference proceedings. E-mail: mika.kirveennummi@cs.utu.fi

Vesa Torvinen is full-time researcher at the Turku Centre for Computer Science, which is one of the centres of excellence in research in Finland. He received a master's degree from the University of Kuopio and is presently finishing his doctoral thesis. His research interests include information systems development methodologies and evaluation of information systems in organisations. He has published articles in several international conference proceedings. E-mail: torvinen@cs.utu.fi

Setting Up to Fail: The Case of Midwest MBA

Andrew Urbaczewski
Indiana University, USA

Jo Ellen Moore
Southern Illinois University - Edwardsville, USA

EXECUTIVE SUMMARY

A mantra of experienced project managers is "failing to plan = planning to fail." In the case of Midwest MBA, a user group is not satisfied with the progress made by the central computing staff on the development of a much-needed information system. In a well-intentioned effort to help the users, the IS staff of the end user area (who had been acting in a liaison role between the user area and central computing) decided to take on the completion of the system. However, the resources needed to absorb this additional project were never accurately estimated, obtained, or allocated. Moreover, tasks and target dates were never firmly established.

Although the decision to take on the completion of the system was a noble one, the IS staff ended up providing a demonstration of the project manager mantra. Not only was the system not completed, but the IS staff's regular duties suffered as well. Animosity between the IS staff and users became rampant and both parties considered the system to be a failure.

Put simply, the IS staff was not dealing in reality. And, as a result, they set themselves up to fail. This case is based on actual events in a real organization. We have, however, changed the identities of the parties involved and other key information to preserve anonymity.

BACKGROUND

Midwest MBA is the administrative division responsible for managing the MBA program in a large Midwestern U.S. public university. The MBA program has existed for many years and is nationally renowned. Midwest MBA attracts students from throughout the United States as well as students from around the world. Currently, about three hundred of approximately seven hundred fifty applicants are accepted into the two-year program annually.

The university currently employs a hybrid IS organizational structure. A central computing group provides an enterprise-wide database and technology infrastructure, and decentralized IS staffs are located within many business units of the university. Midwest MBA is one of the business units that has its own IS staff.

Midwest MBA is chaired by Harry Headhoncho, a professor from the School of Business faculty. He receives release time from most teaching and research duties to manage the program. Headhoncho is primarily involved in fund-raising, public relations, and the general promotion of Midwest MBA, as opposed to details associated with the daily operation of the program. Much of Headhoncho's time is spent away from the office on local and worldwide travel.

Figure 1: Organizational Chart of MBA, Inc.

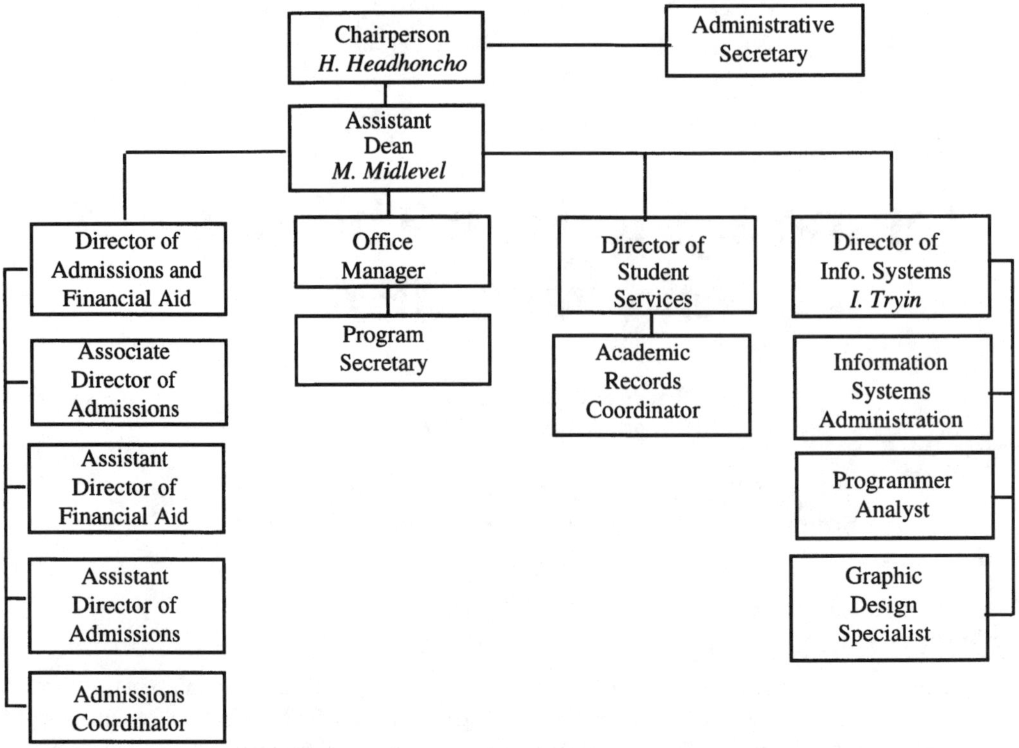

Reporting directly to this chair is the assistant dean of the program, Melvin Midlevel, who manages the day-to-day operations of Midwest MBA. The assistant dean has a staff of thirteen people, divided into four functional areas: Admissions/Financial Aid, Student Services, Information Systems, and Office Management. The mix of professional and clerical staff is approximately equal. Figure 1 shows the organizational chart, minus several 12 hour/week graduate assistants allocated to the department.

The staff of Midwest MBA takes pride in knowing their students. One of the major goals of Midwest MBA is to track every relevant piece of information about a student, from the time the student first requests an information packet to the time he or she leaves the program and obtains employment. This information helps the staff to provide personal, high-quality service to prospective and enrolled students, as well as alumni of the program. The staff also uses this information to respond to queries by various internal and external organizations, such as accreditation groups and the graduate school of the university. The information is vital in recruiting new students and working with the many groups that rank MBA programs. A large coordination effort is required of the functional areas of Midwest MBA to collect this data and facilitate needed access to it.

As one can imagine, it can be quite time-consuming to amalgamate the volume of information needed when dealing with about six hundred enrolled students at any given time. The Student Services area alone is required to produce over 30 reports per year. The admissions report is typically

Table 1: Sample of reports needed by Midwest MBA

Examples of reports:

- Admissions Report (weekly)
- Academic Progress Report (Jan)
- Problem Enrollment Report (Apr)
- Parent Mailing of Hotel Information (May)
- U.S. Citizens by Ethnic Group Report (Aug)

over thirty pages long and is run weekly for use by the Midwest MBA chairperson and assistant dean. A representative sample of required reports is provided in Table 1. Simply stated, Midwest MBA is an information intensive operation.

SETTING THE STAGE

Until the late 1980's, almost all student data were kept in paper folders within Midwest MBA's offices. Each student folder contained information such as demographic data, application materials, and grade reports. Some information was collected directly from the student, whereas other data were collected from outside sources such as employers and other university offices. Due to the sheer volume of data, it was difficult to assemble meaningful reports. However, in the increasingly competitive market of MBA programs, the staff felt the need to collect even more student information for Midwest MBA to gain an advantage in student recruitment and placement. To accomplish this, Midwest MBA recognized the need to computerize information processing as much as possible.

The Silver Bullet

At about the same time Midwest MBA was acknowledging its need for an extensive computer information system, the university's central computing services group (CCS) approached Midwest MBA with an offer to create a student tracking system for their use. CCS hoped to create a model system that they could then sell to similar programs across campus and to other universities. CCS agreed to design and implement *gratis* a system to meet the tracking and reporting needs of Midwest MBA. In other words, CCS had walked in to Midwest MBA and offered to provide the solution to their problems — for free! After meeting several times to discuss requirements, CCS was commissioned to create the Student Tracking System (STS) in June, 1991, with delivery due in nine months.

Midwest MBA did not have regular dealings with CCS in the past, and little was known about what they should expect from CCS for product and service. Midwest MBA's director of information systems, Ima Tryin, had worked for CCS many years prior, but was unfamiliar with CCS' current personnel and initiatives. Midwest MBA's computing services were primarily supported by their own IS staff.

CCS had several problems developing the system. Some problems were technical, some were managerial, and some resulted from a lack of qualified personnel. Regardless of the source of the problems, CCS failed miserably in delivering the system on time to Midwest MBA. Certain parts were delivered but had obviously not been thoroughly tested. Some of the programs did not compile correctly, some had problems getting data from the database, and others had significant logic errors. Well into 1994, CCS had made no significant progress towards a completed STS.

Unfortunately, the demands of the marketplace and regulators do not wait for computer systems to catch up. Midwest MBA's support staff was overwhelmed in trying to keep up with information needs. Makeshift systems were created with spreadsheets and word processors by individual workers. Of course, such uncoordinated end user computing caused terrific problems in data consistency as different people made changes at different times to different files. Reports issued by people in Midwest MBA used data sets that were so different that one wondered whether the individuals were working in the same organization. This was clearly not acceptable to the regulators and other customers of information from Midwest MBA. The Director of Information Systems, Ima Tryin, decided something must be done to remedy the situation.

Biting the Bullet

In an attempt to move Midwest MBA closer to the implementation of the central data source inherent in the STS application, Ima Tryin began working with the parts of STS that had been delivered. The parts of the system that the Admissions group needed were further along than those needed for Student Services, so Ima decided (with Midlevel's approval) to roll the system out to the users in stages. Admissions would be brought online first, and Student Services would follow later. This would make sense for two reasons: students need to be admitted before they can be serviced, and the Admissions parts of STS were much closer to completion than the Student Services modules.

A major problem that plagued Ima in phasing in parts of STS was the inaccessibility of the source code. CCS, protecting its intellectual property, refused to release the source code for STS to Midwest MBA until they signed off that the system was complete. However, Midwest MBA was not willing to accept the rather shoddy work of CCS as final. Ima, an adequate programmer herself, was trying to write patch programs to fix errors in the delivered STS modules so they would be useable, while waiting for CCS to debug the original STS code. She was mostly successful at writing patches, but it became increasingly difficult to maintain the patches when new module versions were frequently received from CCS. Writing the patches was also time consuming, as Ima had to guess at parts of the code because she didn't have the source code to work from.

In the meantime, users were becoming disgruntled. For over three years, they had been told that a completed STS was "just a few months away" from implementation. Their jobs were becoming increasingly data-driven, but they did not have the electronic information support they so desperately needed. Because many of the users didn't know the whole story of who was actually creating the system, tempers flared toward anyone whose job was tied to computer support. This obviously included Ima and her staff. Ima was clearly on the front line as she was writing patches to get the Admissions staff up and running electronically. This, to most users, made her appear to be responsible for the STS development and thus liable for any and all delays which had occurred.

The maintenance and patches to STS had become a monstrosity. In desperation, Melvin Midlevel and Ima Tryin decided to sign off and accept whatever had been completed on STS to date. This would enable them to get the actual source code and make the necessary modifications. In July 1995, Midwest MBA relieved CCS of any further responsibility for the system by accepting delivery of the entire system "as is." Though STS was by no means ready, Midlevel and Tryin decided it was worth the sacrifice to Midwest MBA to obtain the source code and have the implementation of STS under their own control.

CASE DESCRIPTION

The IS staff of Midwest MBA began to rewrite code and fix many of the flaws within STS. This was a major undertaking, one to which no additional resources were allocated. Not surprisingly, other responsibilities of Midwest MBA's IS staff began to be neglected or postponed. Users who depended on the IS staff for technical support became increasingly upset with the inability of the IS staff to perform their maintenance and support duties in a timely and effective manner, as they had always done in the past. Furthermore, the IS staff was nowhere near completing all of the work to be done on STS and fixing all of the problems (some of which were still being discovered).

Adding insult to injury, Midwest MBA's employees attributed all problems with STS to their own IS staff because the IS staff was now completely responsible for the system. Whereas the users and IS staff of Midwest MBA had enjoyed a good working relationship prior to taking on this project, the relationship was now strained and showing distinct signs of hostility. This is reflected in comments from the Director of Student Services regarding the IS staff:

> *"Sometimes I wonder what they do all day. Requests get turned in, but I never see any results. It's like a big black hole. Things have been promised to us for years, but we have learned to accept that they won't get done. This system is more of an impediment to our job than a help."*

The IS staff members of Midwest MBA were also upset with the situation. Night and weekend work became commonplace in an effort to get the STS application up and running as soon as possible. The normal full-time workload of the IS staff was now augmented by the repair work needed to make STS a working reality. While the IS department was eager to make STS work correctly and function for Midwest MBA, there were only so many hours in a day. The IS staff became somewhat testy in their own right when dealing with the users. According to a member of Midwest MBA's IS staff:

> *"We have been going above and beyond the call of duty, working nights and weekends, to create a useable product and instead of thanking us they're griping about us."*

Things were so bad that sometimes when IS staff members were working on the STS project at home, they were accused by some users of slacking off and avoiding work. Likewise, several members of the IS staff were quite critical of the roles played by some members of the user group and openly questioned their team commitment versus desire for personal gain. Although users were asked to identify high priority pieces of STS and critical need-by dates, no additional IS resources were allocated to the effort. Hence, the IS staff became expert performers in the "juggling of burning sticks," delaying the inevitable fire.

In the beginning, CCS had been guilty of not meeting established deadlines and quality expectations. Now, the IS staff of Midwest MBA was becoming the guilty party because they failed to develop and communicate a realistic plan for the completion of STS. The project was never broken into pieces, no time estimates were generated for the tasks to be done, individuals were not given direct responsibility nor allocated time to complete this work, and target dates and deadlines were not established. Melvin Midlevel did not help the situation. He continued to tell the users "success is just a few months away," while telling the IS department "take as much time as you need," "be sure you get it right," and "you'll be getting new staff members shortly."

Unsatisfied with the progress being made by the IS staff, in March 1996, the directors of the Admissions/Financial Aid and the Student Services departments (along with the Office Management director) decided to complain to higher authorities. They went to Ivana Beking, the associate dean of the School of Business, to complain about the lack of progress on the STS implementation. They complained bitterly about the system they had been promised for over five years and wanted someone in power to do something about it. Ivana Beking responded by hiring a consultant to study the problem and determine what should be done about the STS application.

Surprisingly, when the consultant interviewed members of the IS staff and user areas in Midwest MBA, neither party reported a problem with the other party. That is, the users did not indicate the IS staff was at fault and the IS staff did not indicate that the users were at fault. Rather, at this point in time, every member of the IS staff and user areas felt that Melvin Midlevel (the direct superior of the directors of IS, Admissions/Financial Aid, and Student Services) was to blame for the failure of the STS project. They indicated that his tendency to "manage by problem avoidance" resulted in unclear direction and provided no help in problem resolution. In the words of the Director of Student Services:

"This office is allowed to run like a ship without a rudder. We need a rudder."

Indeed, by the end of his investigation, it was evident to the consultant that this really was not a technical system problem; it was a management problem.

The consultant met with Ivana Beking and Melvin Midlevel in July 1996 to discuss his findings. All agreed that the IS staff had the expertise needed to complete the STS implementation, but that Midlevel would need to take a more aggressive approach to managing the effort and allocating resources to it. No official sanctions were levied against anyone.

Upon speaking with members of both user and IS staffs several months later (in November 1996), it appeared nothing had changed. The IS department was still understaffed for the STS project and in need of official direction from Midlevel. And, although the STS was slowly getting better, it was not yet completed. User and IS staffs were both still frustrated with the situation.

SUMMARY

Midwest MBA is an organization with problems on many fronts. Technology in this case is simply a medium for the fronts, a battleground for the organization's members. Identifying this as a technology problem is akin to calling poker a card game; while technically correct, it misses much of the substance at hand. While the silver bullet solution proposed by CCS might work in some organizations, the structure here was not established to support it. Resources were few, tempers were short, and a strong leader, a rudder, was absent from the organization. What would be next?

APPENDIX A - CHRONOLOGY OF THE CASE OF MIDWEST MBA

June 1991	- CCS commissioned to create STS, due date March 1992.
March 1992	- STS not yet ready.
March 1993	- STS still not ready, just "a few more months away".
July 1994	- Some modules of STS delivered, not all working. Patches created by Tryin.
July 1995	- STS accepted "as-is", source code received by Tryin.
March 1996	- Group of Midwest MBA employees seeks help from Beking.
June 1996	- Consultant brought in by Beking to assess current situation.

BIOGRAPHICAL SKETCHES

Andrew Urbaczewski is a doctoral candidate in Management Information Systems at the Kelley School of Business at Indiana University. He holds an MBA from West Virginia University and a B.S. in Finance from the University of Tennessee, Knoxville. Prior to returning to academia, he worked as a programmer/analyst and project control specialist for several U.S. Department of Energy IS projects. His research interests include the organizational impact of IT and electronic commerce.

Jo Ellen Moore is Assistant Professor of Computer Management and Information Systems at Southern Illinois University - Edwardsville. She holds a Ph.D. from the Indiana University School of Business and a master's degree in psychology from Illinois State University. In the corporate environment, she has served as an Information Systems manager, project leader, and programmer/analyst. Her research interests include the management of information systems professionals and technology.

ENI Company

Ook Lee
Hansung University, Seoul, Korea

EXECUTIVE SUMMARY

ENI Company is an electronic commerce firm in South Korea. ENI Company provides English news items and English lessons to the subscribers through daily e-mail service that includes free English news-related question and answer sessions via e-mail. This case study deals with the struggle of this firm to establish and sustain its business in a less-developed national information infrastructure. Information on national information infrastructure and the Internet in South Korea is provided in order to facilitate understanding of the difficulty that ENI Company faces while conducting e-commerce in South Korea. The chronology of ENI Company evolution is described and the organizational structure of ENI Company is also presented. The marketing of ENI Company's products that is the biggest challenge for the firm is also discussed. This case is a good example of how to conduct an e-commerce in a county where national IT infrastructure is not ready for it.

BACKGROUND

Literature Survey

We can easily suggest that doing an e-commerce business in a developed country should be different from doing it in a less-developed country; here "development" refers mainly to the level of national IT infrastructure development. National IT infrastructure can be defined as the vision of broadband communications that are interoperable as though a single network, easily accessible and widely distributed to all groups within society bringing business, education, and government services directly to households and facilitating peer to peer communication throughout society(Kraemer, et al., 1996). But this idealistic vision is hard to achieve for countries with less economic resources. By conducting a literature survey on e-commerce in developing countries, the following factors in addition to having a well-developed national IT infrastructure, were found to be necessary for a country in order to provide a fertile ground for e-commerce.

1. **Active use of credit cards in the Internet**: countries such as Philippines, India, and China, where credit cards are not widely used, can not find a workable payment method for e-commerce (Asuncion, 1997; Bhatnagar, 1997; Liu, 1997; Rao, 1998). In a way, South Korea is similar to these countries even though credit cards are widely used since many consumers on the Internet are very much reluctant to give out their credit card number. In the cultural point of view, another reason can be the fact that Koreans love to use cash in most business

transactions even for a large sum just like people in Hong Kong (Westland, et al., 1997). As for the payment method in e-commerce in Korea, many companies use money transfer through a bank account, i.e., the buyer sends money to the bank account of the seller and the seller will send the goods to the buyer. This kind of pay-first-get-the goods-later payment method is obviously an obstacle for e-commerce to grow since many consumers feel insecure. Thus it is clear that without the full trust between consumers and e-commerce companies, even a country with widespread credit card use can not facilitate the growth of e-commerce.

2. **Fixed fee option of the unlimited use of local telephone lines:** Unlike the U.S., the South Korean telephone company is a government-owned monopoly. Even though it is trying to upgrade the communication lines with fiber-optic cables, its business policy has not changed, i.e., there is no concept of separate billing for the unlimited local call option. In other words, if a person makes a local call to the Internet Service Provider and surfs the net for a long time, the person will get a very expensive phone bill. With this kind of environment, it is not easy to do e-commerce.

South Korean Internet Statistics

It is estimated that the number of Internet users in Korea is about 350,000 in 1996 as shown in Table 1(NCA, 1997).

Table 1. Number of Internet Users in Korea

	Commercial ISP users	Non-Commercial ISP users
Institution	51,850	69,000
Individual	231,226	221
Total	283,076	69,221

In Figure 1 (NCA, 1997), we can see the rapid growth in the number of users of dedicated lines for their Internet use in Korea. The users are, in general, institutions which then provide Internet access to employees or students in the institution.

In Figure 2 (NCA, 1997), we can also see the rapid growth in the number of owners of PPP/Shell accounts for their Internet use in Korea. The owners are normally individual subscribers of commercial Internet Service Providers.

Using a web-based survey(NCA, 1997), National Computerization Agency(NCA) of Korea was able to collect important statistics on the demographics of the Internet users in Korea. The total number of respondents who visited the survey web site and completed the questionnaire was 1725 and following is the result of the survey.

Figure 1. Number of Dedicated Line Users(Institutions) in Korea

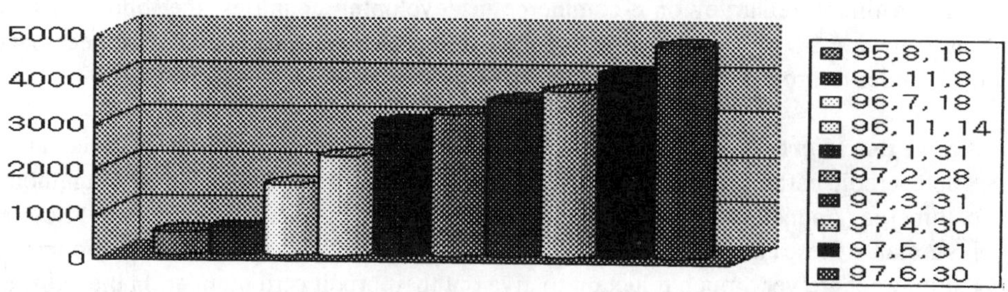

Figure 2. Number of Owners of PPP/Shell Account in Korea

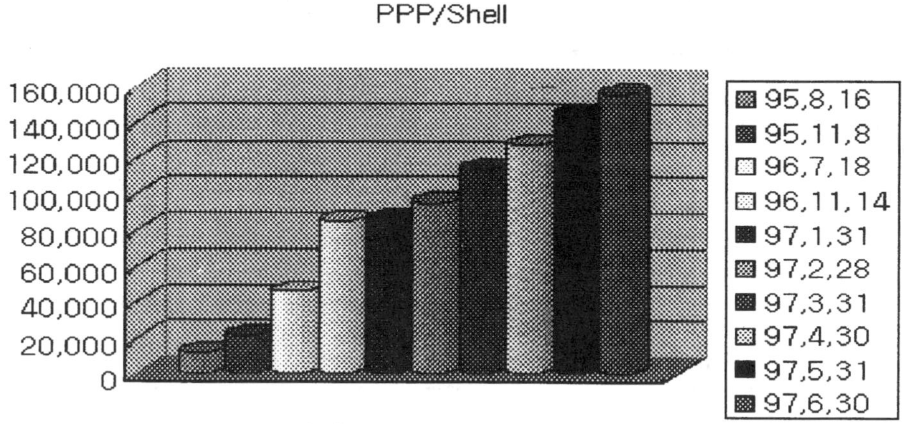

- SEX: male (1463), female (262).
- AGE: under 15 (26), 15-20 (459), 21-25 (453), 26-30 (457), 31-35 (173), 36-40 (75), 41-45 (55), 46-50 (18), over 50 (9).
- OCCUPATION: college student (952), worker-non computer or Internet industry (305), worker-computer industry (125), primary and secondary school student (73), researcher (67), worker-Internet industry (52), teacher and professor (40), government (28), business executive (8), miscellaneous (75).
- EDUCATION: primary school (39), middle school (34), high school (754), 2-year college (261), 4-year college (471), master's (145), doctoral (21).
- PURPOSE OF USING INTERNET: information collection (1301), research (135), entertainment (151), business (82), miscellaneous (56).
- YEARS USING INTERNET: less than 6 months (147), 6 months to one year (854), one-two years (410), two-three years (161), over 3 years (153).

The result shows that the Internet in Korea is a new medium used mainly by young college students and professionals for non-business-related activity; this also means that user population is not very active in utilizing e-commerce as of 1997.

NCA also reports that the total number of e-commerce web sites in Korea is 140 where only 50 have all the mechanisms for automated shopping, i.e., the payment method using credit cards is workable for these sites as of March, 1998 (NCA, 1998). In the same report, NCA estimates the volume of e-commerce in 1998 to be around 9.4 billion won (7 million US dollar). Thus it is clear that e-commerce is in its infancy in Korea.

ENI Company Introduction

ENI Company (pseudonym is being used to protect the anonymity of the firm) is a South Korean firm that specializes delivering overseas news in English and its translation with lessons on English language to the subscribers via e-mail. The primary goal of this business is to educate the public in English reading comprehension as well as inform through up-to-date foreign current affair items. The business was formed in 1997 by a university professor who had extensive knowledge in world affairs. The business charges the subscribers a monthly fixed fee that was approximately US$10. The English news and its translation with lessons on English expressions are delivered everyday via e-mail to the subscribers. The subscribers can also ask questions regarding the content of the news item and its explanation and the firm answers them as soon as possible.

SETTING THE STAGE

Before we embark on presenting our case, we need to address following issues. Since our subject firm is an e-commerce firm in South Korea, it is bounded by the level of development of South Korean national IT infrastructure. Thus information on South Korean national IT infrastructure is

Table 2: Size of Domestic PC Market

Year	1991	1992	1993	1994
Quantity	614	665	773	1200

*In terms of PC's sold(thousand sets).

Table 3: Telephone Lines per 100 Population (1993)

Korea	US	Japan	Germany	UK	France
38	53	47	46	47	53

Table 4: Subscribers of Mobile Communications

	1993	1994
Mobile Telephones	472	960

*Units: thousand sets.

Table 5: Multimedia Industry (1994)

Type	Sales (Billion Won)
Multimedia PC	94.5
CD-ROM drive	37.5
CD-ROM Title	38.0
Sound card	51.4
Image card	25.0
Tools	5.4
Total	251.8

given. Our subject firm also needs to conduct marketing in cyberspace and in order to facilitate understanding of the concept of cyberspace marketing, we introduce some background theories of cyberspace marketing.

South Korean National IT Infrastructure

South Korea embarked on building its national IT infrastructure in 1994 which was officially called "Korea Information Infrastructure (KII)" project (Jeong and King, 1996). The Korean government committed itself to promoting industries such as computer makers, telecommunication network builders and value-added service providers, multimedia firms, cable TV industries, and Internet-related companies. Major aspects of Korean national IT infrastructure are described in following tables (Jeong and King, 1996).

E-commerce firms in South Korea have to struggle in a national IT infrastructure which is not adequate for effective commercial activity yet. This case shows that despite this kind of hardship, it is possible to create and run an e-commerce firm even though it will be difficult to generate a big financial bonanza.

Cyberspace Marketing

As many firms throughout the world try to conduct business on the Internet, the importance of marketing in the Internet has become an important issue to IS scholars and practitioners alike (Copfer, 1998). Some businesses are interested in setting up a WWW site to expand their reach to customers in addition to the physical entity in real world, whereas others try to establish a presence in the Internet without having any physical entity in the real world. In both cases, firms need to conduct marketing in order to solve vexing questions such as who the customers are, how to advertise effectively, etc. (Mosley, 1998).

In this case study we define those consumers who are using the Internet and/or an on-line information provider such as America Online as consumers of cyberspace. Now that cyberspace is born and all these users of cyberspace are reachable by the Internet and they, therefore, have become attractive consumers to the companies that always try to get any individual consumer's attention. For example, a consumer spend their awake time in watching TV or listening to the radio; this is the reason that advertisements exist in those mediums. Thus, similarly, more and more people are spending their time surfing the Internet, which means that companies should advertise in this medium too, since the consumers' precious attention span is being used here just as with TV and radio.

Thus how to advertise effectively in cyberspace has become a very important issue. To advertise effectively, a business should choose the right kind of tool in order to reach enough customers with a reasonable amount of time and money. Many businesses now have web pages which were made for variety of purposes. For example, big corporations created web pages to promote corporate image, thus these web sites are not for trading goods and services. On the other hand, some firms created web sites for the purpose of selling goods and services such as books, computers, flowers, etc.. Among them, some exist only in web sites, i.e., no physical entity exists, while others have physical entities with web sites used for additional business from cyberspace consumers. Except for corporations which created web sites for public relations purposes, all other web sites are engaged in profit-making ventures, which consequently need sophisticated marketing strategies and tools (Hansen, 1998). Thus, Internet marketing can be defined as marketing strategies and tools that are designed to enhance product purchases on the Internet. In other words, Internet marketing is geared toward consumers of cyberspace.

These consumers of cyberspace have some distinctive characteristics compared to real-world consumers. The main differences are as follows. The consumers of cyberspace choose only the web sites which are of interest to them, which means that unlike TV, radio or print advertisement, it is difficult to promote a product to unsuspecting mass customers. In other words, people who are interested in adult-related products will go to those web sites without wandering into some other web sites, such as those selling flowers. Thus, making people aware of the existence of particular web sites even though they are not of interest to the consumer at the moment is not easy. There are some technological breakthroughs such as PUSH technology which does provide web site information to the consumer who seems to have potential interest in the particular area (Burke, 1997).

But even in PUSH technology, one can only push things after the information regarding the customers' behavior is gathered. Thus, with millions newly signing-up to the Internet everyday, the PUSH technology has its limits. Nowadays, Internet advertising is often done in banner ads which take up small space on a web page, are supposedly noticeable and, hopefully, actually being read. But banner ads can appear only in web pages which a particular consumer reads, i.e., if the web site is not visited by consumers, the ad becomes simply obsolete. There has to be a better way to market a product in the Internet in order to reach more people possibly in mass numbers.

Direct e-mail advertisement can be a good answer to solve the above stated problem in Internet marketing. One can not reach mass customers if only banner ads are used. Direct and bulk e-mailing literally means sending bulk or mass number of e-mails directly to unsuspecting users of cyberspace (Gustavson, 1997). The product that is advertised in a direct e-mail advertisement can be anything, which means that unlike PUSH technology which advertises only products considered to be of interest to consumers whose on-line browsing behavior is known, the direct e-mail advertisement can send promotional messages of any product to almost anybody who has an e-mail address. Thus, direct and bulk e-mailing is more or less similar to the direct marketing in real-world shopping. In direct marketing, shoppers are either called or sent a so-called junk-mail by the direct marketer.

Culnan (1993) investigated the consumer attitudes toward direct mail advertisement. In her paper, she claims that strategic uses of information technology based on personal information may raise privacy concerns among consumers if these applications do not reflect a common set of values. Her study addresses what differentiates consumers who object to certain uses of personal information from those who do not object. Data collected by questionnaire from young consumers are used to identify a research approach for investigating attitudes toward the secondary use of personal information for direct marketing. Secondary information use occurs when personal information collected for one purpose is subsequently used for a different purpose. While secondary information use is both widespread and legal, it may be viewed as an invasion of privacy when it occurs without the knowledge or consent of the consumer.

But in the case of direct e-mail advertisement, there is no prior knowledge or consent of the consumer, therefore there can be serious problems using direct e-mailing as a marketing tool. The setting for her study is the use of point-of-sale data from a supermarket frequent shopper program to generate direct mail solicitations. Control emerges as a clear theme in differentiating individuals with positive overall attitudes toward secondary information use from those with negative attitudes. Study participants with positive attitudes are less concerned about privacy (measured as control over personal information, i.e., some people don't like to try hard to keep every aspect of his/her life under tight control while some people like to be secretive about almost everything; like to be in total control over one's life), perceive shopping by mail as beneficial, and have coping strategies for dealing with unwanted mail.

Notice that this result happens when the use of secondary information was already consented. Thus, in direct e-mailing advertisement, even people who have a tendency to favor less control over his/her life and are more tolerant of junk surface mail, can react negatively to those direct e-mail ads since there was no consent of their address for information use, whatsoever. But people who have tendency to tolerate junk surface mail, might still react positively since those e-mails contain information about products that can be useful or cheaper for the consumer.

Some merits of direct e-mail advertisement are as following.

a) Unlike direct marketing which can cost heavily, the cost of bulk e-mailing is very low since sending e-mail letters to any number of people in the world does not cost a dime, at least for the use of the communication line, i.e., Internet communication is free. All the firm has to pay is a local phone bill and, if the firm does not own a server, to pay the Internet service provider which does not charge by the number of e-mail letters, but by the time spent on-line (sometimes, just a flat fee per a certain period time, say, a month).
b) E-mail can cause a very direct response from the consumer since unlike junk letters in real world marketing, people tend to read e-mails even though those are advertisement (Martin, 1998). Junk-mails can be thrown out without even considering what the content could be. But junk or spam e-mails are harder to delete away because people tend to read any e-mail before they delete it. Once they read it, the information which is contained in the ad e-mail can be very effective in terms of getting attention from the customer.

CASE DESCRIPTION

Chronology of ENI Company

1997 January: ENI Company was founded by a marketing professor who spent 10 years studying and working in US and Australia.

1997 February: With 108 subscribers, ENI started to send English news via e-mail.

1997 May: The number of subscribers: 220. ENI Company started an advertising campaign in the conventional media as well as cyberspace.

1998 May: The number of subscribers: 1,002. The amount of projected annual sales revenue: 120,000 dollars. The amount of projected annual net profit: 30,000 dollars.

Organization of ENI Company

The founder, who is also the CEO, wanted to minimize bureaucracy which he regarded as a major cost center. Thus the organization of ENI Company is very flat, with primary workers being teachers whose main tasks include translating English news items and answering questions from subscribers. Currently ENI Company has 10 teachers who work in an autonomous environment where each teacher has the authority to work, i.e., teachers have no management interference on how

Figure 3: Organizational Structure of ENI Company

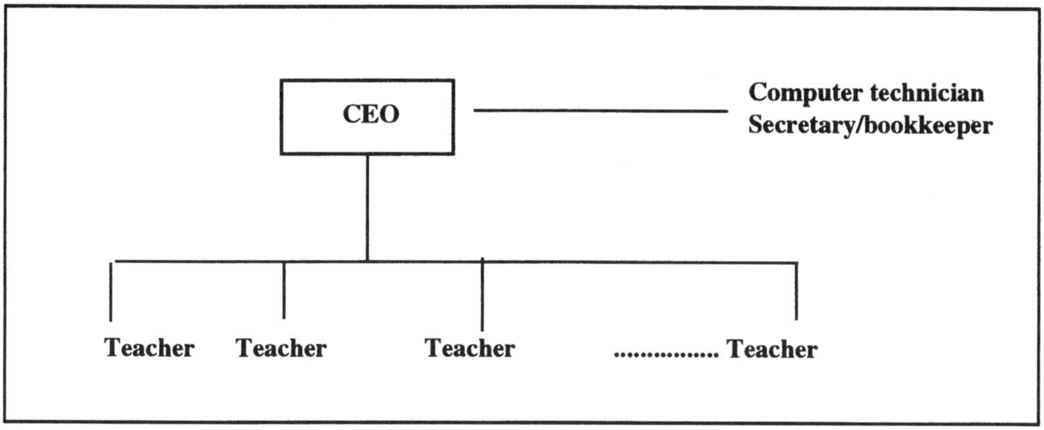

they write English news items and how they answer questions from subscribers. They are given a PC and an e-mail account for their work. Thus translating English news from sources such as AP or Reuters which provide up-to-date news in the Internet, and sending the translated English news plus explanations on English expressions in the news via e-mail, an'd answering e-mail questions from subscribers, all fall into each teacher's responsibility. With the current 1,000 subscribers, each teacher deals with 100 subscribers personally. Other staff include a computer technician and a secretary who also works as a bookkeeper. Figure 3 shows the organizational structure of ENI Company.

Marketing Strategy of ENI Company

The marketing strategy of ENI Company can be analyzed through marketing mix components as following.

1. **Product**: ENI Company tried to differentiate its products from its competitors by giving a personal touch. In other words, by answering questions from subscribers promptly and paying attention to each subscriber's needs in English learning, ENI Company was successful in attracting loyal followers.
2. **Price**: ENI Company tried to differentiate its price from its competitors by offering a fixed monthly fee regardless of the number of questions answered. Other services charged customers by the length of time they spent on their service, since other services were available on the on-line information providers unlike ENI Company which supplies via e-mail.
3. **Place**: ENI Company did not see a presence in the web as a viable option due to the small number of Internet surfers. Instead, ENI Company noticed that even though there were relatively small number of people surfing the Internet, many more people have e-mail accounts either from commercial Internet Service Providers or from non-commercial Internet Service Providers such as one's own schools and companies. Thus ENI Company decided to exist only in the e-mail form until there will be more web surfers in Korea.
4. **Promotion**: ENI Company tried cyberspace advertising medium such as banner ads which failed due to the small number of web surfers, and bulk e-mailing ads which failed due to the high telephone bill for modem-based mailing and the law against such bulk e-mailing ads for non-modem-based mailings. ENI Company relied more on conventional advertising medium such as magazine and newspaper ads.

Table 6: Marketing Strategy of ENI Company

Product	Personal Touch
Price	Fixed Fee
Place	E-mail Form
Promotion	Conventional Medium

In summary, Table 6 outlines the marketing strategy of ENI Company.

In CEO's Own Words

The CEO of ENI Company provided detailed information on the firm. Following conversations are taken from the interview with the CEO.

Q: When and how was ENI created?

A: "In the beginning of 1997 I created the firm since I thought that I could use my extensive knowledge on world affairs in helping Koreans understand foreign news and at the same time, teach them on English lessons using expressions used in the news items. The Koreans are very interested in improving their English skill, especially young students in high school or college since it is vital to acquire a good English skill in order to get into a good college and to land in a better job. My idea was that if ENI Company could provide attentive answers to questions regarding English promptly, many people would sign on to our service. It turned out to be true. But there was a serious difficulty in getting our company known to the general public."

Q: What are its short and long term objectives?

A: "In short term, our objective is to create a critical mass of loyal subscribers whose size should be at least 1,000 people, which we achieved in a year. In long term, our objective is to move to the web eventually and be able to serve multimedia English news and lessons when the IT infrastructure of Korea improves and the number of Internet users increases significantly."

Q: Who are ENI Company's competitors?

A: "There are several similar services in the on-line information providers such as "Chollian" which is similar to AOL in US. But they don't send out individual e-mails; they put English news and lessons in the directory of on-line information providers. Thus they can not give the personal touch as ENI Company does. ENI Company is the only one that provides daily e-mail English news service to the individual subscribers in Korea."

Q: What were the challenges ahead? And how has ENI addressed the challenges?

A: "The challenge was marketing, especially advertising. In order to get more subscribers, it was critical to do effective advertising campaign. We tried conventional advertising medium such as magazines, radio, and newspapers as well as cyberspace advertising medium such as banner ads in the web page and direct e-mail ads. Besides conventional advertising medium whose effect was obvious, we thought that cyberspace ads could be effective too. At first, the banner ad in the web page looked attractive but soon we realized that the number of people who browsed the web pages in South Korea was still not big enough due to the low development of national IT infrastructure. We finally looked at mass mailing technique and it looked workable. We sent out direct e-mail ads and found out that modem-based bulk mailing used up so much telephone time that it was not financially feasible since there was no fixed fee option for local phone call in Korea. Non-modem-based bulk mailing through the Internet could not be used either since the Internet bulk mailing was already against the law in Korea. Thus we relied more on conventional medium such as magazine and newspaper ads."

Q: What were the critical events and outcomes in the evolution of ENI Company?

A: "The critical event happened after we installed the mass mailer and used it for sending advertisement mails for ENI Company. We didn't realize how big the telephone bill could be when the dial-up connection was used for a long time. In other words, during the peak hours the performance of the mass mailer got downgraded, which resulted in taking much longer time to finish sending out all mails. This meant a disaster for ENI Company since the increased revenue from the bigger number of subscribers thanks to the mass mail advertisement was wiped out because of astronomical amount of the local phone bill. This event forced us to give up advertising through the mass mailer."

CURRENT CHALLENGES/PROBLEMS FACING THE ORGANIZATION

Following matters can be considered to be current challenges/problems for ENI Company.

First, how to sustain an e-commerce venture in a less-developed national IT infrastructure is a big concern for ENI Company. Effective e-commerce activity requires well-developed national information technology infrastructure such as well-connected fiber optic computer networks that cover the entire country, and easy availability of computers and affordable cost of network use among ordinary citizens of the country.

Second, how to conduct marketing for the venture is another big concern for ENI Company. Not only is it very difficult to establish a functioning e-commerce firm but it is even more difficult to do marketing on the Internet. The product of an e-commerce firm is usually advertised in banner-style ads in the web page, which is a common practice and is considered to be a working solution in developed countries. However, in countries such as South Korea, not many people are surfing the Internet, i.e., the number of people who can afford to surf the Internet for quite a long time are very limited due to the outdated billing practice of the government-owned monopolistic phone company. Thus, in this environment an e-commerce company which wants to sell a tangible product such as computer hardware components or software items has no choice but to create a web page that sells those products and faces the problem of how to make its web page noticeable to the small-number of Web surfers in Korea. Furthermore, those e-commerce companies who sell intangible goods face even more difficult challenge in advertising their products. The company which is the subject of this case sells English news and English lessons as products; they deliver these products to the individual's e-mail account. In short, any e-commerce company in Korea faces the same problem in marketing, which is how to advertise effectively in an environment where there are only small number of people who are surfing World Wide Web pages.

Third, how to establish a workable payment method is the biggest concern for ENI Company. This matter is a serious obstacle in the development of e-commerce even in developed countries such as the U.S. since consumers are worried that their credit card numbers might be stolen and might not get the ordered goods after paying with credit cards. In less developed countries such as South Korea, the consumers are, of course, very much worried when they order a product on-line or through the Internet, not because of the possibility that their credit card numbers might be stolen, but mostly because of the fact that the cash they have already paid might be robbed, i.e., the goods never arrive. The reason is rather simple. In a less-developed country like South Korea, not many people have credit cards and even for those with credit cards, tend to prefer cash when the price of the good is not very high. For this reason, most on-line payment is made as follows: First, the customer goes to the bank and does a wire-transfer of cash to the designated bank account of the seller which is an e-commerce firm. Second, the e-commerce firm's employee goes to the bank and checks if a certain customer's cash was transferred. Last, if the money-transfer is confirmed then the firm sends out the good that was ordered by the customer. With this kind of payment method, the consumers are naturally worried, since in actual incidents some firms took the money and ran away, i.e., no goods arrived.

Last, how to handle customer relations is also a concern for ENI Company. The subject of our case is a firm that claims to be attentive to the needs of the customers, i.e., it promises to answer all questions regarding English news items and lessons using those news items that are delivered to the subscribers. But in reality, it is very difficult to give satisfying answers to every question from the customer. Thus, quality control of customer service is a challenging puzzle to solve.

FURTHER READING

Kalakota, R. and Whinston, A.B. *Electronic Commerce: A Manager's Guide*, Addison-Wesley Publications, 1996.

Kalakota, R. and Whinston, A.B. *Frontiers of Electronic Commerce*, Addison-Wesley Publications, 1996.

Keen, P.G.W. and Ballance, C. *On-Line Profits: A Manager's Guide to Electronic Commerce*,

Harvard Business School Press, 1997.

Meeker, M. and Stanley, M. *The Internet Advertising Report*, HarperCollins, 1997.

O'Keefe, S. *Publicity on the Internet: Creating Successful Publicity Campaigns on the Internet and the Commercial Online Services*, John Wiley & Sons, 1996.

Sterne, J. *What makes people click: Advertising on the Web,* Que Education & Training, 1997.

Whinston, A.B., Stahl, D.O. and Choi, S.Y. *The Economics of Electronic Commerce*, Macmillan Technical Publishing, 1997.

Wong, P.K. (1996). Implementing the NII Vision: Singapore's Experience and Future Challenges. *Information Infrastructure and Policy*, 5(2), 95-117.

Zeff, R.L., Aronson, B. and Zeff, R. *Advertising on the Internet*, John Wiley & Sons, 1997.

REFERENCES

Asuncion, R.M. (1997). Potentials of Electronic Markets in the Philippines. *Electronic Markets*, 7(2), 34-37.

Bhatnagar, S. (1997). Electronic Commerce in India: The Untapped Potential. *Electronic Markets*, 7(2), 22-24.

Burke, R. (1997). Do you see what I see? The future of virtual shopping. *Journal of the Academy of Marketing Science*, 25(4), 352-360.

Copfer, R. (1998). Marketing in an online world. *American Salesman*, 43(3), 19-21.

Culnan, M. (1993). 'How did they get my name?': An exploratory investigation of consumer attitudes toward secondary information use. *MIS Quarterly*, 17(3), 341-361.

Gustavson, J. (11/10/97). Netiquette and DM. *Marketing*, 102(42), 46.

Hansen, G. (01/19/98). Smaller may be better for web marketing. *Marketing News*, 32(2), 10-11.

Jeong, K.H. & King, J. (1996). National Information Infrastructure Initiatives in Korea: Vision and Policy Issues. *Information Infrastructure And Policy*, 5(2), 119-133.

Kraemer, K., Dedrick J., Jeong, K.H., King, J., Thierry, V., West, J., & Wong, P.K. (1996). National Information Infrastructure: A Cross-Country Comparison. *Information Infrastructure And Policy*, 5(2), 81-93.

Liu, Z. (1997). China's Information Super Highway: Its Goal, Architecture and Problems. *Electronic Markets*, 7(4), 45-50.

Martin, J. (1998). You've got junk mail. *PC World*, 16(4), 45-46.

Mosley, J. (01/19/98). Deck us all in on-line shopping. *Marketing News*, 32(2), 6.

NCA(National Computerization Agency of Korea). (1997). A Study on the Technical Trend of the NIC(Network Information Center) and Internet Technologies. http://ncalib.nca.or.kr/HTML/1997/97066/f97066.htm.

NCA (National Computerization Agency of Korea). (1998). The White Paper on Informatization of 1998. http://calsec.nca.or.kr/knowledgebase.htm.

Rao, M. (1998). Ad Convention: Indian Advertising Agencies Urged to Harness Internet Technologies. *Electronic Markets*, 8(1), 48-49.

Westland, J.C., Kwok, M., Shu, J., Kwok, T., & Ho, H. (1997). Electronic Cash in Hong Kong. *Electronic Markets*, 7(2), 3-6.

BIOGRAPHICAL SKETCH

Ook Lee is a professor of information systems in the Department of Business Administration at Hansung University in Seoul, Korea. Previously, he worked as a project director at Information Resources, Inc, in Chicago, Illinois, and as a senior information research scientist at Korea Research Information Center in Seoul, Korea. His main research interests include electronic commerce, digital libraries, expert systems, neural networks, and critical social theory. He holds a B.S. in Computer Science and Statistics from Seoul National University in Seoul, Korea, and an M.S. in Computer Science from Northwestern University in Evanston, Illinois. He also earned an M.S. and Ph.D. in Management Information Systems from Claremont Graduate University in Claremont, California. His email address is leeo@hansung.ac.kr.

Remote Management of a Province-Wide Youth Employment Program Using Internet Technologies

Bruce Dienes
University College of Cape Breton, Canada

Michael Gurstein
University College of Cape Breton, Canada

EXECUTIVE SUMMARY

A province-wide network of Community Access Internet sites was supported during the summers of 1996 and 1997 by Wire Nova Scotia (WiNS), a government funded program to provide staffing, training and technical support for these centres. The program was managed remotely from an office in Sydney, Nova Scotia (Canada) using a variety of Internet-based technologies, including email, a web site, conference boards, real-time chat, and mailing lists. Remote management enabled the efficient and low-cost operation of a program involving 67 sites with field placements, plus six regional coordinators and the technical and administrative staff at the hub in Sydney. Effectiveness of remote management was enhanced when employees participated in an initial face-to-face regional training workshop. This training not only familiarized the employees with the communications technologies, but, perhaps more importantly, put a human face and personality to the messages that later came electronically over the Intranet.

[Note: For the benefit of those readers who may not be familiar with Internet technical terms, a brief glossary (Appendix A) is provided at the end of this case.]

BACKGROUND

Remote management as a key strategy for enabling new kinds of enterprise and other cooperative endeavours has emerged from the necessity of creating networked organizations to deal with the complexity and fluidity of the Information Age society. In the business world, the companies who are able to create effective virtual teams, particularly if they are able to collaborate with other companies that have complementary resources, are the ones who are thriving. For example, in 1995, three "arch-competitors" in the magazine business, *Men's Health, Esquire,* and *Rolling Stone* discovered an opportunity to collaborate on a bid for a significant advertising contract, and in three short weeks managed, via rapid and effective use of electronic communication and virtual team building, to put together a bid that beat out the media giant, Time, Inc. for the contract. In Denmark, government policy encouraged small enterprises to create collaborative networks, and this has resulted in a significant increase in the success of these businesses (Lipnack & Stamps, 1997).

Note that it was not just that electronic media created more efficiency or enabled existing projects to be done more effectively. It created the possibility for an entirely different way of doing business. As Marshall McLuhan (1964) has told us, "The medium is the message." Even if our initial motivation in using electronic media is to simplify or speed up existing routines, the nature of the medium inevitably generates secondary effects (Grundy & Metes, 1997), changing the way we work, the way we think, and the way we relate to other people and organizations. One impact of working virtually is that the old hierarchical models of the Industrial Age become transformed into a more interdependent networking model, where access to information and participation in decision-making is not limited to the upper echelons.

> *Instead of asking, "What is the information that matters and how do we most effectively manage it?" companies must start asking, "What are the relationships that matter and how can the technology most effectively support them?" — Michael Schrage, The Wall Street Journal, March 19, 1990 (cited in Johnson-Lenz & Johnson-Lenz, 1995).*

This creates a need to develop new management and leadership styles, new ways of training workers to collaborate effectively, and a readiness to work across organizational boundaries, whether in commerce (e.g., Sieber & Griese, 1998) or in not-for-profit organizations (e.g., James & Rykert, 1997). Moreover, the network that is built must be constructed in anticipation of rapid change, able to respond immediately to new realities and new opportunities (Metes, Grundy, & Bradish, 1998).

In the case described below, the initial motivation for using remote management via use of an intranet was to create a cost-effective way of coordinating a province-wide program, delivering technical support, and receiving reports. However, once the network was created, it became clear that it also had potential to enable types of collaboration that would not have been practical with conventional organizational structures. Despite the fact that there was a nominal hierarchical structure in place, with both central and regional levels of coordination, the universal access to information and to communications tools enabled both fieldworkers and regional coordinators to operate with a high degree of independence, and were able to use the intranet communications to brainstorm, plan and implement collaborative activities involving multiple sites. The existence of the communications network encouraged and enabled formation of temporary regional action networks, which have the potential to evolve into more permanent Community Enterprise Networks (CENs), described below (see also Gurstein & Dienes, 1998). Such secondary effects of remote management have great potential and should be considered in both the planning and evaluation stages of a project.

Governments, educational institutions and non-profit community organizations are increasingly becoming involved in partnerships for sharing information systems and intranets for collaboration and for delivering services. One excellent example is the Missouri Research and Education Network (MORENet). "The primary mission of MOREnet is to provide collaborative networked information services to its members and customers in support of education, research, public service, economic development and government" (http://www.more.net). Resources on their web site include technical support, discussion lists, and access to tools such as online databases. Its link to the Missouri Express project for supporting Community Information Networks (CINs) provides an online newsletter as well as contact information, instructions and templates for creating a CIN.

The Wire Nova Scotia (WiNS) case described below incorporates these ideas into its intranet, and adds the element of remote management of staff, including monitoring, reporting, documentation of the project, collaboration on inter-site projects, and technical support.

SETTING THE STAGE

In the summer of 1996, the associate chair in the Management of Technological Change (MOTC) at the University College of Cape Breton (UCCB) in Sydney, Nova Scotia, Canada opened the Centre for Community and Enterprise Networking (C\CEN). C\CEN's function was to research, develop and incubate ways of using the emerging Information and Communications Technologies (ICT) to promote local economic development in non-metropolitan regions, linking university, government and private resources with local needs. C\CEN followed the ongoing traditions of community-based economic development of the Antigonish Movement, which began in the Cape Breton area, and the mission of UCCB, which is to provide support to the economic and social development of the Cape Breton region. (See also the C\CEN web site, located at <http://ccen.uccb.ns.ca>.)

C\CEN had no core funding, so staffing and facilities would grow and shrink based on the particular projects it was funded to run at any given time. Because of this, its organizational structure was continually in flux. While this was at times confusing, it did create a climate of openness to experimentation with different management and organizational strategies. Base staff included the director, administrator, administrative assistant, systems administrator, and a technical support officer. A project leader would be hired for each new project, along with additional secretarial and technical support as needed, and staffing specific to the project. Project coordinators were given latitude to innovate and implement creative solutions to issues as they arose, generally reporting directly to the Centre's director. As the number of projects grew, a technology coordinator was appointed to oversee the incubation of new projects. Later this function was taken over by two assistant directors, one responsible for project development, the other for project implementation. Often staff roles and job descriptions would change to adapt flexibly to the needs of a new configuration of projects. The ability to share common technical and administrative expertise and resources between projects enabled a more cost-effective solution to staffing needs.

In its first year of operation (June 1996 to June 1997), leveraging an initial $165,000 grant into almost one million dollars via various partnerships and sponsorships, C\CEN was responsible for the creation of 140 person-years of skilled technically intensive work in non-metropolitan Nova Scotia, primarily in Cape Breton.

Cape Breton is suffering from the effects of a depleted resource-based economy, with the former mainstays of fishing, coal and steel no longer able to provide sufficient employment for the local population, resulting in outmigration and increased poverty. C\CEN piloted several information-based employment-generating projects with a view to encouraging the formation of a cluster of activity in this sector in the region, which, once it reached a "critical mass", would be self-sustaining. In particular, C\CEN encouraged the development of the network of Community Access Internet sites initiated by Industry Canada's Community Access Program (CAP), which provides a one-time-only grant on a competitive basis to organizations in up to 10,000 communities across Canada to support public Internet access. (For more information on CAP, see <http://cap.unb.ca>.)

Each CAP site is operated autonomously by a local community board. In isolation, such sites are unlikely to be economically viable, much less stimulate economic development in their area. However, as a coordinated network of centres (a Community Enterprise Network), they can serve as a distribution system for value-added facilitated access to various government and commercial information services (e.g., Job Banks, health information, on-line banking) and also as a coordination network for distributed production and joint marketing of local products and services to the global marketplace, using the flexible networking model (Piore & Sable, 1984; Gurstein, 1998a).

In order to stimulate the development of the CAP network in Nova Scotia, and to encourage communication and cooperation between the CAP sites, C\CEN developed a project called Wire Nova Scotia (WiNS) to coordinate and support summer staffing of the sites. (For a description of the WiNS program see the Case Description below.)

One key aspect of WiNS was to demonstrate that a province-wide program such as this could be administered from Cape Breton using appropriate communications technologies, as easily as it could from the central provincial administrative hub in the capital, Halifax. Instrumental to this was the development of an effective, inexpensive system of remote management. It is this system that is the subject of this case study.

Prior to developing the WiNS program, the C\CEN office was working with a small Windows95

Local Area Network (LAN), connected to the Internet via dial-up modem using WinGate software, plus a Unix server, located at UCCB, which hosted a web server as well as SMTP, POP-mail and Majordomo mailing list services. Internet connectivity for the Unix server was through UCCB's pipeline. Dial-up connectivity for the LAN was donated by MT&T Sympatico. All CPUs were Pentium 100 or 133mhz.

With these simple tools, the Centre managed an impressive array of projects, including creating an on-line bilingual real-time interface to the Access'96 conference held in Sydney, NS in November, 1996. Participants on the Internet could view real-time summarized transcriptions of ongoing presentations and seminars, simultaneously translated into English and French. Questions and comments from Internet participants would be directed to the speakers for immediate reaction.

Some examples of other projects include:

- **Environmental Web Project**: Web-based index of all organizations in Cape Breton with the capacity to provide services in the Environmental sector.
- **The Cape Breton Music Centre On-line**: Web site for the promotion and preservation of Cape Breton music, with an events calendar, resource directory, links to musicians' sites, etc.
- **Technical Support Group**: Offering on-line technical support to Community Access sites, First Nations sites, etc.
- **Electronic Democracy Project**: Providing a web site and email lists for open discussion of election issues in Nova Scotia.
- **Student Connection Program**: This was a province-wide program in which students were hired to train small businesses in use of the Internet.
- **ANGONET**: A Government of Canada project, "NetCorps" is being designed to link Canadian youth with technology-based projects in developing countries. As a pilot, C\CEN sent two young Cape Bretoners to provide technical support for ANGONET, Angola's Internet Access Network.

Many of these projects, particularly Student Connection, ANGONET, and the Technical Support Group included a remote management aspect, and experiences with various communications modalities led to a better understanding of what worked and what did not. This experience enabled C\CEN to design an effective remote management system for WiNS.

CASE DESCRIPTION

WiNS'97 Project Description

Industry Canada's Community Access Program (CAP) has supported Community Access sites by providing startup funds as well as funding for summer youth workers. These workers help staff the sites and also encourage involvement of local businesses, government and community groups. The Wire Nova Scotia Project (WiNS), developed and piloted in 1996 by C\CEN, in collaboration with the Nova Scotia Community Access Committee, was an enhancement to this program. WiNS assisted Community Access sites and summer youth workers in Nova Scotia to reach their potential through coordinated training, management and administration, including on-line technical support. The program was open to any site providing community access to the Internet, not only the CAP sites.

The WiNS pilot was subsequently adopted by Industry Canada as a national model. The following summer, C\CEN was asked to implement WiNS '97, which served 67 access sites across Nova Scotia directly, and many others indirectly. This effort included helping sites to strategize with school boards, libraries, regional development authorities, federal agencies and corporate sponsors to work towards the creation of local collaborative networks, with a view to creating long term sustainability.

In addition to the site workers, there were six regional coordinators with management skills. These staff helped to coordinate the efforts of sites in their area, facilitated provision of training and technical support and provided regular supervision or intervention where necessary.

WiNS is a medium for community capacity building. Community Access and a trained workforce are both critical to the infrastructure for other initiatives that will follow. They provide the groundwork for a new commercial sector—Rural Informatics, and the structures to enable rural enterprises to be both competitive with, and appropriate collaborators for, their urban counterparts.

The development of WiNS involved the investment of many partners including:

- Industry Canada
- Enterprise Cape Breton Corporation
- Human Resources Development Canada
- Nova Scotia Department of Education & Culture
- Nova Scotia Department of Economic Development and Tourism
- University College of Cape Breton
- MT&T / Sympatico
- Nova Scotia Regional Libraries
- Regional Development Agencies
- Acadia Centre for Small Business and Entrepreneurship
- Strait East Nova Community Enterprise Network
- Chebucto Community Network

In order to satisfy the various reporting requirements to these bodies, detailed records of fieldworker activity and site usage had to be kept.

One fieldworker was assigned to each participating Access Site. Precise job descriptions would vary with the specific needs of each location, but in general, the fieldworkers' tasks would include:

- staffing the Access Site in collaboration with local volunteers to maximize hours open.
- promoting the Community Access site to the community at large and marketing its services.
- providing training on Internet and related services to representatives of community organizations, business people and individual citizens.
- providing technical support to the Community Access site.
- outreach to neighboring communities to promote the Community Access concept and encourage others to develop proposals for the CAP program.

Funding for the fieldworkers' salaries was designated for the 16-30 age group, so most workers were young students with limited experience. This created a greater need for supervision and technical support. Complicating the management process was the fact that the workers had a dual-reporting relationship. The workers were interviewed and hired by the local CAP site committee, and a local site supervisor was assigned to them. Depending on the site, these supervisors were more or less accessible during the summer months. Since payroll was managed directly through C\CEN-UCCB, and the funders had specific expectations for outputs during the work term, the fieldworkers also reported to their WiNS regional coordinator, each of whom supervised between 10 and 25 sites. When the agendas of the site committees and the funders clashed, it was up to the WiNS coordinators to mediate the situation.

Overview of the Remote Management System

The newest meaning of "virtual" attests to forces that are fast moving teams into an altogether different realm of existence—virtual reality—or more precisely, digital reality. Electronic media together with computers enable the creation of spaces that are real to the groups that inhabit them yet are not the same as physical places. The eruption of the World Wide Web in the last decade of the millennium has allowed virtual teams to create private electronic homes. These interactive intranets—protected members-only islands within the Internet—signal a sharp up-tick in the human capability to function in teams. (Lipnack & Stamps, 1997, Chapter 1, Section 2).

Human resource management is a critical component to a successful project. Limited financial resources and a large geographical area require that traditional methods be adapted to adequately monitor, administer and manage a large and widely dispersed group. Using a storehouse of available knowledge and skills, developed from the projects described above, C\CEN was able to institute a form of Virtual Management. Using a Web site, O'Reilly's WebBoard™ software, e-mail, electronic mailing lists and IRC chat, the project was managed "on-line," with a formal reporting structure implemented and a variety of support functions available to assist the fieldworkers in their responsibilities. Five regional coordinators and one community network developer were placed across the province. These staff used the virtual management process to coordinate supervision of field workers in their region, as well as communications and cooperation between the regions and with technical and administrative support staff from C\CEN.

Needs

To be an effective system, the following management functions had to be implemented online:

- **Reporting**
  Weekly work summaries of fieldworkers
  Weekly site usage statistics
  Weekly regional reports from regional coordinators
- **Documentation/Archiving**
  Permanent record of reports accessible online
- **Conferencing**
  Several distinct online asynchronous discussions, threaded by subject
- **Announcements**
  Read-only announcement board
- **Problem Solving**
  Ability to form virtual task forces and workgroups to solve emerging problems
- **Tech Support**
  Online technical support for CAP sites, fieldworkers and regional coordinators.
- **Shared insights/experience/idea building**
  Free flow of information between all workers
- **Regional collaboration/planning**
  Coordinating local multi-site projects and efforts
- **On-line real-time meetings**
  Real-time chat (IRC) in a secure environment allowing workers to meet "virtually" to brainstorm projects and plan joint activities

In addition, the communication media had to have the following characteristics:

- **Flexible participation**
  Multi-modal access to communication services (via email, web and IRC).
  Both synchronous and asynchronous communication modalities.
- **Access control**
  Access to resources contingent on status, function and regional location within the organization.
  Signal to noise ratio (i.e., information versus socializing) appropriate to each channel of communication.
- **Cost-effectiveness**
  Resources had to be available at a cost easily affordable in terms of both central server and connectivity requirements and local workstation specifications.
- **Ease of use/User-friendliness**
  Even non-technical users should be able to use the system with a minimum of training.

Solutions

When the Web first got started, threaded discussions and live chat were interesting diversions *on Internet sites and corporate intranets, but little more. Not any longer. Today online conferencing has moved to the forefront of many web sites. Some of the Web's most highly trafficked sites are devoted to live discussion areas. And workgroups in small and large companies alike have discovered the productivity advantages offered through Web-based collaboration. (<http://webboard.oreilly.com/wb3/product_info/index.html>, [July 12, 1998]).*

The interactive capacities of the World Wide Web give rise to applications that enable a radical shift in communications options for organizations. Initially, the Web was used primarily as a giant bulletin board, but now is fast becoming a set of interconnected virtual communities. Secure interfaces allow for controlled access to information and communication resources, at the same time as allowing access from any terminal connected to the Web, provided the user has the appropriate access codes.

What this means is that organizations with small budgets can take advantage of the existing infrastructure of the Information Highway with only a modest investment in hardware and connectivity costs. Fieldworkers can access the organization's intranet from home, from a work site, or from a public access terminal. Regional coordinators with a notebook computer and a cell phone can travel anywhere in their region and do not necessarily require the expense of a branch office. The head office can be located anywhere, freed of the high rents of urban locations and able to bring jobs to non-metropolitan areas. These are the underlying concepts of the WiNS intranet: flexibility, accessibility, low cost, and effectiveness.

To manage the WiNS'97 project, two new P-150 NT servers were added to C\CEN's network: one to run an expanded LAN at the central office, and one to host the WebBoard and NT-Mail (to provide email service for staff). The main components of the WiNS Intranet were:

1) **WiNS Web Site** (<http://ccen.uccb.ns.ca/wins97>) housed on the C\CEN Unix server.
This was the "front end" of the Intranet, providing a gateway to the other features and a public face to the world. There were six main sections to the site:

a) Project Information Area

This was the public interface, which gave an overview of the project and links to the other areas on the site, including contact information. The WiNS general information flyer was available. Prospective employees could find out requirements for qualified applicants and sites could learn how to apply for a summer worker. Success stories from the previous year were highlighted. Press releases and updates about program innovations were posted as they became available. Links to other relevant sites provided information about the broader context of the project. These included links to the Nova Scotia Community Access Committee, the Community Access Program, the Strait East Nova Community Enterprise Network, the main C\CEN page and all other funders and partners of the project.

b) Technical Support Area

This area gave a description of services that were offered (as well as clear statements about services that were NOT offered!), a Frequently Asked Questions (FAQ) file, based on the highest volume queries dealt with by the technical staff, links to other technical resources on the Web, and, most importantly, a link to the Technical Support WebBoard conference, described below in the section on the WebBoard.

c) Regional Resources

A clickable map of the province, colour-coded by regions, gave quick access to resources of relevance to each particular region. These included a map of the region, showing all the access sites, and contact information for each site, each fieldworker, and the regional coordinator. These coordinators were encouraged to develop their own information resources for their region to add to this area. Graphic components were made available via FTP so that the regional sites thus developed could have a similar "look and feel" to the central site. Some regional sites were hosted on regional servers (such as local Community Network servers) rather than on the C\CEN server. Decentralizing development and hosting of regional web sites enabled local creative control over content and design, thus giving local workers more "ownership" of the site and making it more likely for them to use it. It also decreased the workload of the central technical team, enabling them to put more attention on technical support rather than on web design and maintenance.

d) Fieldworker Resources

This area gave an overview of fieldworker responsibilities, examples of specific tasks that workers would undertake, information on how to get help or technical support, and instructions on using their personal Web conferences for regular reporting purposes.

e) Conference Area (WebBoard) (See description of WebBoard below.)

f) Contact Information

Phone, fax, email and mail addresses for the central office, regional coordinators and technical support were posted. Links also were provided by region to a list of all participating sites, their WiNS fieldworkers and their contact information. This gave a simple, accessible method of reaching anyone in the organization.

2) E-Mail

For those staff who did not already have email accounts, C\CEN provided this service. [Note: With the new availability of free web-based email services, such as HotMail or EudoraMail, this would no longer be required.] Whereas the WebBoard was used for reporting, announcing and group discussions, email provided a medium for one-to-one correspondence, particularly for arranging meetings, dealing with personnel problems, or getting quick feedback. It also provided a mode for social interaction between the workers, a key aspect of maintaining high morale, particularly in situations where workers were staffing sites alone much of the time.

3) WebBoard

While the web site was the matrix that integrated the various communications modalities, the WebBoard was the "heart" of the system. (See Peck & York, 1998 or <http://www.oreilly.com/webboard> for a detailed description of the WebBoard software.) WebBoard provides web-based on-line discussion conferences that are accessible to anyone with a web browser and Internet access. A key feature of the WebBoard is the ability to restrict access to each individual conference depending on the status of the user within the organization.

WiNS used four types of conferences: general, regional, staff and personal, incorporating the functions of reporting, planning and coordination, problem-solving, documenting, socializing, idea generation and technical support.

a) General conferences were open to all employees and served a variety of purposes.

i) The *Announcements* conference was a read-only conference to which only senior staff could post but all could read. This created a zero-noise board that everyone would check regularly for important updates without having to plow through any chaff.

ii) The *Business Development* conference was an idea-generating and sharing medium for people to brainstorm methods of using the access sites to generate income, to support local business, or to create new businesses, using the ICT tools.

iii) The *Technical Support* conference provided answers to a variety of technical questions on hardware, software, networking, training, etc. Technical support was made available to any access site worker or supervisor, not just the WiNS worker. Anyone could sign themselves up to the WebBoard and automatically be given access only to the Technical Support and Business Development conferences. A particularly useful aspect of using a web conference rather than one-to-one email was that, in addition to getting information from the technical support officers, participants could assist each other and share expertise they developed as they solved problems locally. This created a broader pool of expertise and reduced the workload for the technical support staff. It also reduced the tendency for field workers to develop too much dependence on the central technical support, encouraging them to develop their own skills and share them with others. The technical staff would of course monitor the answers provided by other employees to ensure that they were not giving inappropriate advice.

iv) The *General Discussion* conference provided a medium for anyone to raise issues, talk to groups from other regions, chat, etc. This was an informal environment that facilitated socializing, networking, and resource sharing across the whole project. Providing a forum specifically catering to informal interaction helped to keep the other fora more focussed and productive. It is important to provide a variety of communication venues so that people of a wide variety of personalities and communication styles will feel comfortable using the on-line medium.

b) Regional Conferences served each region, enabling planning of joint ventures between nearby access sites, region-wide projects, coordination of promotion and advertising campaigns, announcements of regional on-line real-time chat meetings, and sharing of resources and expertise between people who had met at regional training sessions. Regional conferences were readable by anyone involved in the WiNS project across the province so that others could learn from innovations in other areas, but generally only participants from that region would post to the regional conference.

c) The **Staff Conference** was restricted to the senior staff: the WiNS coordinator, the WiNS administrator and the regional coordinators. It was used for weekly regional reports and for discussion of any problems involving sites, fieldworkers, or policies. Having this as a web conference rather than a one-way report to the WiNS coordinator enabled regional coordinators to learn from one another's experiences, and to offer support as appropriate. Coordinators had a variety of distinct expertise. For example, one may have training in community development, another in technical issues and another in business development. The staff conference made these skills available to all as problems and solutions were brainstormed.

d) Personal Conferences were the primary reporting method for all fieldworkers. Each fieldworker had his or her own personal conference that was used to file weekly reports and site usage statistics. This conference was viewable only by the fieldworker, their regional coordinator, and the WiNS coordinator. This gave the WiNS coordinator a log of activities and problems encountered by all staff. He could get a "snapshot" of any given site for any period by simply reviewing the weekly logs of the site worker. The absence of a log would point immediately to a potential trouble spot, and this could be addressed before the problem was exacerbated. These logs (in addition to the regional weekly reports) provided a rich source of data for compiling success stories and for writing the final report on the project.

4) On-Line Mailing Lists

C\CEN provided three email lists of relevance to this project. One was a general discussion list ("centre") for those working at the central hub in Sydney, allowing exchange of ideas between people working on different projects; another was a general discussion list for all the Community Access sites in Nova Scotia ("CA-NS"), providing a medium for sharing problems and solutions, new grant or income generating possibilities, upcoming workshops, conferences and training opportunities, etc. In addition, there was a list for general discussion of community access issues across Canada, with a particular focus on universal access to the Information Highway ("UA-C"). Originally run on majordomo software, the latter two are now functioning using LISTSERV. Archives of these lists can be found at <http://ccen.uccb.ns.ca/archives>.

5) IRC Chat

Internet Relay Chat (IRC) enables multi-user real-time text-based interactive communication. Individual users' typing is identified by an alias name at the beginning of each line they send, and each participant sees comments from all others as they scroll up the monitor screen. IRC-based meetings work best when there is a designated moderator and a predetermined agenda or set of questions about which participants are prepared to dialogue.

Chat functions were available for WiNS through the WebBoard or via IRC client software. Most sites used the MIRC shareware software for this purpose, as the chat capabilities in version 2 of WebBoard were limited. (Note: This limitation has been addressed in version 3 and private chat lines can operate within the WebBoard environment.) Use of chat for regular regional meetings was at the discretion of the regional coordinators. Fieldworkers could also set up chat meetings at any time, for example to coordinate multi-site activities (such as creating content for a booth representing area Community Access sites at the local county fair).

There were mixed opinions as to the usefulness of the medium. Some felt that it was important for fieldworkers to interact in real-time, as this built a sense of community and provided an environment for brainstorming or for sharing immediate issues. Others found the medium somewhat ponderous, especially for those without fast typing skills, preferring the asynchronous modalities of email or web conferencing, which allow participants to reflect on their messages before posting and to respond at whatever time they happen to be free.

6) **Face-to-face regional orientations and training sessions**

a) Prior to the fieldworkers' start date, the WiNS coordinator held an orientation and training session attended by all regional coordinators. In addition to familiarizing these staff with the policies, procedures and communications technologies of the WiNS project, the session afforded the opportunity to get to know one another, identify one another's skills and resources and, perhaps most importantly, put a human face to the email addresses that would be the primary source of communication throughout the project.

b) Each regional coordinator later held a one or two day orientation and training workshop for their fieldworkers, covering issues parallel to those described above. Those who for whatever reason were unable to attend the training session were harder to integrate into the WiNS team, regardless of the amount of technical training they were given. The lesson seems to be that, for this population at least, it is difficult to build a sense of teamwork and community, or to create a desire for on-line socialization, without some prior face-to-face contact. Whether this would be true for people who have much more experience working virtually is an open question.

CURRENT CHALLENGES/PROBLEMS FACING THE ORGANIZATION

Funding cutbacks

Funding for C\CEN expired in March, 1998, and at this point it seems unlikely that it will be resurrected. The 1998 summer staffing program for Community Access sites was operated directly from the MOTC office with reduced staff and for a shorter time. While there were insufficient funds to engage in development work to create new opportunities and modalities for rural Community Access, the remote management system (now using WebBoard™ v.3) enabled implementation of a similar program despite the much-depleted central resources.

Community Enterprise Networks (CENs)

One challenge facing emerging CENs is to adapt the remote management model for business purposes. Issues include higher security (encrypted email, secure servers, etc.) and secure web interfaces to dynamically updated databases such as inventory, contact lists, etc. Depending upon the nature of the particular enterprises fostered by the CEN, unique solutions would have to be designed. The needs of a distributed production network are quite different from that of an information service or technical support delivery network, for example.

SUMMARY

The coordination of the program using remote management via the Internet worked well. The web-based, password-protected Intranet enabled workers and staff to keep in touch with one another, track activities in other regions, collaborate on group projects, and create an automatic archive of all reports and discussion for future analysis. The on-line technical support service was particularly useful, as it allowed not only for rapid response from the technical staff at C\CEN, but created an open forum where WiNS participants could answer one another's technical questions and learn from watching the solutions unfold. The WebBoard™ allowed for management of the information flow such that, for example, fieldworkers' reports were accessible only to themselves, their regional coordinator and the WiNS coordinator, but regional discussion conferences were accessible to all staff. The WiNS Staff conference was accessible only by coordinators, creating a secure forum where confidential issues (such as personnel problems, etc.) could be discussed. One challenge with the system is that it required more training than anticipated, and a more user-friendly interface may be sought for future projects. Overall, the system saved considerable amounts of money on travel and phone costs, and it leaves an archive of all correspondence and worker logs, enabling future programs to learn from previous experience.

APPENDIX A: GLOSSARY OF TERMS

These definitions were extracted from the *ILC Glossary of Internet Terms*, Copyright © 1994-97 by Internet Literacy Consultants™. Used by permission.The URL of the full document is: <http://www.matisse.net/files/glossary.html>, which is where you can look for the latest, most complete version. Permission is granted to use this glossary, with credit to Internet Literacy Consultants, for non-commercial educational purposes, provided that the content is not altered including the retention of the copyright notice and this statement. ILC regards any use by a commercial entity as "commercial use". For permission to use it in other ways or to suggest changes and additions, please contact us by e-mail.: matisse@matisse.net +1.415.575.1156. Last update: 01/28/98

Browser

A Client program (software) that is used to look at various kinds of Internet resources.

Client

A software program that is used to contact and obtain data from a Server software program on another computer, often across a great distance. Each Client program is designed to work with one or more specific kinds of Server programs, and each Server requires a specific kind of Client. A Web Browser is a specific kind of Client.

E-mail

(Electronic Mail) — Messages, usually text, sent from one person to another via computer. E-mail can also be sent automatically to a large number of addresses (Mailing List).

FTP

(File Transfer Protocol) — A very common method of moving files between two Internet sites. FTP is a special way to login to another Internet site for the purposes of retrieving and/or sending files. There are many Internet sites that have established publicly accessible repositories of material that can be obtained using FTP, by logging in using the account name anonymous, thus these sites are called anonymous ftp servers.

Home Page (or Homepage)

Several meanings. Originally, the web page that your browser is set to use when it starts up. The more common meaning refers to the main web page for a business, organization, person or simply the main page out of a collection of web pages, e.g. "Check out so-and-so's new Home Page."

Another sloppier use of the term refers to practically any web page as a "homepage," e.g. "That web site has 65 homepages and none of them are interesting."

Host

Any computer on a network that is a repository for services available to other computers on the network. It is quite common to have one host machine provide several services, such as WWW and USENET.

HTML

(HyperText Markup Language) — The coding language used to create Hypertext documents for use on the World Wide Web. HTML looks a lot like old-fashioned typesetting code, where you surround a block of text with codes that indicate how it should appear, additionally, in HTML you can specify that a block of text, or a word, is linked to another file on the Internet. HTML files are meant to be viewed using a World Wide Web Client Program, such as Netscape or Mosaic.

Internet

(Upper case I) The vast collection of inter-connected networks that all use the TCP/IP protocols and that evolved from the ARPANET of the late 60's and early 70's. The Internet now (July 1995) connects roughly 60,000 independent networks into a vast global internet.

Intranet

A private network inside a company or organization that uses the same kinds of software that you would find on the public Internet, but that is only for internal use. As the Internet has become more popular many of the tools used on the Internet are being used in private networks, for example, many companies have web servers that are available only to employees. Note that an Intranet may not actually be an internet — it may simply be a network.

IRC

(Internet Relay Chat) — Basically a huge multi-user live chat facility. There are a number of major IRC servers around the world which are linked to each other. Anyone can create a channel and anything that anyone types in a given channel is seen by all others in the channel. Private channels can be (and are) created for multi-person conference calls.

LAN

(Local Area Network) — A computer network limited to the immediate area, usually the same building or floor of a building.

LISTSERV®

The most common kind of maillist, "LISTSERV" is a registered trademark of L-Soft international, Inc. Listservs originated on BITNET but they are now common on the Internet.

Login

Noun or a verb. Noun: The account name used to gain access to a computer system. Not a secret (contrast with Password).

Verb: The act of entering into a computer system, e.g. Login to the WELL and then go to the GBN conference.

Maillist

(or Mailing List) A (usually automated) system that allows people to send e-mail to one address, whereupon their message is copied and sent to all of the other subscribers to the maillist. In this way, people who have many different kinds of e-mail access can participate in discussions together.

Network

Any time you connect 2 or more computers together so that they can share resources, you have a computer network. Connect 2 or more networks together and you have an internet.

Node

Any single computer connected to a network.

Password

A code used to gain access to a locked system. Good passwords contain letters and non-letters and are not simple combinations such as virtue7. A good password might be: Hot$1-6

POP

(Point of Presence, also Post Office Protocol) — Two commonly used meanings: Point of Presence and Post Office Protocol. A Point of Presence usually means a city or location where a network can be connected to, often with dial up phone lines. So if an Internet company says they will soon have a POP in Belgrade, it means that they will soon have a local phone number in Belgrade and/or a place where leased lines can connect to their network. A second meaning, Post Office Protocol refers to the way e-mail software such as Eudora gets mail from a mail server. When you obtain a SLIP, PPP, or shell account you almost always get a POP account with it, and it is this POP account that you tell your e-mail software to use to get your mail.

Posting

A single message entered into a network communications system. E.g. A single message posted to a newsgroup or message board.

PPP

(Point to Point Protocol) — Most well known as a protocol that allows a computer to use a regular telephone line and a modem to make TCP/IP connections and thus be really and truly on the Internet.

Server

A computer, or a software package, that provides a specific kind of service to client software running on other computers. The term can refer to a particular piece of software, such as a WWW server, or to the machine on which the software is running, e.g. Our mail server is down today, that's why e-mail isn't getting out. A single server machine could have several different server software packages running on it, thus providing many different servers to clients on the network.

SMTP

(Simple Mail Transport Protocol) — The main protocol used to send electronic mail on the Internet. SMTP consists of a set of rules for how a program sending mail and a program receiving mail should interact. Almost all Internet email is sent and received by clients and servers using SMTP, thus if one wanted to set up an email server on the Internet one would look for email server software that supports SMTP.

Terminal

A device that allows you to send commands to a computer somewhere else. At a minimum, this usually means a keyboard and a display screen and some simple circuitry. Usually you will use terminal software in a personal computer - the software pretends to be (emulates) a physical terminal and allows you to type commands to a computer somewhere else.

UNIX

A computer operating system (the basic software running on a computer, underneath things like word processors and spreadsheets). UNIX is designed to be used by many people at the same time (it is multi-user) and has TCP/IP built-in. It is the most common operating system for servers on the Internet.

URL

(Uniform Resource Locator) — The standard way to give the address of any resource on the Internet that is part of the World Wide Web (WWW). A URL looks like this:
http://www.matisse.net/seminars.html or telnet://well.sf.ca.us or news:new.newusers.questions

etc.

The most common way to use a URL is to enter into a WWW browser program, such as Netscape, or Lynx.

WWW

(World Wide Web) — Two meanings - First, loosely used: the whole constellation of resources that can be accessed using Gopher, FTP, HTTP, telnet, USENET, WAIS and some other tools. Second, the universe of hypertext servers (HTTP servers) which are the servers that allow text, graphics, sound files, etc. to be mixed together.

REFERENCES / FURTHER READING

Dienes, B. (1997). *WiNS '97 final report.* [Report]. Sydney, NS: Centre for Community and Enterprise Networking, University College of Cape Breton. (Executive summary available online: <http://ccen.uccb.ns.ca/articles/winsum.html>.)

Grundy, J, & Metes, G. (1997, June). *Intranet challenges: Online work and communication.* [Online] http://www.knowab.co.uk/wbwintra.html [1998, October 8].

Gurstein, M. (1998a). *Flexible networking, information and communications technology and local economic development.* [Online] http://ccen.uccb.ns.ca/flexnet [1998, May 31].

Gurstein, M. (1998b). Information and communications technology and local economic development: Towards a new local economy. In G. A. MacIntyre (Ed.), *Perspectives on communities: A community economic development roundtable.* Sydney, NS: UCCB Press. (Also available online: <http://ccen.uccb.ns.ca/articles/ict4led-chapter.html>.)

Gurstein, M., & Andrews, K. (1996, October 31). *Wire Nova Scotia (WiNS '96): Final report.* [Report]. Sydney, NS: Centre for Community and Enterprise Networking, University College of Cape Breton.

Gurstein, M., & Dienes, B. (June, 1998). *Community enterprise networks: Partnerships for local economic development.* Paper presented at "Libraries as Leaders in Community Economic Development" conference, Victoria, BC. [Online] http://ccen.uccb.ns.ca/flexnet/CENs.html [1998, June 28].

Gurstein, M., Lerner, S., & MacKay, M. (1996, November 15). *The initial WiNS round: Added value and lessons learned.* [Report]. Sydney, NS: Centre for Community and Enterprise Networking, University College of Cape Breton.

James, M., & Rykert, L. (1997). *Working together online.* Toronto: Web Networks.

Johnson-Lenz, P., & Johnson-Lenz, T. (1995, March) Humanizing distributed electronic meetings. Paper presented at Groupware '95, Boston. [Online] http://www.awaken.com/Awaketech/AWAKEN1.NSF/d4cbbb795713bdee882564640074729d/8441cffd01b521d2882564a9008390dc?OpenDocument [1998, October 6].

Lipnack, J, & Stamps, J. (1997). *Virtual teams: Reaching across space, time and organizations with technology.* New York: Wiley. (See also <http://www.netage.com/index_vt.html>.)

Lipnack, J, & Stamps, J. (1994). *The age of the network.* New York: Wiley. (See also <http://www.netage.com/Knowledge/NetAge_top.html>.)

Lipnack, J, & Stamps, J. (1993, August 1) *Networking: Not just for big business.* New York Times. (Also available on-line: <http://www.netage.com/TNI/Publications/Articles/article_nyt.html>.)

McLuhan, M. (1964). *Understanding media: The extensions of man.* New York: McGraw-Hill.

Metes, G., Grundy, J., & Bradish, P. (1998). *Agile networking: Competing through the Internet and intranets.* Upper Saddle River, NJ: Prentice-Hall. (See also <http://www.agility.co.uk/an1>.)

Peck, S. B., & York, J. (1998). *Conferencing with WebBoard* (2nd. Ed.). Cambridge, MA: O'Reilly Software. (See also <http://webboard.oreilly.com>.)

Piore, M. & Sable, C, (1984). *The second industrial divide: Possibilities for prosperity.* New York: Basic Books.

Sieber, P.; Griese, J. (Eds.). (1998). *"Organizational Virtualness", Proceedings of the VoNet-Workshop, April 27-28.* Berne: Simowa-Verlag. (See also <http://www.virtual-organization.net>).

BIOGRAPHICAL SKETCHES

Bruce Dienes received his doctorate in psychology from the University of Illinois at Urbana-Champaign. He coordinated the Campus Support Groups Program for five years, developing training models and initiating a local network of on-line support groups. This program was designated a national model by the American Self-Help Clearinghouse. He has been involved in community development as an organizer, as a trainer and as a consultant. He worked for the Centre for Community and Enterprise Networking (C\CEN) at the University College of Cape Breton as program coordinator for Wire Nova Scotia (WiNS) and later as an assistant director of the Centre.

Michael Gurstein has a bachelor's degree from the University of Saskatchewan and a doctorate in sociology from the University of Cambridge. Dr. Gurstein worked as a senior public servant in British Columbia and Saskatchewan, managed a consulting firm concerned with the human aspects of advanced technologies and served as a management adviser with the United Nations Secretariat in New York. He is currently the ECBC/NSERC/SSHRC Associate Chair in the Management of Technological Change at the University College of Cape Breton and the founder/director of the Centre for Community and Enterprise Networking (C\CEN) of that institution.

Managing the NICS Project at the Royal Canadian University

Charalambos L. Iacovou
Georgetown University, USA

EXECUTIVE SUMMARY

This case describes the installation of an IBM mainframe computer at the Royal Canadian University. The goal of the described project was to establish a Numerically Intensive Computing Service (NICS) in order to provide "first-class" computing facilities to the researchers. Due to a number of factors, NICS failed to meet its objectives and the university abandoned the project within the first two years of its operations. The factors that contributed to its failure include: advancements in computing technology and changes in the computing style of end users; political and other non-technical considerations in selecting the system; and the weak and adversarial relationship between the computer center staff and the senior university administrators. These factors, with a special emphasis on organizational issues, are discussed throughout the case. At the end of the case, the reader is invited to provide solutions for managing the current failure situation and minimizing its negative consequences.

BACKGROUND

The University

The Royal Canadian University (RCU)[1] was established over 70 years ago. It is currently one of the largest universities in North America and employs about 2,000 faculty members in more than 100 academic departments, schools, and research centers. More than 30,000 students are currently enrolled at RCU. RCU's annual revenue exceeds $300 million. Provincial government subsidies and research grants account for about 85 percent of RCU's revenues and student tuition constitutes the remaining 15 percent. RCU considers itself one of the premier research institutions in North America. Currently, the university receives about $100 million annually in research grants and contracts. About 100 spin-off companies, with more than $700 million in annual revenues, have been established by RCU to market technology and know-how generated by its researchers.

RCU's administration structure includes the president, the chancellor, the board of governors and the university senate. The president of the university is RCU's chief executive officer and is responsible for overseeing its entire operations. The chancellor is elected by the university community and represents the university on official occasions. The 12 appointed and elected members of the board of governors are responsible for the administration of RCU's property and revenue. The senate, which has more than 60 appointed and elected members, is responsible for the academic governance of the university.

The daily operations of the university are managed by the president, five vice-presidents and twelve deans (see Figure 1). The Vice-President (VP) of Academic and Provost oversees the operations of the academic units of the University. The VP of Administration and Finance oversees

Figure 1: RCU's Organizational Chart

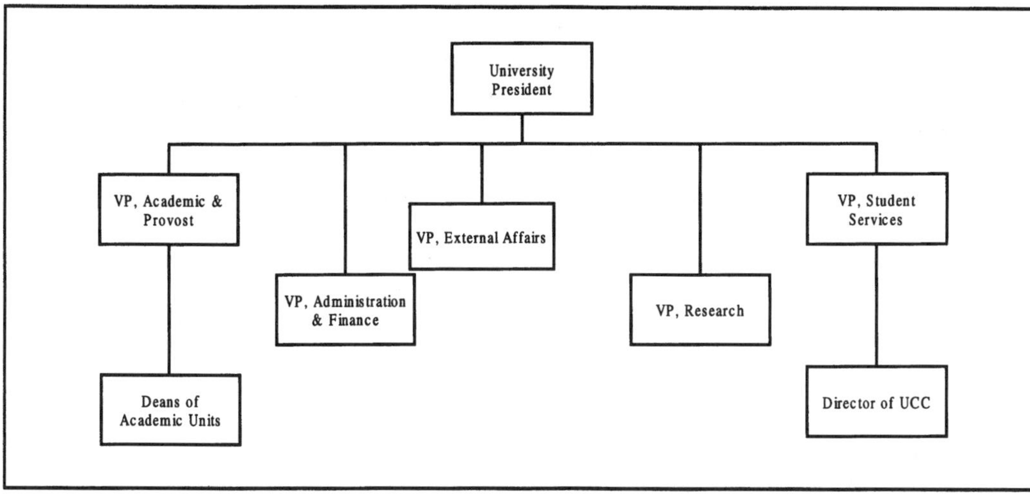

many of the administrative departments of the University, including Finance, Human Resources, Plant Operations, Security, the Bookstores, Planning and Development, and Purchasing. The VP of External Affairs is responsible for all external university relations, fundraising and development. The VP of Research oversees the research activities of the University and manages the relationships with grant agencies and private research organizations. The VP of Student Services oversees many of the support operations of the University, including the Registrar, Athletics, Computing, Telecommunications, Housing, Libraries, and Student Services.

History of Computing at RCU

In the mid-1950s, the university president established a committee to assess the university's interest in "computing machines and the study of automation in general." After a review of RCU's needs, the committee recommended the purchase of a computer for academic use. Contributions from local organizations in exchange for future computer usage were sought to help pay for its cost. A number of local firms declined the university's request because they did not see a reason for using such a machine! One of them even replied that "with reference to your letter of August 20th, I confess that I am unfamiliar with the electronic computer and its possible uses." Despite this lack of awareness in the business community, RCU managed to raise $20,000 in contributions from local organizations and acquired its first computer for $60,000 in 1957. This computer was an Alwac III E, a first generation, single-user computer capable of performing 250 instructions per second. This was among the first installations of computers in Canada. As expected, the Alwac III E became very successful soon after its installation. In its first two years of operation, it was used by more than 25 university departments and 16 outside organizations.

Due to the increased demand for computing services and the introduction of newer, more powerful machines, the Alwac III E was replaced in 1961 by an IBM 1620 computer. These two trends, the introduction of more powerful machines in the marketplace and the increasing demand for computing services, continued to play a key role in the university's computer purchasing decisions for a number of years. By the early 1990s, RCU had acquired ten new machines, each providing two to 15 times the computing power of its predecessor. These frequent computer purchases were made necessary by RCU's annual increase in mainframe usage, which is estimated to be about 20 percent.

The arrival of the first computer in 1957 was accompanied by the creation of RCU's first computing center (CC). The center was developed to support the computing needs of the academic community. A director (who later became the president of the university) and two computer programmers were hired to staff the center. The responsibility for overseeing the center's operations was assigned to the Provost. This center continued to operate as the sole computing facility at RCU until the mid-1960s when the university acquired its first computer to support its administrative services. At the time of the purchase, RCU established a separate data processing center (DPC) to support its non-academic staff. DPC was placed under the responsibilities of VP of Finance.

Since the establishment of DPC in the 1960s, the organization of computing services at RCU went through a number of structural changes. In 1980, the two centers were merged. In the Fall of 1984, the university decided to put renewed emphasis on administrative computing and the academic and administrative computing operations were again reorganized. The administrative systems staff were moved to a new department, Information Systems Management (ISM), under the supervision of VP of Finance; the academic computing services were moved under the supervision of a newly appointed VP of Student Services (who received a mandate for improving the university's computing and networking facilities). In 1990, after an internal review, RCU's administration again restructured its computing services. The CC and ISM were integrated into one department, University Computing Center (UCC). At the same time, all network related services (data networking, cable plant and telephone services) were moved to a newly established Data Network and Telecommunications (DNET) department.

By the early 1990s, UCC grew to include over 100 staff (three management staff, 52 programmers and analysts, 42 operations staff and seven administrative clerks). The staff was organized in several groups: Office of the Director, Academic Operating Systems, Administrative Operating Systems, Educational Services, Computer Operations, Statistics and Numerical Analysis, and Applications Support. The annual budget for UCC was about $6 million; $3.2 million were spent on salaries. Until recently, these funds were directly allocated to UCC as a line item in the university's overall budget.

UCC received input and direction from both the users and senior administrators. User input was received through a committee, the Campus Advisory Board on Computing (CABC). CABC was established in 1968 to "discuss and comment upon future plans and communicate feedback concerning the operations of the center." The interface between UCC and senior management was implemented through a direct reporting relationship between UCC's Director, Bob Lewis, and the VP of Student Services, Dr. John Parker. The relationship between Lewis and Parker was not a close one. This was reflective of the adversarial working relationship and politics between UCC and the university administration in general. The senior administrators viewed UCC as an auxiliary service and treated it as a cost center. Their main concern was to reduce its costs. The management of UCC was assigned to technical people who lacked the political power and leadership abilities to alter these negative perceptions and attitudes towards UCC and its services. A staff member commented on this issue:

> The VPs that we reported to did not have a good understanding of what was involved. They had a very high level overview of what was happening. I don't think they had a good understanding of really what was involved in providing a computing environment for either academic or administrative computing. They were not familiar at all with the center's operations. I think this low level of involvement was typical in industries that the computing side was still seen as black magic. Computing services were really only understood by people doing it and by key user groups because they were very aware of what was involved. I don't think senior management ever understood our operations. Indeed, the planning, the execution, the operation of computing services was all done by the technical people alone.

SETTING THE STAGE

Until the late 1980's, UCC enjoyed a somewhat monopolist power as it was the sole provider of computing services at RCU. All computer-related funds were centrally allocated to UCC which decided which systems to implement and which services to offer to the users. This monopolist advantage of UCC, however, was eliminated by two recent changes.

The first threat to the power of UCC was caused by changes in the hardware industry. Specifically, the introduction of RISC-based[2] computers into the marketplace made it possible for smaller organizational units to acquire their own relatively inexpensive and powerful computers. The introduction of these new computers created quite a concern in the computing industry, which up to that point was mostly dominated by large mainframes centrally located and administered in corporate IT units. For the first time, serious departmental users of intensive computing were able to gain access to powerful computing facilities without having to purchase a mainframe or subscribe to a centralized organizational computing service. However, as this new technology represented a dramatic shift from the traditional, well-accepted "big iron" approach and the RISC technology was not yet proven as a robust alternative to mainframe-based computing services, there was some uncertainty about its eventual success in the marketplace. Some computer experts felt that mainframes and supercomputers would continue to dominate the intensive-computing niche of the market; others felt that the newer RISC-based machines would be able to erode the virtual monopoly of the mainframe and push the industry towards an alternative, decentralized-model of computing.

In addition to the challenges created by the introduction of the RISC machines, UCC faced additional pressures from RCU's administrators who unilaterally decided to decentralize the allocation of computer funds. The administrators felt that this decentralization of resources and decision-making authority would improve the quality and reduce the cost of the computing services at RCU. The president's office strongly supported these changes because "a decentralized budgetary model encourages users to make informed choices as to which type of equipment or service is most effective, desirable and affordable for their particular needs." Under this new plan, the various academic units (instead of UCC) would receive annual allocations of computing funds. In turn, these units would have to pay UCC, based on a charge-back policy, for all the computing services they receive from it. To further increase the efficiency of its operations, UCC would be required to recover all computing-related investments through charge-back fees. This decentralization of the computer funds was gradually implemented. During the first year of this model, only ten percent of the funds were allocated to the academic units. Eventually, 100 percent of all computer funds were allocated to them. One of RCU's vice presidents explained the rationale behind the move to this decentralized model of computing:

> It was a time of change and it was not easy for anybody but with the president being as sort of strong-willed as he is, he felt, and I agreed with him, that decentralization in the longer term was the best bet. It gave individual units choice of what they wanted to do. And the argument that I used to hear was that there would be a lot of unused MIPS sitting on people's desks if you decentralized it. So, if you take the global view of the university there is a lot of redundant capacity and therefore you can have economies of scale by having a central machine and that's a traditional argument. But then my response to it was that if I drive on a highway I see a lot of redundancy with cars having only one passenger. And the reason why we tolerate that one passenger in a car is the individual freedom, flexibility of the people. We felt that we needed to give a similar type of flexibility to our computer users.

UCC personnel expressed strong opposition to this decentralization policy partly because of its potential limiting effects on their discretion and partly because of its lack of involvement in the policy's planning and implementation. One UCC manager commented:

> The decentralization process has been proceeding on an ad hoc basis. Our organization was so uncertain and the critical thing was that the fee for service transition was never outlined in a planned manner. It wasn't clear whether the university wanted us to become fee for service, whether they were going to force us to or not. What was definitely clear was that none of our customers like the idea and the idea was never ever promoted within the university. There was no process by which the university community was involved and could buy into the idea.

Overall, the transition to a decentralized, charge-back system, coupled with the availability of more powerful, smaller computers, which were being acquired by the users independently, had a negative impact on the perceived power of UCC. This is reflected in the following comment by one of RCU's deans:

> The computing center was technically a very good organization that kind of lost its way in the early 1990s. At that time they were providing less and less of a service. They were becoming less and less relevant to what was going on in our department because the administrative systems were in place and people were using workstations. Computing services has not been a powerful department within the university since they started to decentralize its funding.

CASE DESCRIPTION

Project Initiation

After the introduction of the charge-back rates, a number of science researchers in Chemistry, Physics, Engineering and other disciplines began lobbying the administration of the university to increase the level of computing support that was provided to them. This group of researchers, led by a Chemistry professor, demanded that the established mainframe CPU usage rates be reduced for off-peak use so that researchers with intensive computing needs could perform their computing tasks in the evening hours without draining their research budgets. In addition, they requested that the university seriously consider the purchase of a numerically intensive supercomputer for its researchers. At the time, only a couple of supercomputing facilities existed in Canada. Researchers with a need for such facilities had to independently arrange and pay for access to them.

The president of RCU responded favorably to these initiatives. Off-peaks rates were drastically reduced. Most importantly, the president, who felt that "a first-class university should have first class computing facilities available to its researchers," established a committee to evaluate the intensive computational needs of the researchers. Several researchers and CABC members were appointed to the committee, which was chaired by the VP of Student Services. After considering the needs of the researchers, the committee concluded that the university should indeed develop a large, numerically intensive computer service (NICS). The committee felt that such a facility would be pivotal in ensuring the future success of the university's research endeavors. In response to this committee's finding, the president created a university-wide vendor selection committee, composed by researchers and UCC staff members and chaired by two senior science professors, to identify and review candidate systems for this service and recommend a specific solution to the VP of Student Services for purchase.

Vendor Selection

Due to the uncertainty caused by the introduction of the RISC-based machines and powerful personal computers in the marketplace, the members of the vendor selection committee had difficulties agreeing on a specific system configuration for NICS while preparing the request for proposal (RFP). Some members of the committee believed that the proposed facility should consist of a single supercomputer. Others felt that the university should acquire a number of powerful workstations and connect them using a network. As a committee member pointed out, the selection of an optimal configuration was a difficult task due to the diversity of alternatives and preferences:

> The workstation technology was changing very rapidly and throughout the discussions there were proponents of the workstation solution, the clustered workstation solution, the multi-processor approach as well as proponents of the very expensive supercomputer, CRAY approach. Some felt that the only solution was the purchase of a "big iron." Others were making the decision between doing their intensive computing on a central machine and their own personal computers. They would have runs that would take perhaps days to do on a personal computer but there was no problem in terms of cost once you bought the machine—the cost is fixed. There are no problems in terms of scheduling— you didn't have to worry about anyone else's workload. And because the style of computing was changing, you could, for example, break a problem up into small pieces that they could run a piece overnight and come back the next morning and look at the results and continue from that point. The change in computing at the time created a big question for us.

Because the committee was unable to specify the exact architectural configuration of the potential service, it decided to let interested vendor recommend specific configurations and products. However, all members of the committee agreed (and so indicated in the RFP) that the proposed service should be a UNIX-based[3] one and should be scalable so it could act as "the beginning of a more comprehensive network-based large scale computing." The RFP was sent to 35 vendors. Thirteen of them responded to it.

The vendor proposals varied greatly, both in terms of computer architectures and processing power. The proposed solutions included super-workstations, mainframes, mini-supercomputers, near-supercomputers, supercomputers and various combinations of these. For about three months, the selection committee met to discuss the submitted proposals and review their technical and financial feasibility. However, the committee was unable to reach a consensus in selecting the "best" proposal due to two controversial issues. These issues related to the scale of the facility (whether the service should be based on a large supercomputer or a smaller machine) and its management (whether it should be managed by UCC or the science department). Due to these disagreements, the chairpersons approached the VP of Student Services and described the difficulties faced by the committee. During their consultations with him, they indicated that two proposals received considerable support but neither of them received the unanimous approval of the committee members. The first proposal was by Cray and it was recommending the acquisition of a Cray XMP-14 supercomputer. This solution represented the most powerful computer among the proposed systems and received the support of a few researchers who believed that a supercomputer was the only way to implement a numerically intensive service. The second proposal was by Convex and it was recommending the purchase of a Convex C220 vector computer. This solution was endorsed by UCC staff members and many researchers who felt that even though C220 was not a supercomputer, it would offer adequate computing power to the users at a significantly lower cost than that of Cray.

During its discussions with the senior administration, the committee was informed that the university was not in a position to purchase the Cray supercomputer due its high capital cost and operating expenses. The committee was also asked to reconsider a proposal by IBM that had rejected during its early deliberations. The proposal recommended the purchase of an IBM 3090 mainframe computer using the AIX [4] operating system. The cost of IBM's proposed solution was about $4 million.

After discussing the feedback of the senior administration, the selection committee reconsidered and rejected IBM's proposal again. The director of UCC and the chair of the selection committee wrote memos to the VP of Student Services indicating that the Convex solution was preferable to IBM's proposal, because the cost of the Convex C220 system was significantly lower than the cost of an IBM3090 even though both systems had comparable levels of performance. The following extract from the UCC's director memo illustrates the reasoning behind his decision:

> The Computing Center recommends that the University purchase the proposed Convex C220 system. The Convex is the superior choice because of its price, software maturity, performance, and ease of installation and operation. The Cray proposal is operationally very expensive and carries too much risk in terms of future cost and installation difficulty. The IBM proposal is too expensive relative to the performance of the computer. In addition, industry observers currently caution against purchase of low-end IBM 3090 computer for economic reasons.

Despite the negative feedback received from both selection committee members and UCC staff, the senior administration of the university continued to express a strong interest in IBM's proposal and engaged in discussions with the vendor to refine it. According to a senior administrator, because of RCU's relationship with IBM, the administration felt that it could significantly influence IBM in improving the terms of the proposal. Indeed, the relationship with IBM was a multifaceted one. RCU was among the recipients of the largest IBM donations in Canada. Also, at the time, IBM was contemplating the implementation of an air traffic control project (as part of a governmental contract) and RCU was being considered as a candidate partner for this project. A UCC staff commented on the attitudes of the senior administration:

> In the minds of many people, the NICS project was part of a bigger plan. It was part of RCU and IBM's existing and future relationship. At the time, I think the university was relatively predisposed to work with IBM. I recall that at the time we were talking about having a major computer lab out here and RCU was one of the candidate partners for it. The sad part is that lab was tied to some air traffic control bids and it never happened.

The apparent support for IBM's solution was met with strong resistance by both researchers and UCC staff members. They were concerned with the high cost and poor performance of the IBM 3090 computer and the immaturity and instability of the AIX software. As the following comment indicates, they attributed the administration's support towards the IBM proposal to non-technical, political reasons:

> IBM's proposal was not included in the short list of the selection committee, let alone be the top choice. Certainly the opinion of most of us was that [the selection of IBM's proposal] was a decision made at very high levels of the university for political reasons that had to do nothing whatever with the technical suitability of the solution —but had do to with a relationship with IBM. That remains my opinion to this day.

To address the financial and technological concerns raised by faculty members and UCC staff, IBM modified its initial proposal to make it more economically attractive for RCU. According to involved IBM managers, IBM was willing to reduce the cost and risk to the university because it was under a lot of pressure to maintain its market share (which was being attacked by manufacturers of smaller machines) and it wanted to establish itself in the growing UNIX market. IBM's modified proposal recommended a three year large scale computing software joint study with the university. As part of this study, IBM was willing to give RCU free use of an IBM vector facility for the duration of the study, and transfer title of the machine to the university upon its conclusion. It would also waive any license fees for the use of AIX for three years, and offer at least a 50% discount for such fees after the completion of the project. IBM was also willing to provide RCU with an early support plan (ESP) for its pre-release version of AIX for a few months until the product became more robust and commercially available. In addition, IBM indicated that it was willing to assist RCU in securing external funding for the lease payments.

Despite the significant improvements in IBM's proposal, the resistance among the UCC staff remained strong. In an eight-page report to the VP-Student Services, one of the senior programmers strongly opposed the acceptance of the IBM proposal because of its risk. The report listed a number of unsuccessful installations (and some "disastrous" ones) of the IBM 3090 vector facility and stated that the proposed solution offered low performance and high level of software unreliability at a too high price. Other UCC employees voiced their strong opposition to the revised proposal as well. Unfortunately, as the following comment indicates, their opinions did not carry significant weight:

> It didn't do any good saying that we had problems with this system because the decisions were made outside UCC. I don't think any of us had the ear of the president' office. Those of us who said things developed the reputation for being troublemakers. A couple of technical people who had said things at the VP's level were at the point where they were being shut out. They had the reputation as being negative towards IBM and therefore anything said was not being taken seriously. People were basically jeopardizing their own careers by arguing levels about the UCC director. In fact, both of these individuals left UCC because of they way they were treated by the administration.

After considering the new proposal and the recommendations of the committee and UCC, the VP of Student Services selected IBM as the winning vendor. Many UCC staff members felt ignored and were angered by this decision:

> The original decision to purchase an AIX system was viewed with total astonishment by most of the people on the technical side. They couldn't understand why the decision was made. [That this was a mistake] was extremely clear to the technical people. The trend of moving to smaller machines was not a new trend. It was clear that advances in RISC based systems would have dominated the mainframe market.

Many of the opponents of the IBM solution attributed the selection of the administration to political, non-technical reasons:

> IBM interacts with the University at many levels. It interacts with the university on different kinds of computers and different faculties and different kinds of uses so presumably all these different interactions were taken into account when the final decision was made. The result in the end was that the university decided that when factors other than the price performance ratio that we were looking at narrowly on this project were taken into account, IBM could make a proposal that would benefit the university better overall. That may well be the case. I mean there are other things that come into play when you look at this from the president's office. But I think it is true that the technical assessment was not that we buy an IBM computer.

According to the joint study agreement that was signed between IBM and RCU, the university was to receive an IBM 3090/150S mainframe with a vector facility operating the AIX operating system under ESP.[5] The primary of objective of this study was "to convert the major scientific applications from MTS (a non-IBM environment) to an IBM environment using AIX/370 and the Vector Facility." In a related agreement, RCU was to acquire the hardware through a four-year lease. According to the lease, RCU was to make the following payments to IBM: $400,000 during the first year of the project; $650,000 during the second year; $1,275,000 during the third year; and $680,000 during the last year. According to an IBM executive, this transaction was structured as a lease (instead of an outright sale) because of RCU's concerns about the eventual viability of the system:

> From our point of view as well as RCU's point of view there was a bit of anxiety about whether this solution was really going to make sense given the changes in the industry. Certainly some RCU people questioned whether this was going to be the right technology in the long term, but many others thought it would be a good solution. IBM did everything it could to adjust to the new environment. But, as you may know, when IBM was getting into the Unix area, it kind of stumbled a couple times in terms of the basic boxes it was making.

Finally, IBM agreed to purchase RCU's Amdahl 5860 computer for $165,000. These agreements were supplemented by a standard Government Term Lease Agreement as RCU was a provincial university. This agreement included a "non-appropriation system return clause" that would allow RCU to return the machine if it was not able to receive appropriations of sufficient funds to make the lease payments after making bona fide requests for such funds.

NICS Installation

When the IBM mainframe was delivered to RCU, an early release version of the AIX operating system was installed. UCC staff members tested the system and quickly identified a number of issues related to the performance and reliability of the system. UCC asked IBM to guarantee that it would address 18 specific issues that were identified by its staff during its early testing of the mainframe and software. Included in these issues was the need for a usage tracking software module that would enable RCU to track usage and bill its users.

Six UCC staff were assigned to maintain and support NICS during its pre-release stage. Also, 25 faculty members and their graduate students participated in the AIX pre-release testing. These researchers used the computer for free and reported problems to the UCC staff. A number of bugs in the system were identified by the users:

> There were a number of problems with the pre-release version of the software, including wrong answers. The Pentium problem all over again. So there would be a situation where they would simply get the wrong answer in very simple situations so it wasn't a complex one in a billion chance that the Pentium was. It was common and it was to the point that one plus one didn't equal two. It was that simple.

Despite these problems, UCC spent a significant amount of energy and resources on getting the NICS service operational. The attitude of UCC's management towards the implementation of NICS is reflected in the following comment:

> Once the decision was made it was our baby. And it doesn't matter if you don't like the baby to begin with, you still have to work with what you have, you still have to offer the services based on resources we have. So I think we took the point of view of doing everything we possibly could to make that system work. We worked with IBM in solving a whole number of technical issues.

Due to bugs in the system, the announcement of NICS was delayed several times. As a result, IBM extended the ESP indefinitely until the system was put into production. To compensate RCU for some of the additional costs associated with these delays, IBM paid $300 thousand to the university. Two hundred thousand dollars were allocated to the science department and $100 thousand to UCC. This allocation upset many UCC staff who felt that the UCC should be the only recipient of these funds (as it was now operating on a cost-recovery basis).

After a number of delays, the usage tracking software was installed and the system was finally put in production about a year after its delivery. According to the usage policy, interested researchers had to apply to the VP of Student Services to receive approval for an account. The usage rates that were established for the service were: $4.50 per minute for CPU usage; $0.50 per MB-minute for memory usage; and $0.08 per MB-day for disk usage. As part of the newly implemented charge-back policy, these charges had to be paid using distributed computing funds or research grants. A project participant described the effects of this policy on usage:

> There were a number of technical problems that meant we couldn't charge for the service initially but those eventually got resolved and at the point they turned charging on, usage dropped dramatically. Basically the system was being used by a whole number of people who simply didn't have the money to pay for computing. Graduate students were using it; some researchers were using it. The assumption that they made going into the project was that there would be funds available for this style of computing. However, as it turned out, the university budgets were being cut, researchers were not getting access to large amounts of grant money, and nobody had the money to pay for the service. At the same time, personal computing was becoming more powerful and affordable. And so from their point of view, the decision people were making was: do I run my programs on my PC or do I do it on NICS? If I can do it on the mainframe for free, then I'll do that. Then I use my PC for word processing too. If I have to pay for it however, I'm going to bring it back and put it on my PC. So, when we actually got the charging operational on NICS, on that particular date, the usage dropped from 100% to less than 7%. In the first day of the service we generated something like $3.87!

NICS continued to be operational with disappointing results. The average monthly CPU utilization for the first six months of its operation was about 480 hours (which is equivalent to about two thirds of the system's theoretical capacity). While the users continued to express their increased need for UNIX-based computing services, very few of them were willing to use NICS. Due to the wide availability of inexpensive RISC-based UNIX workstations and powerful personal computers in the marketplace, most users felt that the NICS rates were too high and a few of them began purchasing their own workstations and other computers using their research funds.

While NICS was being put in operation, the administration of the university established a new senior administrative position to improve the relationship between UCC and the university (which was further deteriorated by the selection of IBM as the NICS' vendor) and better coordinate the various technology units on-campus. To fill this position, Dr. David Williamson was hired as an Associate Vice-President (AVP) of Information and Computer Systems. As a result of this administrative change, the reporting relationship between UCC and the senior administration was altered. The director of UCC began reporting to Dr. Williamson instead of the VP of Student Services. Dr. Williamson commented on the responsibilities of this newly created position:

> I think the university computing services has always had at RCU, and in fact across the country, a very good reputation as a first class service. The other departments in my portfolio were relatively small and less significant at the beginning, when I started. We sort of expanded their role in a way that made it more integrated over computing and communications. Even though the computing center was quite well respected for what it did, due to changes within the university and in the computing environment in general, the administration saw a need for reorganization and direction and probably for getting on with a different role for the 90s than its role in the past.

Soon after his arrival to RCU, Dr. Williamson became aware of the issues related to NICS. Overall, he was concerned that UCC was losing both political capital and revenue by not taking advantage of the inexpensive RISC technology to offer UNIX computing services to the users (other than the expensive NICS service). More importantly, he was concerned with the ability of UCC to raise sufficient funds from usage charges to meet the next lease payment to IBM. As Dr. Williamson commented, this was a significant concern for him as the UCC budget was severely limited (due to the decentralization of the computer funds):

> My initial awareness [with NICS] had to do what I think was about 300 thousand dollars, or something of that neighborhood, of IBM donations. From what I was told, essentially IBM gave the university that amount because it really hadn't delivered what we had anticipated. I had a discussion with the VP about this. As I dug deeper to better understand the situation, I was getting the feeling that this project was not going to take off. In fact, because we were moving towards a cost-recovery model I was worried that when we eventually put the service in production it would not generate enough money for the next lease payment that was coming up. UCC, which was part of my portfolio, was expected to cost-recover all of its investments and I didn't think that was possible with this system. So, I began examining the issue in more detail and kept the administrators closely informed and involved with all the decision-making.

To address the first concern, the lack of inexpensive UNIX service, UCC decided to offer a second UNIX service using inexpensive RISC workstations to users who could not afford to purchase their own workstations or use the NICS service. To implement this service, UCC acquired a Sun SPARCstation 2 and two Silicon Graphics computers. After considering the acquisition and maintenance costs of these computers, UCC set the general UNIX usage rates. These rates were significantly lower than those of NICS. Specifically, the rates for the general UNIX service were $0.50 per minute for CPU usage; $0.063 per MB-minute for memory usage; and $0.04 per MB-day for disk usage. Shortly after its introduction, the UNIX service became very popular: there were more than 900 accounts on this service, with an average CPU utilization of 670 hours per month. Due to the high performance and low cost of the new UNIX systems, many NICS users moved their accounts to the new service. However, about 75 of them, who needed to use the vector facility and certain AIX-based software on the IBM system, continued to use NICS.

To address the issue related to the financial viability of NICS, Dr. Williamson spent many hours consulting with UCC staff and senior administrators. This was a critical issue for RCU which was facing the deadline of the third lease payment. At the time, the outstanding lease payments totaled almost $2 million while the market value of the IBM computer was estimated to be about $50,000 (according to the Computer Merchant's Price Guide). The low demand for NICS and the depreciated value of the hardware itself made the continuation of NICS an extremely difficult choice for Dr. Williamson.

CHALLENGES FACING RCU

Issues

Two years after the inception of the agreement and about a year after the introduction of NICS, RCU was faced with the decision of whether or not to continue the service. The timing of the decision was also affected by the preparation of the university's annual budgets, which needed to take into consideration the lease payments to IBM.

If RCU decided to continue to operate NICS, Dr. Williamson knew that UCC's budget would be severely impacted by the lease payments and the lack of revenue. Although abandoning the project could reduce the financial loss to the university, he knew that it would be difficult to convince IBM to accept the system's return. Also, the abandonment of NICS could create a potential public predicament for RCU. Abandoning the service after having paid about $2 million for it and having spent over a year to eliminate software bugs could be a major embarrassment for the university. On the other hand, if the university continued to operate NICS, there would be no available funds for the acquisition of additional RISC computers, further limiting the ability of UCC to respond to the needs of its users and increasing its credibility liability. Furthermore, Dr. Williamson needed to decide whether it was indeed wise for UCC to continue offering centralized UNIX services as RISC machines were becoming less and less expensive enabling academic departments and even individual researchers to acquire them on their own (and therefore reducing the demand for a centralized service).

In summary, Dr. Williamson needs to develop a plan to manage the issues related to the future of NICS while (1) ensuring that the university avoids additional financial losses and a public embarrassment and (2) ensuring that the needs of users for UNIX services are appropriately satisfied. What should he do?

ACKNOWLEDGMENT

The author would like to express his appreciation to the individuals who participated in this case study and acknowledge the insightful comments provided by Albert S. Dexter and the anonymous reviewers.

ENDNOTES

[1] Certain names and other information have been altered to protect the identity of the organization and individuals involved in this case. In all other respects, the case provides an accurate account of the facts. The data presented in the case are based on structured interviews and an analysis of numerous documents (meeting minutes, agreements, memorandums, electronic messages, etc).

[2] RISC stands for Reduced Instruction Set Computer. RISC is a computer processor containing a small set of simple instructions. Such processors are capable of performing faster processing through the use of the limited instruction set, uniform encoding, homogeneous register sets, and simple addressing modes.

[3] UNIX is an interactive, time-sharing open operating system.

[4] AIX stands for Advance Interactive eXecutive, which is IBM's version of UNIX. Even though IBM had just announced the development of AIX at the time, the software was not ready for commercial release and use yet.

[5] This was a single processor IBM 370 Enterprise System Architecture (ESA) machine and was rated at 12MIPS. The vector facility was rated at 10 MFLOPS. It had 64 megabytes of central storage (the maximum available on this model). Two IBM PS/2-70s were used as front-end processors. AIX's Transparent Computing Facility (TCF) was used to connect these machines so that they appear

as a single computer to the end users. Initially, the following software was installed on the NICS: AIX/370 operating system, FORTRAN VS compiler, IBM's Engineering and Scientific Subroutine Library (ESSL), International Mathematical and Statistical Libraries (IMSL), Numerical Algorithms Group (NAG) Library, U.S. Department of Energy Laboratories' SLATEC library, and standard UNIX utilities.

BIOGRAPHICAL SKETCH

Charalambos L. Iacovou teaches information systems at the McDonough School of Business at Georgetown University. His research focuses on the management of project failures, the adoption of information technology by small organizations, and the role of trust in electronic commerce. His papers have appeared in Management Information Systems Quarterly and the proceedings of conferences in Canada, Europe and United States.

Business Process Reengineering for the Use of Distance Learning at Bell Canada

Tammy Whalen
Bell Learning Solutions, Canada

David Wright
University of Ottawa, Canada

EXECUTIVE SUMMARY

The Bell Online Institute (BOLI) represents a radical change to the way Bell Canada provides internal training to its 27,000 employees. BOLI specializes in Web-based training, one type of technology enabled (distance) learning. Web-based training is a significant departure from the more traditional classroom-based practices at the Bell Institute for Professional Development, which is the organization that oversees all employee training at Bell Canada. This case study examines the use of Web-based training at Bell Canada in the context of business process reengineering. We present a theoretical context and a practical guide to how technology enabled learning changes the business processes in an organization. The study defines the processes that are required to deliver Web-based training, the value to the internal and external business practices of the organization, and the costs for each process. The wider applications of this case study are identified and will be of interest to those in other organizations that are moving from classroom delivered training to distance delivery.

This case study describes changes in the organization that result from reengineering, including the impact Web-based learning has on training plans, student needs assessments, the ability to provide specialized curricula, training students and instructors in using new technologies, and establishing a principle of continuous improvement. Alternative ways of achieving project objectives are presented, along with organizational impact, technology alternatives, and cost-benefits.

BACKGROUND

The telecommunications industry in Canada has undergone dramatic changes within the past few years with deregulation, rapidly changing technologies, and the globalization of business practices. A downsizing effort at Bell Canada in response to the changing nature of the telephone industry has affected almost every area of the company, with the internal training function being no exception. The Bell Institute for Professional Development (BIPD) is the business unit that oversees all internal training activity at Bell Canada. BIPD has been re-invented over the past 3 years, and changes are still on-going. The first change occurred in 1995, when all training that had once been conducted in-house by BIPD was outsourced to 4 companies known as the "training partners." BIPD's role changed from that of training provider to strategic planner. The second change occurred in 1997, when a decision was made at the corporate level to increase the use of distance learning for training delivery to Bell Canada employees.

Methodology

In a work which predates the concept of business process reengineering, Jim Stewart (1991), sums up the role of training in achieving organizational change when he states that organizational change can happen only when individual change occurs through learning. While there are a great number of published methodologies for applying BPR concepts, one model that emphasizes organizational transformation is the "6 R's" of business process reengineering by Johnson A. Edosomwan in *Organizational Transformation and Process Reengineering* (St. Lucie Press, 1996).

Edosomwan's 6-R's methodology draws upon previous well-known works on business process reengineering. Michael Hammer, the originator of the business process reengineering concept emphasizes the need for complete redesign of the organization to accomplish reengineering objectives. In *Reengineering the Corporation* (Hammer and Champy, 1993), identify innovation, speed, customer service, and quality as key success factors in the process. Other authors have also applied the original concepts of business process reengineering to specific situations such as training. In *Reengineering the Training Function* (1996), Donald Shandler describes preparation, process identification, vision development, solution formulation, and transformation as major reengineering elements and clearly shows how they apply to corporate training.

The advantage of the 6-R's approach is that the methodology is concise, yet well-defined. Below is a summary of the 6-R's methodology:

I. **Realization**
Realization is the understanding that dramatic changes are needed to meet the needs of organizations, teams and individuals and support the changes in the organization that occur due to rapidly evolving economies and technologies; and the re-design practices to fit with the organization's business objectives.

II. **Requirements**
The identification of strategic direction for the reengineering effort requires an analysis of customer needs. Decisions may then be made about the products and services that will fill the gap between the desired outcome and the processes, products, and services that are needed.

III. **Rethink**
New structures and systems must be put in place to accomplish the objectives of the reengineering program. Here, new procedures, processes, and technologies are identified.

IV. **Redesign**
At the redesign stage, specific targets are set for implementing the reengineering process.

V. **Retool**
At this stage, technologies or other more competitive systems are evaluated and adapted to the reengineering project.

VI. **Reevaluate**
During reevaluation, performance indicators are measured to determine how well the reengineering effort met its goals.

SETTING THE STAGE

Distance learning is an example of a changing business practice that Bell Canada hopes will improve both the bottom line and the quality of training delivery. Training is a significant employee benefit and employees at Bell have been very interested in the changes to training delivery that are occurring in the organization. The Bell Online Institute was created in order to explore the possibilities for using technology enabled learning in the organization. Web servers were procured for this purpose, and a core team was established with expertise in corporate training delivery, technology, and business development. Other interested parties were consulted as necessary, such as representatives from end-user groups. Distance learning has several obvious advantages over classroom training that needed to be explored:

- convenience and flexibility in the time, geographic location, and pace of training delivery, and
- cost savings over current classroom based training based on reduced training time and travel savings.

A related change to the organization was the establishment of Bell Learning Solutions, a business unit that provides distance learning services to customers through consulting, reselling third party products and services, and creating and marketing technology based course content.

CASE DESCRIPTION: 6-R Analysis

I. Realization: Establishing the Need for Business Process Reengineering

In the past year, there has been a great deal of interest in distance learning within Bell Canada. One of the reasons that BOLI was established was that Bell Canada believes that technology enabled training adds value to the training program. Employees find attending classes increasingly difficult to coordinate with their work schedules and believe that distance learning will offer them more convenience and flexibility than classroom based training. Classroom training is an "event," requiring pre-registration and courses offered at fixed periods of time, at a location apart from an employee's normal place of work. Distance training delivery to the desktop is one alternative that can provide continuous learning to employees. From a management perspective, distance learning offers cost savings over current classroom based training and enables employees to acquire and retain the skills necessary for Bell Canada to complete effectively in a global marketplace.

Outsourcing

Before 1995, BIPD conducted all employee training at Bell Canada. BIPD had a $40 million budget, classroom space, and 426 employees working in course design and delivery, as well as in administrative support functions such as project management, technology support, registration and course logistics. In 1995, training at Bell Canada was outsourced to 4 training partners. One company provides telecommunications technology related training, a second company provides information technology related training, a third company provides "soft" skills training in areas such as leadership and management, and the Bell Online Institute provides support for distance learning. Courses once offered by BIPD were on the topics of telecommunications, information systems and technology and general business skills, the same subject areas now covered by the training partner companies.

After outsourcing, the BIPD budget was $9 million and staff was reduced to 60 employees. Training partners also now provide classroom space as one of their outsourcing services along with all required course materials. The demand for employee training remains strong. In 1997, 14,000 students, equivalent to 43,300 student days, were taught at BIPD through the training partners.

Current Assessment

The current situation points to several possible directions for BIPD. These include:

- the need to reduce cost of training delivery
- the need to update course curriculum frequently so that topics of current interest are provided as they are required by the workforce to carry out their jobs, and reflect the evolving computer and telecommunications technologies used in the organization
- the need to provide high quality training
- the need to give end-users flexibility in the location where they take training
- the need to reduce training time by focusing training very specifically on topics needed by clients to carry out their jobs, and

- the need to align training with profit-making business ventures of the company, i.e., the support of Bell Learning Solutions, a business unit that designs and delivers tailored solutions to customers interested in implementing distance learning in their own organizations.

II. Requirements: Defining Customer Needs

BIPD's Strategic Role

BIPD now functions primarily in a strategic role to set the direction for training within Bell Canada and is the liaison with the outsource training partners. Liaison activities include building business relationships, identifying pricing and contract issues, and ensuring that internal customer needs are met. A customer services team of BIPD learning professionals has specialists dedicated to specific groups of Bell Canada employees to ensure that employee training needs are addressed. For example, a specialist may do an evaluation of an employee group to identify performance gaps and then put the employees in touch with one of the training partners to help fill those gaps through appropriate courses.

There is no transfer pricing done between the business units of employees taking courses and BIPD. However, the training partners do charge Bell Canada business units for courses employees take. Business units are given their own training budgets which they may use where they wish, and are not limited to the courses offered by training partner companies only. If an employee does takes a course through one of the training partners, payment is made directly to the partner company and BIPD gets no part of the payment. Tuition fees vary from course to course, depending upon course features such as the length of the course, the venue, and whether there is a need for equipment rental.

Both supervisors and employees themselves may make the decision to take a course at BIPD and balancing the needs of these two client groups is important. On the one hand, employees need the opportunity for personal development, while on the other, the needs of the corporation must be at the heart of BIPD's program. There is shared accountability between the employee and the employer when an employee takes a course; when an employee signs up for a course, a copy of the registration form goes back to the employee's supervisor.

The idea for a new course often comes from the training "customers" themselves. For example, a call center supervisor may feel that there are problems with his staff that a formal training session could remedy. BIPD would respond by using a set of diagnostic tools to determine whether the supervisor is correct and a course is the best solution to the problem. If the development of a new course is indicated, the next step would be for BIPD to bring in the appropriate training partner, who would use a standard instructional design systems process to move from the "idea" stage to developing an actual course. If a suitable instructor is not already on the staff of the training partner, the partner company would find an expert in the field and, if necessary, train him to be an effective trainer.

Before outsourcing, BIPD operated at a tactical level, performing all aspects of employee training in-house from instructional design to course delivery. Now, BIPD spends much more time identifying how training can address the needs associated with supporting Bell Canada's key business issues, for example, changes in the technologies the company sells, or even change management issues themselves. Apart from planning training strategy and management of outsourcing partnerships, BIPD is also still responsible for executive education, employee orientation, career management, technology support, and administrative support.

Included in BIPD's strategy for the coming year is the exploration of training options with the highest learning value to ensure a competitive advantage for Bell Canada. Within that strategy, the organization wants to evolve the technology-based learning capability within Bell. BIPD's key success factors in implementing this strategy are:

- the availability of human and financial resources
- the shared commitment of the training partners
- the availability of suitable computer systems to end-users and the ability of end-users to adapt to a technology enabled training environment

Table 1: Analysis of BIPD Before Distance Training Delivery

STRENGTHS • outsourcing has created opportunity for flexibility in training provision • outsourcing has meant that BIPD functions on a strategic rather than an operational level • senior management committed to using new technologies for training delivery and for new profit-making opportunities	WEAKNESSES • classroom training has high cost • course curriculum does not always reflect end-user needs for current topics of interest due to evolving job requirements and changing technologies
OPPORTUNITIES • end-users requesting training at their desktops • needs to be alignment with BIPD and Bell Learning Solutions, a business unit which delivers distance learning solutions to customers	THREATS • deregulation of telephone industry has increased need for broader business opportunities • changing technologies and globalization has increased need for current information

- the ability to move quickly with a shorter cycle time, and
- the availability of learning applications that meet the needs of customers and the business needs of Bell Canada.

Table 1 presents an analysis of the strengths, weaknesses, opportunities, and threats at BIPD.

Recommendations

From the analysis above, several next steps can be identified:

- research costs of Web-based training
- research Web-based learning platforms
- test usability of platforms and acceptance of distance training through a pilot project
- identify further courses for distance delivery within the coming year
- get commitment of training partners to support a distance learning strategy, and
- increase use of Internet / intranet delivery of training.

III. Rethink: Integration of Appropriate Technologies

Distance Learning Initiatives at Bell Canada

The challenges of deregulation, globalization, and constantly evolving technologies represent both threats and opportunities for the company. On the one hand, increased competition has necessitated drastic re-organization and painful cost-cutting, as well as high levels of stress for employees due to the new, and often unknown, demands the current reality brings. However, employees are rising to the challenge and evolving the business into new areas that draw on many strengths in the company. One of the new ventures is the provision of consulting expertise and distance learning outsourcing to customers through Bell Learning Solutions, an end-to-end business unit. Bell Canada has 20 years of experience in the distance education field. Bell has explored the internal use of various distance training methods and has also provided customers with distance learning technologies such as video-conferencing and business television. The many business activities in the company have allowed for the re-deployment of much expertise related to distance learning, for example a wide variety of Internet and electronic commerce applications.

Distance learning, like the current classroom training, is outsourced to the training partner companies. In-house, BIPD is developing expertise in instructional design related to multimedia materials in order to work effectively with the training partners. Partnership with the Bell Online Institute (BOLI) is BIPD's first initiative in Web-based training. The Bell Online Institute (BOLI) is Bell Canada's business unit that delivers technology-enabled training to Bell Canada employees. BOLI also functions as a testing ground for various Internet-based learning platforms used for training delivery at Bell Canada.

The Bell Online Institute has no direct reporting relationship with BIPD. Because BOLI has been independent of the existing training infrastructure, there has been more opportunity for creativity in that organization. BIPD uses the products that BOLI develops to provide services to the end-users. Courses that are developed by, or for, BOLI become part of the BIPD curriculum that is updated and managed by the training partners.

The use of distance learning has encountered no resistance from BIPD staff because it represents an interesting challenge rather than a threat. The real changes to training delivery at Bell Canada happened when outsourcing occured several years before distance learning was introduced. To BIPD, distance learning is simply a service supplied by the training partners in response to "customer" demand.

Profit-making opportunities for Bell Canada have increased from the establishment of BOLI. Bell Learning Solutions is a business unit created to re-sell Bell Canada expertise in the area of distance learning to external customers. Basic products of Bell Learning Solutions are consulting services, course production in a distance format, and re-seller services for course content from many different sources. The Bell Online Institute acts as the technical resource to fulfill customer needs for course production and technical expertise in consulting projects. While the Bell Institute for Professional Development, Bell Learning Solutions, the Bell Online Institute, and the other training partner companies are independent organizations, they are also able to work together to meet a wide variety of internal and external customer needs. For example, the Bell Online Institute, the Bell Institute for Professional Development, and Bell Learning Solutions have partnered with an SAP implementation partner to design and deliver SAP training internally to Bell Canada employees.

One of the other benefits that Bell Canada expects will come with the use of distance learning is that courses will be available to employees without a formal registration and approval process. As well, employees will not need to leave their normal place of work. The result may be an increase in the amount of training delivered to employees, making employees well-equipped with knowledge that they can use to deal with constantly changing job demands, right at their fingertips.

Continuous Improvement and Alternative Delivery Options

In choosing an appropriate technology to deliver the courses in the distance learning pilot, the advantages and disadvantages of various modes of delivery were analyzed.

Web-based course delivery is one of the alternatives to classroom delivery. Other technology options include business television, video-conferencing, audio-graphic conferencing, and CD-ROM or other type of computer based training (CBT). Depending on instructional design considerations that in turn depend on the needs of users and the type of course material being presented, these technologies could be used as alternatives to Web-based delivery or as an extension of the capabilities of the Web. For example, CD-ROMs may contain hot-links to Internet or intranet sites, and video-conferencing and business television can broadcast Internet applications to students.

A major part of business process reengineering is continuous improvement. In Bell Canada's case, this means continuously evaluating which technologies are most appropriate for the training required. Tables 2 and 3 below give a comparison of the advantages and disadvantages of distance delivery using CD-ROM, intranet, Internet, audiographic or video-conferencing, and business television compared to classroom delivery. Comparisons are made on the basis of organizational impact and costs.

IV. Redesign: Setting Goals for Performance Improvement

BIPD has established a 1998 goal of moving 20 percent of the current curriculum to a Web-based distance learning platform. This represents a four percent increase in the materials currently

Table 2: Comparison of Classroom to Distance Delivery Using CD-ROM, Intranet or Internet

	Classroom	CD-ROM	Intranet (Web)	Internet (Web)
Mode	Live	Asynchronous	Synchronous / Asynchronous	Synchronous / Asynchronous
Availability to Bell Canada Employees	High (although some employees must travel)	High	High	High
Possible Methods of Communication	Live lectures, Q&A	Telephone Fax E-mail	E-mail Discussion groups Live text chat Synchronous audio or video	E-mail Discussion groups Live text chat Synchronous audio
Student and Instructor Interaction	High	Low	Medium (potential for real-time text, audio and video)	Medium (potential for real-time text and audio)
Interactive Exercises	No (although breakout possible)	Yes	Yes	Yes
Effective Course Presentation	Excellent (interaction with instructor possible)	Good (limited communication)	Excellent (interaction with instructor possible)	Good (graphics, video and real-time communication slow)
Easy Access to External Resources	No (although may access Internet from separate application)	Yes (if use hotlinks to connect to Internet/intranet sites)	Yes (if have Internet connection)	Yes (to other Web sites)
Easy to Use	Yes	Yes (with initial instruction)	Yes (with initial instruction)	Yes (with initial instruction)
Easy to Update Course Information	No (if use printed course materials)	No (need to press and redistribute)	Yes	Yes
Distribution Cost	Medium/high depending on travel required	Low/medium (depending on # of updates)	Low	Low

available in a distance learning format, and is equal to an additional 32 courses. Most of the 32 courses will be developed and delivered as Web-based training, as BIPD feels that offering courses via the Bell Intranet is a solution that will best meet the needs of most employees.

The Web offers significant advantages over CD-ROM based training, such as the ability to update materials quickly and inexpensively, and the ability to reach virtually all employees without the need to physically distribute course materials. There is also the benefit of the larger storage capacity of a server as opposed to a CD-ROM, allowing the use of more multimedia elements such as animation and video. The primary means of distance delivery is expected to be through the corporate intranet or the Internet. Due to relatively high costs and limited flexibility, little use of video-conferencing or business television is expected.

BIPD presently has 150 independent study courses on CD-ROM or diskettes. It is assumed that all of these courses could potentially be taught using any of the asynchronous platforms. There are presently 700 classroom-based courses offered through BIPD. It is estimated that 10 percent of these could be potentially offered using the Centra Symposium synchronous platform.

In order to judge the cost-effectiveness of a distance training solution, economic analysis is required and a positive return on investment is determined. To test the economic benefits and feasibility of using Web-based training in the organization, a pilot project was designed.

V. Retool: Testing of the Concept

Bell Online Institute Pilot Project

The Bell Online Institute piloted three Web-based courses in 1997 to measure the cost, and

Table 3: Comparison of Classroom to Distance Delivery Using Audio-graphic or Video-conferencing, or Business Television

	Classroom	Audiographic Conferencing	Video Conferencing	Business Television
Mode	Live	Synchronous	Synchronous	Synchronous
Availability to Bell Canada Employees	High (although some employees must travel)	Medium (requires multimedia PC or equipped room)	Medium (requires multimedia PC or equipped room)	Low (requires broadcast studio and receiving sites)
Possible Methods of Communication	Live lectures, Q&A	Live voice White board	Live audio and video Can include White board	Live audio and video Can include White board
Student and Instructor Interaction	High	High	High	Medium (potential for real-time communication)
Interactive Exercises	No (although breakout possible)	No (although breakout possible)	No (although breakout possible)	No (although student polling possible)
Effective Course Presentation	Excellent (interaction with instructor possible)	Excellent (interaction with instructor possible)	Excellent (interaction with instructor possible)	Excellent (interaction with instructor possible)
Easy Access to External Resources	No (although may access Internet from separate application)	No (although may access Internet from separate application)	No (although may access Internet from separate application)	No (although may access Internet from separate application)
Easy to Use	Yes	Yes (with initial instruction)	Yes (with initial instruction)	No (requires scripting and tele-vision production)
Easy to Update Course Information	No (if use printed course materials)	No (if use printed course materials)	No (if use printed course materials)	No (if use printed course materials)
Distribution Cost	Medium/high depending on travel required	Medium	High	High

evaluate the effectiveness, of training delivered on four different Web-based learning platforms:

- WebCT (http://www.webct.com/) developed at the University of British Columbia
- Mentys (http://www.globalknowledge.com/) from Global Knowledge Network
- Pebblesoft (http://www.pebblesoft.com/), and
- Centra Symposium (http://www.centra.com/).

In order to obtain an accurate picture of Web-based course production in the marketplace, the three courses designed for delivery on these learning platforms were produced by three independent vendors. Before these three Web-based courses were developed, 16 percent of BIPD's curriculum could be offered in a distance format, but these course materials were on paper or audio-video cassettes. All of the courses were on telecommunications topics: TCP/IP, frame relay, and routing, and all of the courses were estimated to be equivalent to 2-day classroom courses. The pilot courses were delivered to engineers working in Bell Canada's Advanced Communication Systems group.

Three of the learning platforms, WebCT, Mentys, and Pebblesoft present course materials asynchronously. Asynchronous course delivery uses network communication that is not done in real time; for example, online discussion groups, e-mail, and the use of HTML pages. In other words, course materials reside on a server on an on-going basis and may be accessed at the student's convenience. The frame relay course on the Pebblesoft learning platform was authored in French. Both asynchronous courses took students approximately two hours to complete.

The fourth platform, Symposium, is a synchronous learning platform using 28.8 Kbps delivery over the Internet or an intranet. Synchronous course delivery involves real-time network communication such as audio or video conferencing, or text-based chat groups. The Symposium system supports text, graphics, and animation to present course materials. Shared features for system users include audio communication among the course participants and the instructor, an electronic white board, an Internet browser, and a live text chat room.

The course on the topic of routing was delivered on the Centra Symposium platform. The pilot course was delivered to 20 Bell Canada engineers from Quebec City, Montreal, Ottawa, Toronto, and Hamilton. The course was offered in two parts, one in the morning and one in the afternoon. The morning session lasted two hours and the afternoon session lasted one hour, for a total of four hours. All the engineers were present for the entire course.

The course was highly successful with participants reporting that they felt their needs were being met with flexible, convenient course delivery on topics that were of current interest for their job activities. Testing indicated that learning for all courses was as effective as classroom training.

VI. Reevaluate: Evaluating the New Process

Course Production Costs

Cost is one of the key performance measures in business process reengineering. In their report *Methodology for Cost-Benefit Analysis of Web-Based Telelearning: Case Study of the Bell Online Institute* (in press), Whalen and Wright examine the costs for the business process needed to produce

Figure 1

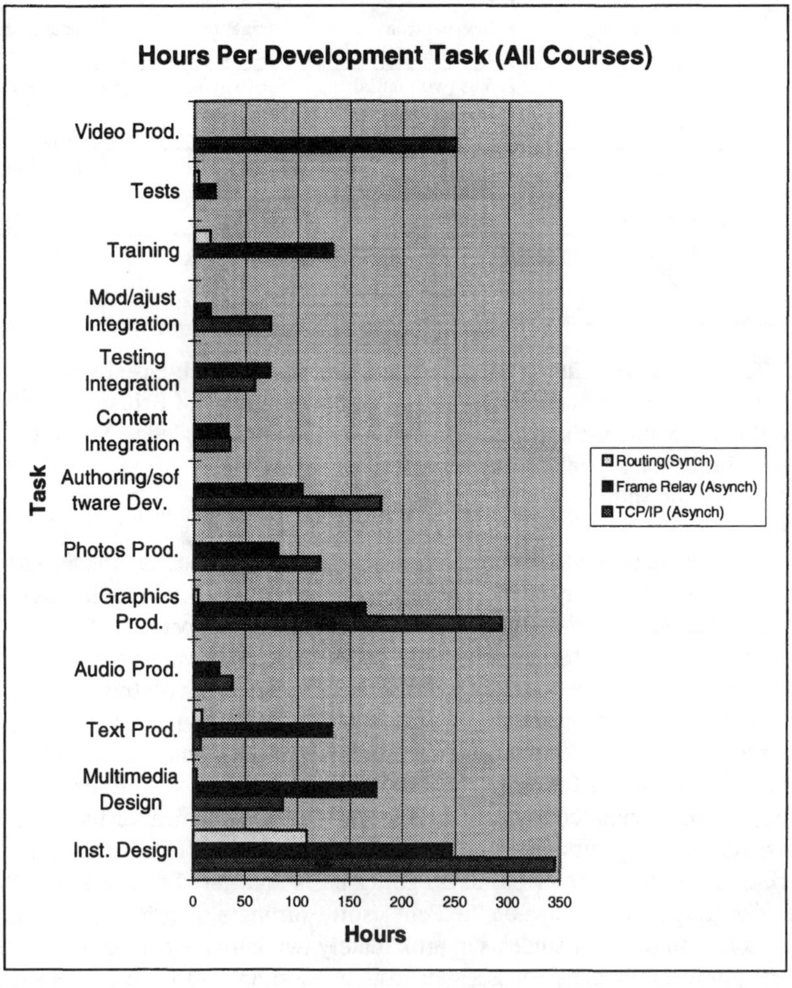

Figure 2

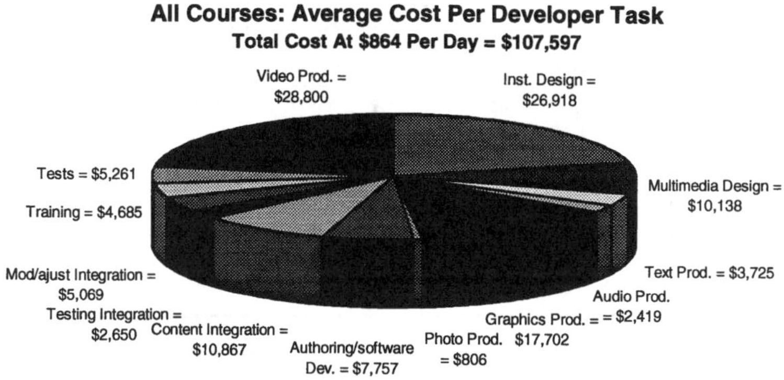

a Web-based course. These processes include: instructional design, multimedia design, the production of text, audio, graphics, photographic and video elements, authoring and software development, content integration and modification, and training and testing design and delivery.

Figure 1 compares each costing element across all courses. There was some variation in the use of any particular element among the courses, although the variations were small, and some elements were consistently used more or less frequently for all of the courses. None of the costing elements could be discounted as unimportant, and none of the costing elements clearly dominated the others in importance.

The cost for each of these processes varies with the number of development hours required. As can be seen from Figure 1, there were some variations in multimedia elements used and in the length of development time for each. To get a better picture of costs for the development of each element, an average of the time spent on developing each element and an average of the hourly rates charged by the three production companies were used and results summarized in Figure 2.

Economic Return

Like all large corporations, Bell Canada routinely analyzes new business opportunities to determine economic attractiveness. However, this study is the first to calculate the economic return on an internal training initiative. Using classroom course delivery costs at BIPD as a baseline, the cost savings of using Web-based training were analyzed. Course development costs for Web-based training are higher than those for classroom courses, but savings result from lower delivery costs. After calculating the savings in Web-based delivery per student, the return on investment was determined as well as the number of students that must take the course before the high development costs were recovered through savings in course delivery over the client-server architecture. The case study analyzed the breakeven number of students required to recover Web-based course development costs and the return on investment over a 5-year period. All of the measures of financial performance indicated that the business case for Web-based training is strong.

The breakeven point of an investment project is the point at which either cost savings or revenue generated equals the cost of investment. Realizing savings for Web based courses requires a sufficient number of students to recover course development costs. In this case, the breakeven point is the number of students that must be trained for the fixed costs of Web-based course development to be offset by the reduced delivery costs of Web-based training. The breakeven points for the three Web-based courses in this pilot ranged from 4 to 112 students.

Return on investment (ROI) is the financial gain (or loss) that results from an investment project. The returns on investment were positive for all of the Web-based courses in the study. Returns on investment ranged from $3 saved for every $1 spent on Web-based training, to $33 saved to every $1 spent. The average savings per student was $702 for the asynchronous courses and $1,103 for the synchronous course. Of the asynchronous courses, the TCP/IP course using the Mentys learning platform had the least cost savings, at an average of $625 per student. The asynchronous frame relay course saved an average of $850 per student.

Table 4: Summary of Multimedia, Breakeven, and ROI

Course	Multimedia Development Hours	Breakeven Number of Students	ROI Over 5 Years
Routing	144 hours	4 students	3283 percent
Frame Relay	1,156 hours	51 students	697 percent
TCP/IP on WebCT	1,487 hours	111 students	288 percent
TCP/IP on Mentys	1,487 hours	112 students	283 percent

Table 5: Levels of Training Evaluation

Level 1	Reaction Evaluation: How well participants liked the training.
Level 2	Learning Evaluation: Principles, facts and techniques that were understood and absorbed by the participants.
Level 3	Transfer-of-Learning Evaluation: Transfer of training skills to knowledge on the job.
Level 4 (BIPD Goal)	Results Evaluation: Impact of training on the organization.

The most critical factor in the cost-benefit of using Web-based training was found to be the amount of multimedia content in a course. Courses with a higher amount of multimedia content, such as the TCP/IP course which contained a 5-minute video segment showed a correspondingly lower ROI. This was the only course which used video and, as a result, the number of hours spent in multimedia production was far higher than for the other courses. The synchronous course, which was the most cost effective course in the pilot, contained a few graphics and live audio only. The limited amount of multimedia content in the synchronous course offset the higher costs of course delivery which resulted from having a live instructor present during delivery, as well as greater student salary costs due to the extra hours required to deliver the course. Table 4 summarizes the number of hours spent developing each course, the resulting ROI, and the breakeven number of students.

Training Quality

BIPD is involved in the continuous improvement of training quality. To accomplish this continuous improvement, training effectiveness must have a means of being measured. Table 5 summarizes the four levels of training evaluation as defined by Kirkpatrick (1979), an industry standard for evaluating training and a system currently used at BIPD.

The evaluation of training quality is on-going in the organization. Course effectiveness is measured at "Level 1" evaluation, through student surveys about their satisfaction with the instructor, course materials, and venue. This year, BIPD intends to move to "Level 3" evaluations, in order to determine post-course effectiveness. Participants, supervisors, peers, and possibly customers will be surveyed to evaluate whether an employee's training has made a difference to his performance in the workforce. Also in the near future, BIPD will use a "Level 4" type of evaluation, where return on investment will be done for selected courses. This Level 4 evaluation will be done because of demand from senior management at Bell Canada to be informed about training effectiveness and the ROI for training.

A substantial amount of the expertise that was available within BIPD before outsourcing is still available to the organization. In fact, some former employees of BIPD now work for the training partner companies and other former BIPD employees found new jobs within Bell Canada. Quality control is an ongoing issue in the outsourcing relationship. BIPD works very closely with the training partners to ensure that standards are maintained. The use of "Level 3" evaluation techniques may help reduce some of the need for this close relationship.

The curriculum has improved since outsourcing. Outsourcing was the impetus needed to review the existing curriculum, and many courses were found to be outdated. In some cases, courses were updated, in others the courses were dropped from the curriculum. As well, some courses did not have

Table 6: Application of 6-R's to Bell Canada and General Applicability to Distance Learning

Step in Process	Aim	Bell Canada	Generalization to Other BPR Effort
I. Realization	• understanding of the need for reengineering effort to better meet the needs of the organization	• classroom training not meeting the needs of employees or the company • management and employee interest in distance learning • outsourcing has created opportunity to try new methods	• review current training strategy to evaluate opportunities for improvement • evaluate opportunities for change using internal and external resources
II. Requirements	• identification of strategic direction so BPR effort meets customer needs	• identification of employee and organizational training requirements • identification of appropriate training partner to deliver specific courses • increase use of Web-based training	• review employee and management training goals • identify new technologies or competitive business system • identify whether best met with internal or external resources
III. Rethink	• procedures, processes and technologies required are identified and put in place	• ensure that organization is structured to accommodate reengineering effort • draw on strengths in organization to take advantage of existing technologies and expertise already in place	• ensure that organization can meet reengineering effort required by evaluating organizational structure, existing technologies and processes required • identify resource gaps
IV. Redesign	• specific targets set for reengineering implementation	• reaffirm chosen technology • set goals for use of Web-based training	• plan reengineering implementation by setting targets for new methods of training design and delivery using specific technology
V. Retool	• technologies evaluated and adapted to re-engineering project	• design and delivery Web-based training pilot using several different platforms for evaluation • collect data on time and costs for course development and delivery	• design and delivery pilot of technology-based training to test feasibility • collect data necessary to evaluate pilot
VI. Reevaluate	• performance indicators measured to evaluate reengineering effort	• conduct cost-benefit analysis of pilot • evaluate alternative delivery options to ensure continuous improvement	• conduct cost-benefit analysis of pilot • evaluate alternative delivery options to identify those that best meet needs of organization

detailed written documentation and outsourcing was an opportunity to ensure that documentation was brought up to date.

CURRENT CHALLENGES

Today, the Bell Online Institute is steadily expanding the number of Web-based courses available to employees and has also been working very closely with a company that is developing a competency and training management system for Web-based course delivery. A record for each Bell Canada employee has been created and current efforts are focused on developing suggested online training curricula based on each employee's job. Links will also be made to the enterprise-wide computer system used to manage human resources, in order to track courses employees have completed. BOLI is involved in the delivery of customer projects generated by the sales and marketing staff of Bell Learning Solutions as well. The number of permanent staff assigned to BOLI has been increased and there has also been an increase in the number of outside consultants brought in to work on customer projects. The current challenge to BOLI is to remain focused on projects that will add the greatest value to the organization, as well as effectively manage the many projects that are now underway.

CONCLUSION

The Bell Online Institute pilot project is a reengineering effort that illustrates some of the general principles for reengineering classroom training to distance learning:

- the need for business process reengineering was defined at the Realization stage
- customer requirements were established in the Requirements phase and both employees and senior management supported the proposed changes
- appropriate technologies were integrated into the new business process in the Rethink phase
- the reengineering project had clear goals for performance improvement set in the Redesign stage
- the change was carefully managed at the Retool stage by carrying out a pilot project to test the feasibility of the solution, and
- there was an evaluation of the new process based on clear performance measurements during Reevaluation.

Table 6 summarizes the 6-R's, Bell Canada's reengineering of training, along with generalization to other reengineering efforts in the area of corporate training. The 6-R's methodology was effectively applied to the case of distance training at Bell Canada and could be generally applied to business process reengineering efforts for training in other organizations.

FURTHER READING

Carr, David K and Henry J. Johansson (1994). *Best Practices in Reengineering*. Coopers & Lybrandt.

Guthrie, Edward R. (1995, March). Belief and behavior: Which is the chicken, which is the egg? *Management Review 84*, 62.

Hammer, Michael (1990, July-August). Reengineering work: Don't automate, obliterate. *Harvard Business Review*, 104-113.

Infor (1995, November). *33(4)*, 223-283.

Infor (1996, February). *34(1)*, 1-58.

Koonce, Richard (1996, January). The human face of change. *Training and Development 50*, 23.

ACKNOWLEDGMENT

The research for this chapter was supported in part by Industry Canada's Network of Centres of Excellence in Telelearning.

REFERENCES

Edosomwan, Johnson A. (1996). *Organizational Transformation and Process Reengineering*. Delray Beach, Florida: St. Lucie Press.

Hammer, Michael and James Champy (1993). *Reengineering the Corporation*. New York: Harper Business.

Kirkpatrick, Donald L. (1979). Techniques for evaluating training programs. *Training and Development 33(6)*, 78-92.

Shandler, Donald (1996). *Reengineering the Training Function: How to Align Training With the New Corporate Agenda*. Delray Beach, Florida: St. Lucie Press.

Stewart, Jim (1991). *Managing Change Through Training and Development*. London: Kogan Page.

Sykes, Stephanie (1997, March 11). Director, Bell Institute for Professional Development. Personal Communication.

Whalen, Tammy and David Wright (1998). Methodology for cost-benefit analysis of Web-based telelearning: Case study of the Bell Online Institute. Manuscript submitted for publication.

APPENDIX: Glossary

Asynchronous: network communication which is not done in real time; for example, e-mail or the use of HTML pages.

Audiographic Conferencing: real-time audio and graphics-only (using a device such as a White Board) among groups of people, often in specially equipped rooms.

Breakeven Point: the number of students that must be trained for the fixed costs of Web-based course development to be offset by the reduced delivery costs of Web-based training. Breakeven = (Fixed Costs of Web-Based Course - Fixed Costs of Classroom Course) / (Delivery Cost Per Student for Classroom Course - Delivery Cost Per Student for Web-Based Course).

Browser: a computer program, such as Netscape, which allows users to access and view documents on the Web.

CD-ROM (Compact Disk Read Only Memory): a plastic disk coated with aluminum which stores information digitally as non-reflective pits or bubbles. A 5.25 in CD-ROM is capable of holding about 650 Mbytes of information.

Client: an end-user computer using a browser program (such as Netscape) to receive multimedia documents from a Web server.

HTML (Hypertext Markup Language): the standard document format used on the Web.

Internet: an international network linking millions of computers via telephone lines using the TCP/IP protocol. .

Intranet: an internal network linking corporate computers using the TCP/IP protocol.

Multimedia: any combination of text, graphics, images, audio, and video which have been digitally encoded.

Return on Investment (ROI): the financial gain (or loss) resulting from the use of Web-based training. ROI = (Present Value of Savings From Web-Based Training / Present Value of Costs of Web-Based Training) * 100.

Server: an Internet host computer using a program (such as NCSA's HTTPD) to store and transmit multimedia documents for an information service (such as the World Wide Web).

Synchronous Communication: real-time network communication such as video conferencing.

TCP/IP (Transmission Control Protocol/Internet Protocol): Protocol which enables universal interconnection and addressing of all Internet hosts.

Video Conferencing: real-time audio and video between groups, often in specially equipped rooms.

Web: a network of multimedia documents stored on Internet or intranet servers, which can be seen by using a browser. Documents are formatted with the HTML standard and transmitted with the HTTP standard.

White Board: a shared electronic drawing board which can be displayed on each workstation.

BIOGRAPHICAL SKETCHES

Tammy Whalen is a Senior Consultant at Bell Learning Solutions, a business unit of Bell Canada. She is also a graduate student in the Master of Business Administration program at the University of Ottawa and is writing her thesis on the business case for distance learning.

David Wright is a Full Professor at the University of Ottawa. He has provided distance education using a variety of technologies and is currently sponsored by Industry Canada to investigate cost effective distance education alternatives and their economic impact. He has 25 fully refereed publications in a range of journals, as well as conference papers, and is the author of the book Broadband: Business Services, Technologies and Strategic Impact published by Artech House, Boston. He is cited in "Who's Who in the World" and in "Who's Who in Science and Engineering".

The Integration of Library, Telecommunications, and Computing Services in a University

Susan A. Sherer
Lehigh University, USA

EXECUTIVE SUMMARY

Today many IS departments and individuals are attempting to transform from technical groups and specialists to user oriented functions and customer support personnel. The major responsibility of the traditional IS department has evolved from the development, operation, and support of technology to the management of information. In the university environment, managers of information have traditionally been librarians. Librarians have increasingly become users of electronic information resources. A merger of the library with computing and telecommunications brings together technical expertise with information management skills.

This case study describes the process of integrating the library, computing and telecommunications services in a University. Within the last two years, a new manager in the newly created position of Chief Information Officer merged these diverse organizations. We will describe the techniques used during the first year to foster communication, develop new strategic direction, and create and implement a new organizational structure. We will focus on establishing leadership, the organizational change and operational planning process, and the initial implementation of the new organizational structure. We will describe some of the problems and obstacles that needed to be addressed, including new management's establishment of trust and control, creating an environment for change, managing change amid strong time pressures, human resource issues, and resource constraints.

It is expected that many of the issues that arose during this merger will be addressed by organizations in other industries as they attempt to evolve from technical IS groups to more customer oriented organizations. Today it is imperative that IS functions provide client support, which requires different types of skills from those traditionally nurtured among technical experts in traditional IS departments.

BACKGROUND

The University is an independent, coeducational institution located in northeastern United States. Approximately 5000 undergraduate and 1500 graduate students are enrolled in programs in the arts and humanities, business, education, engineering, natural and social sciences. The University employs approximately 1400 faculty and staff.

The University's mission is to advance learning through the integration of teaching, research and service to others. It is committed to integrated learning, promoting the discovery, integration, and communication of knowledge. The University's mission statement embraces a commitment to the intellectual, physical, social, ethical, and spiritual development of all members of the academic community.

To support the discovery and integration of knowledge, the University built extensive library resources, including a collection of over 1 million volumes. The advent of the campus wide network enabled the library to supply all users with electronic as well as traditional services. In 1995 the university library staff numbered approximately 60, having just been significantly reduced from 75. The organization of the libraries in June 1995 is shown in Appendix A.

The University also built extensive computer and communications resources to support its mission. The computing center served the needs of students, instructors, researchers, and administrative users. A total of approximately 90 people worked in this department, whose structure is shown in Appendix A.

SETTING THE STAGE

The library and computing center at this university, as at most universities, evolved separately, with different roots and cultures. We describe here the evolution of these two groups, similar integration efforts at other schools, and the University's decision to integrate these functions.

Library Services vs. Computing and Communication Services: Evolutionary Differences

Historically the library's primary function has been to provide information to faculty and students who are in the pursuit of knowledge. Initially when computers were introduced into universities, their primary purpose was the storage and dissemination of data. Whereas data are simple observations of states of the world that are easily structured, captured, quantified, and transferred, information is data endowed with relevance and purpose which requires a unit of analysis, needs consensus on meaning, and requires human mediation (Davenport, 1997). The role of the librarian has generally been to aid the user in finding meaningful information that can be transferred to knowledge, information that has been reflected upon. As computers have become more pervasive and user oriented, they have evolved as a source of information rather than simply data. As a result, data processing centers have evolved to encompass information management.

While the library and computing and communications services both provided support for information gathering and processing, there were significant cultural differences within these professions. These differences were embedded in different approaches to problem solving. For the computing people, new technology was often the key to solve problems because it enabled more efficient and effective processing of data and information. For the librarians, the information itself provided answers.

Gender differences, the nature of the educational training, and the problems that were solved by librarians as opposed to computing personnel contributed to cultural differences. The library profession is predominately female while computing is male dominated. At the university, the library's director and four out of six associate directors were female. In Computing and Telecommunications, only one associate director was female and she was two levels down within the hierarchy. The demarcation between exempt and nonexempt staff was much greater in the library as compared to the computing center, most probably because of the greater importance of formal education in the library.

Librarians were generally excellent oral and written communicators. They used their skills in meetings where they joined together to plan and discuss various issues. The computing people had strong analytical skills and analyzed and solved problems often without as much discussion. Whereas the librarians worked best in a very collegial atmosphere, computing people often stressed doing the best individual job, i.e., developing the best program, solving the problem quickest, etc.

The librarians interacted primarily with faculty and students. They were comfortable working with novice users, often providing general solutions to meet the needs of large groups. Computing people entered the profession when communication with external constituents was minimized through the segregation of the computing center. "Many computer people were attracted to the IT profession because they were happier communicating with machines than with people." (Caldwell, 1996) As the computing environment evolved to a more distributed model, computing people generally supported users with some degree of technical literacy, providing one-to-one interactions and customized solutions. Communication between the computing center and telecommunications personnel and faculty and students continued to be fairly limited to help desks.

Not only were the cultures of library and computing different, but both professions had different subcultures as well. In the library, there were differences between public services and technical services librarians. Media specialists had their own background and differences. In computing, subcultures existed between academic and administrative computing, between hardware and software support personnel, and between user services and programmers or technicians (Hirshon, 1998).

These departments also had very different structures. One year prior to her retirement, the Director of Libraries flattened her organization, resulting in the six associate director positions shown in Appendix A. The computing and communications groups remained more hierarchically organized. There was no comparability in the management titles.

Mergers at Other Universities

During the mid to late 1980s a number of universities integrated libraries and computing. Few of these mergers survived. They did not effectively merge the departments below the top level because the technology had not yet advanced to enable full integration. However, in the 1990s, with the advent of the World Wide Web, the need to access large amounts of interactive information, replacement of administrative systems with enterprise information systems, and increased incorporation of technology in the curriculum, many schools revisited the need for integration. Universities pursued several different integration models ranging from total integration to limited integration of one or two key service areas such as help desk, or joint management of some projects. Of the approximately 100 schools that have merged their functions, more than 80% integrated within the last five years and only 17% have attempted full integration (Hirshon, 1998).

The University's Decision to Merge

In 1994, the University's president decided to merge the library with computing and communications services. The vision was to integrate information systems, resources, and services in support of the teaching, learning, and research missions of the university. The decision was based upon the following recent trends in the environment:

- Library resources were increasingly electronic.
- Technical computing personnel were increasingly required to support end user computing as new technology enabled the transformation from centralized to distributed information processing and users became more technically capable.
- The objective was to manage information, not just systems.
- Both were support functions, providing information resources to the academic community.
- Content and technology were merging.

Most members of the community believed that this was a reasonable strategy. There were some who thought that economic considerations were driving the decision as well, suggesting that the ability to reduce administrative costs was a driving factor. The opportunity to merge was created when the Director of Libraries retired and the Assistant Vice President for Computing and Telecommunications returned to the teaching faculty.

CASE DESCRIPTION

This case will focus on the *process* of organizational change. It describes the establishment of new leadership, the six-month process of planning for the new organizational structure, and the initial implementation of that structure, including the importance of training personnel to meet the new requirements (Agarwal et al., 1997; Keen, 1991). A time line of major events is shown in Appendix B.

Leadership

In 1994 a search committee was constituted to select a Chief Information Officer (CIO). Since this was a newly created position, the committee not only selected a candidate, but also formulated the job responsibilities and qualifications detailed in Figure 1. Since this was a new type of position, it was recognized that it would be difficult to find an individual with experience in all areas: computing, telecommunications, and libraries. This was recognized as a potential problem, as members of the group whose background differed from that of the new leader were concerned that the new CIO might not be as supportive of their areas of operation. However, given the fact that few universities had previously combined these areas, candidates with experience in any one of these areas were considered.

Figure 1: Chief Information Officer Responsibilities and Qualifications

Responsibilities

- Integrate library and computing services and resources in support of the teaching, learning, and research missions of the university.
- Create an organizational structure and a working environment that encourage creativity, cost effectiveness, and change.

Qualifications

- Possess a vision for incorporating innovative information technologies in higher education.
- Understand the changing paradigm in scholarly communication.
- Understand the use of information technologies to support instruction, research, and administration.
- Understand the administration and operation of libraries, computing, and telecommunications services.
- Experience and expertise in at least one of the areas: libraries, computing, and telecommunications services
- 7-10 years of increasingly responsible management experience.
- Demonstrated commitment to participative management style and open decision making.
- Excellent interpersonal and communication skills.
- An advanced degree in a relevant field.

After reviewing the qualifications of numerous candidates, an individual was chosen who provided a vision for the use of information technology in higher education. This candidate understood the opportunities to improve support of the teaching, learning, and research missions of the university presented by the merger of the information management resources of the university. The candidate understood the challenges of merging the different cultures and offered a management style that was expected to recognize and deal with these challenges. The successful candidate had previous experience primarily in the area of library management with a background in library technology.

The new Chief Information Officer came on board in late 1995. The first weeks were spent meeting the staff, with scheduled interviews with every professional and offers to meet any individual support staff member. In addition, the CIO visited every department and held group department meetings.

Interim Management

Given the significantly different organizational structures of the two component units, an interim management group was established, consisting of 15 individuals pulled from the management of the existing organizations. Since the library had six direct managers, the CIO went down to the next reporting level to provide similar representation from computing and communications. This resulted in a group of nine managers from the computing and communications group, shown in Figure 2. The purpose of the initial management group was to help the CIO manage solely during the transition for the six-month planning process, after which a new management would be established in a new organizational structure.

Figure 2: Interim Management Team

Associate Director, Library Technical Services
Associate Director, Library Automation
Associate Director, Library Special Collections
Associate Director, Library Reference Services
Associate Director, Library Access Services
Associate Director, Library Media Services
Director, Computing User Services
Director, Administrative Systems and Telecom
Associate Director, Computing and Networking
Associate Director, Computing Facilities
Associate Director, User Services
Manager, Computer Operations
Manager, Network Facilities Support
Manager, Systems Programming
Manager, Academic Programming

The interim management group met twice a month throughout the six-month planning process. Several participants noted that in early meetings, library and computing people typically sat separately and talked among themselves before meetings. Gradually, however, the managers began to mingle and communicate with others in different areas.

Interim Organization: Getting to Know Each Other

The combined organization was composed of approximately 140 people. At its inception, many people did not know the responsibilities of their colleagues in the other component units. Several initiatives were instituted to enable personnel to become better acquainted. First, in January each group was asked to provide an overview presentation of their area. Everyone was obligated to attend these sessions. Some consideration was given to making this a competition, asking groups to compete for the best presentation, to elicit more excitement about the process. However, it was thought that this might be unfair because some groups, by the nature of their jobs, had much higher presentation skills than others, e.g. media productions. Attendees believed that they learned some new things by attending these group orientation meetings. While these sessions provided a forum for presenting information, they did not necessarily engage people in cooperative efforts nor was this its purpose. If the group were to coalesce, they needed to have opportunities to work together. A retreat and cross-functional planning team structure were instituted to accomplish this task.

Planning for Organizational Change: The Retreat

Prior to instituting any organizational change, an intense planning process was begun. This was necessary to create motivation for change or "unfreeze" the set of different assumptions, values, and beliefs in the library and computing cultures (Schein, 1997). A one-day retreat was scheduled in early 1996. The entire organization was invited; approximately one half attended. The focus of the retreat was to develop mission, vision, environment, and values statements, which were discussed in a series of small group sessions.

Planning for Organizational Change: Working Teams

The major goal for the new CIO was to develop a plan for an integrated organization, and thereafter to restructure the functions to implement that plan. To develop the plan, four working teams were set up to focus on technology, information, financial and staff resources, and client services. In addition, a lead team coordinated the activities of all the teams. The lead team was made up of the four team leaders and four at large members. Each of the four working teams included 8-10 people. Each team was managed by a member of the interim management group and included at least one other member of that group. Teams were organized so that there could be a fresh look at the problems and extensive charges were prepared for each team to outline some issues. Figure 3 shows the leadership and composition of these teams.

Figure 3: Planning Team Composition

Lead Team
Leader: Chief Information Officer
Members: Team leaders (listed below), 4 at-large members from computing and libraries

User Services Team
Leader: Associate Director, Computing User Services
Members: Libraries (5), Computing and Telecom (3)

Information Services Team
Leader: Associate Director, Library Technical Services
Members: Libraries (4), Computing and Telecom (4)

Financial and Staff Resources Team
Leader: Associate Director, Library Reference Services
Members: Libraries (3), Computing and Telecom (4)

Technology Services Team
Leader: Associate Director, Computing and Networking
Members: Libraries (2), Computing and Telecom (7)

To identify appropriate people throughout the organization to participate in this planning process, the interim management group was consulted. They were asked to identify non-parochial individuals in the organization who would have the vision to help move the combined organization forward. Teams were to mix individuals who represented the different subcultures within the combined group. The CIO consulted with others concerning team appointments, but ultimately decided who would lead the teams. As unanimity in such cases is impossible, in some cases, the selections were inconsistent with some of the interim management group recommendations.

After the work teams were initially formed, there were some changes to the composition of the teams to incorporate further advice from the interim managers. In some cases, a team lacked experience in an area assigned to that team. Changes were made to add or change team members so that "experts" were assigned to each team, albeit not always the ones the group might have selected. Several people felt that lack of inclusion of a particular person who was considered an expert in a particular area caused a lot of unnecessary time to be spent in team meetings. However, this was purposeful, as change was a primary objective, and teams had the opportunity to call upon additional expertise as needed.

The team structure directly involved approximately 25% of the organization. Some who were involved found that time pressures were very uncomfortable because they could not get their regular jobs done, which raised stress levels. On the other hand, those who were not involved were anxious because they did not know how the outcome would affect them. The team structure seemed to have accomplished several objectives:

- It brought together individuals with little prior relationship and got them to work well together.
- It brought together different and oftentimes new perspectives for looking at issues, getting people working intensively together in a cooperative environment while developing intra-organizational alliances not previously in place.
- It got staff involved in the planning process so that they would "own" the resulting strategy and organizational structure, building a base line of support for the outcome.
- It provided an opportunity for the new CIO to observe the actions of personnel in order to identify future leaders in the new organization, providing a testing ground for personnel.

Communication during the Planning Process

Throughout the planning process, there was an attempt to provide open communication both internally and externally. Traditional channels of communication were altered. The sensitivities of the process coupled with time constraints lead to the structuring of much of the communication.

Teams met from January through April to develop the strategic plan draft. During the process, there was an attempt to make this process more open via listservs and discussion groups providing everyone with input and access to the deliberations. As work progressed, however, information became more filtered. For example, as the teams got closer to defining organizational structure, there was concern that sharing too much raw information might cause panic among individuals. This sensitivity prompted the limitation of team postings to minutes that eliminated some of the detail, but resulted in the promulgation of significant rumors.

As the planning process evolved, traditional flows of communication were no longer satisfactory. With multiple teams working on issues simultaneously, managers could not provide answers to their own staff if they were not on certain teams. Channels of communication were altered. During this period, many individuals missed the camaraderie and felt a loss of community when, for example, librarians no longer met together, but they also bemoaned the significant amount of time they were spending in planning meetings. However, given the ambitious schedule and requirements on individuals, there was little time to mourn this loss. Open meetings and informal brown bag lunches were held periodically throughout this six-month process to share information. In fact, some members of the organization felt that there was too much communication and, in fact, they were overloaded with information.

During this time period, external communication was limited. Focus groups were brought together to gather input from faculty and students. Although these sessions were short (typically 1-1.5 hours) and very structured, many attendees noted that this had been the first time that they had been asked for their opinions. Some participants felt that the structure limited their ability to provide information on real problems but some structure was essential if useful information was to be gathered in a short time period.

Centralization of Internal Business Functions

During this time period, the CIO centralized the internal business functions that were previously in three separate structures into a central location to track and control financial resources. With a large and complex budget, the CIO believed this was necessary to manage effectively. Some managers who were now required to go through a central process to acquire needed items resented this act because they perceived that local support was removed. This caused some hard feelings during a time period when trust was still being established and staff were still very anxious and fearful about any type of change.

The Process of Introducing the New Organizational Structure

In mid April an open meeting was held to discuss the strategic plan draft. After the first draft of the strategic plan was sent out, each lead team member was given a week to propose the rough outlines for a new organizational structure that would advance the strategic plan. Extensive discussions ensued within the lead team, and there were many iterations of the structure. In May, the first draft of a new organization was unveiled at a brown bag lunch. No individuals were named on the chart, only the broad outlines of the organizational structure. The purpose of this meeting was to allow people to ask questions. As anxieties were very high, the meeting was referred to facetiously by some staff members as a "body bag," rather than a "brown bag" lunch. Most individuals were trying to determine whether and how they might fit into this new structure. A concerted effort was made to use new terminology to communicate whenever possible. The intent was to help break away from the old patterns of communication and ideas and create a common language for the new integrated culture (Schein, 1997). This also caused some confusion and discomfort. In fact, one participant commented that "it is a bit like a puzzle to see where you fit in".

Within two weeks after this lunch, the organization chart was revised and the CIO identified the high level managers who would populate it. All others were given a formal opportunity to request in writing the types of jobs they desired and their placement in the organization by completing a job request form developed by University Human Resources. It was recognized that not everyone could be matched with his or her desires, but ultimately a substantial percent (90%) were placed in one of their top three choices. The newly identified managers selected people for their organizations after reviewing these forms. There were some cases where two managers wanted the same person in their group, which resulted in some discussions of what was best for the entire organization.

The New Organizational Structure

Major recommendations from the planning process were the need for improved management of technology in classrooms, and the creation of client service teams. It was recognized that client needs varied broadly across campus, from sophisticated researchers requiring heavy library resources and computing power, to those with very basic needs. The organizational structure that evolved to address these needs combined elements of the old structure with new requirements (Appendix C).

The old "glue" behind the organization that was retained in the new structure was the familiar functional organization or infrastructure group. What was radically different was a new matrix user (client) services organization. Each of the colleges was assigned a team of consultants in four areas: computing and networking, library assistance, instructional technology support, and administrative information systems support. While each consultant's primary assignment was to a College team, s(he) also was to report to a "functional team" leader who headed each of these four functional areas. For some individuals, these secondary assignments created conflicting loyalties.

Personnel and Job Changes

A key issue was finding personnel to fill new positions, particularly in the new user services area. In many cases, these jobs required very different skills from previous job responsibilities. Particularly among computing people, there was a concern that technical expertise was no longer a valued skill, but that management now primarily valued the ability to support and work with users. For some, this change was not welcome. Not unexpectedly, some individuals, particularly computing personnel who were most mobile in the job market, left to find employment elsewhere where they could continue to apply primarily their technical skills. Shortage of personnel also lead to the sharing of consultants between more than one college, which was an additional cause of stress.

All job descriptions were re-written within two months to reduce job ambiguity and insure that no tasks would fall between the cracks. This was particularly important in the creation of the new user services organization. The university Human Resources department audited all of the jobs. As a result, some salary levels were changed. Approximately 25% of all staff saw an increase in salary while a small percentage, about 5% of staff, received decreases in salary because managerial responsibilities were reduced.

Physical Changes

By early the next year, employees began to move their offices so the new teams would be physically housed together. User services teams were consolidated by their college rather than by their functional departments. For some individuals, this physical move was as upsetting as the organizational change. It meant giving up "homes" to which they had become accustomed. Moreover, it meant physically leaving behind traditional sources of information that had supported functional capabilities. Some individuals experienced a loss of identity as they were moved away from their functional colleagues, and were concerned that they would not continue to learn and grow professionally.

Technological Change

Another item that had a big impact on the ability to implement these changes was the need to undertake simultaneously major technological changes. In particular, beginning in the second half of 1996, installation of Windows 95 began across campus. This occurred at the same time that the organizational changes were put into place. Some users blamed the new organization for difficulty in meeting this challenge, but it should also be recognized that in a technology-based organization there is always a major technological change that needs to be implemented.

Implications for Users

When the organizational change went into effect, it took some time for users to learn whom to contact in the new IR organization. Many users had to change old habits, as they were accustomed to contacting specific individuals in the old departments to obtain help. Users now were provided with specific contacts for their colleges. Initial user response was a function of the quality of their assigned user services team. Users who were assigned consultants who were primarily dedicated to their college and who possessed a good customer service ethos saw service improvements. Some users complained because service was not as promised. In some cases the clients may have been getting used to better service. In other cases, they may have been promised but did not receive improved service because individuals in the new structure were either not yet trained to provide it or were attempting to perform several jobs at the same time. The need for providing additional training for the new jobs, particularly in user services, became very clear. It is not possible to just change organizational structure; people need to be trained to work within a new organization (Agarwal et. al., 1997). As staff moved into their new positions and had to keep old operations going at the same time, there was not time to get all people trained at once. Some training occurred, but more was needed.

CHALLENGES/PROBLEMS

We will now review some of the specific problems and challenges that arose through the planning and implementation of the new organizational structure. Throughout the planning process, major challenges were associated with new leadership and the aggressive time schedule. Choices often had to be made that would satisfy some people and address some concerns, but would leave others unsatisfied. The challenges of implementing the new organizational structure included human and financial resource issues.

Leadership: Establishment of Trust

This traumatic change was instituted under the direction of a new leader. New leaders are often brought in from the outside to create change. They often have a "honeymoon" period to do so, but for the change to be accepted new leadership must build trust during this period. Initial respect is based upon the experience and credentials of the new leader. In this case, the computing people, in particular, were skeptical about the new CIO because of the perceived lack of computing experience. Since these people expected that he would not become involved in technical decisions, any efforts by the CIO to do so were considered "micro-management."

To help build trust not only of the CIO but throughout the organization, the collaborative and collegial planning process was put into effect. However, some staff believed that the CIO did not have the opportunity to learn about the capabilities of all managers and staff prior to the constitution of the interim management group and the working teams. Since the management structures in the library and computing organizations were so different, the CIO initially created a larger and more inclusive management group rather than attempting to select a few from the existing structures. Several staff felt that, while this larger group enabled more personnel to have input, the size made it more difficult to build trust. As previously noted, the CIO's decision to centralize the internal business functions prior to the completion of the planning process also contributed to further difficulty in establishing trust.

Aggressive Time Schedule

The aggressive time schedule led to numerous challenges. A very ambitious and strict time line was established, announced, and adhered to. At the January 1996 retreat, the CIO announced that the organization would definitely change, and that all organizational changes would be announced within 6 months. It was felt that the schedule should be fixed not to exceed six months to minimize the stress of the traumatic changes, resolving doubts and ambiguities for staff as quickly as possible. However, the schedule needed to allow time for staff to participate while continuing with the work of the traditional functions. This meant that stress levels for these six months were exceptionally high. Staff were continuing with their own jobs and working very diligently on the new team assignments. There was often not a lot of time for exploration of actions. For example, when the management team was chosen, there was little explanation of its composition. This led to initial feelings that further input from staff might not be considered.

While the communication was intended to be very open, and very often was, the aggressive time schedule and the high stress level made open communication more difficult. Some staff found too much to review and too little time to digest all the material. Moreover, it was felt that there was too much at stake to publish everything in such a short time period. Opportunity for small group discussion was limited because of the time pressures. Some staff felt uncomfortable about their instructions and the level of detail required in the process, while others felt that they wasted time on issues that went away after the new organizational structure was announced. Communication with external constituents primarily occurred through job interactions, structured focus groups, and an open request to the campus for feedback on the published draft plan.

Resources

Other major challenges of the new organization were related to resources, both human and financial. The new organization required major changes in human resources. A change management plan and a human resources program were imperatives. The human resources plan included new job descriptions, performance measurements, and some training, but more training was required. "Changes in roles and performance appraisal criteria have to be supported by individuals who are appropriately trained to fulfil these roles. It is not simply a matter of calling an erstwhile systems analyst a project consultant if she is not equipped to handle the new job." (Agarwal et. al., 1997)

Existing staff had to learn to redefine long standing culture assumptions and develop new behaviors that would be reinforced by the external environment (Schein, 1997). A few technical personnel were immediately moved directly into user services positions. Some personnel did not have the training while a few others did not have the interest to redefine their own values and focus on user services. These staff revered their technical skills over their service skills. With a limited number of staff from which to choose, it was not always possible to place each person into their ideal position, but it was necessary to select those willing to acquire new skills for these new roles, provide training in new skills, and mentor those individuals.

The initial implementation was exacerbated by the lack of personnel. There were some vacancies to be filled when the new organizational structure went into effect, caused in large part by the significant university reduction in staff. These vacancies were not filled as quickly as hoped. In the meantime, some individuals performed more than one job, limiting their abilities to deliver user services and mentor others. There were also some shortages in critical areas with technical people assigned to cover areas for which they did not have all the requisite skills, which created some user frustrations. However, as these positions were filled, the new employees were generally better oriented toward their jobs than the people who were already on staff and had to adjust from their previous to their new positions.

In this case, increasing hardware and software demands had to be evaluated along with demands for new staff in a period of constrained financial resources. This was further exacerbated by a complete turnover of the senior management of the University, including the person to whom the CIO reported. These changes occurred in 1997 just after the new information services organization was implemented, and as individuals were beginning to develop new operating processes to test the effectiveness of an organizational structure that was driven by the strategic vision of the former top management of the university. Transitions at the top levels of management often alter institutional priorities and allocation of financial resources. As this top management turnover occurred sequentially and not at one time, it made it particularly difficult to secure commitments for additional resources.

SUMMARY

While this case illustrates the planning process used by a university to reorganize its library, computing, and telecommunications functions, it has applicability to information systems organizations in many other industries who are developing a new vision to provide end user support while supporting infrastructure investment. Most information systems organizations need to re-deploy human capital to allow for careers in business services and support as well as development support and technical services (Keen, 1991). Organizations must develop human resource strategies to develop new skills while providing opportunities to advance with traditional skills. IS functions who want to evolve from a technical oriented to a customer oriented approach must employ change management processes and provide training for client service functions.

Appendix A: June 1995 Organization Chart

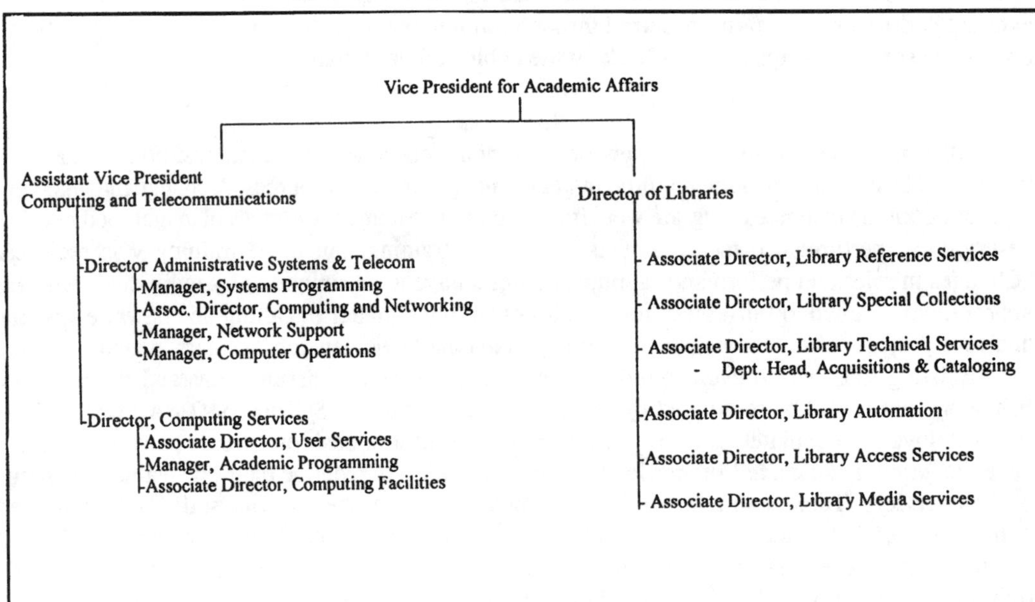

Appendix B: Schedule of Events

1994	Director of Libraries announces retirement and Assistant Vice President of Computing and Telecommunications announces intention to return to faculty Search committee established
Fall 1995	New Chief Information Officer arrives Interim management group established
First quarter 1996	Group orientation meetings The planning retreat Working teams established Working team meetings Client focus groups Centralization of internal business functions
Second quarter 1996	Open meeting to discuss strategic plan draft Discussion of proposed organizations Brown bag lunch unveiling first draft of broad organizational structure Revised organization chart released High level managers identified Staff completes job placement request forms
Third quarter 1996	Implementation of new organizational structure begins Installation of Windows 95 begins
1997	Physical relocation of employees

Appendix C: July 1996 Organization Chart

Vice President for Academic Affairs

Chief Information Officer

- -Associate Director, Technology Infrastructure
  - -Manager, Systems Development
  - -Manager, Systems Administration
  - -Manager, Installation & Maintenance Support
  - -Manager, Computing & Communications Operations
  - -Manager, Instructional Technology Support Services
- -Associate Director, User Services
  - -College and department team leaders*
  - -General Help and Information Desk
- -Associate Director, Library General and Special Collections
- -Associate Director, Library Access Services
  - -Manager, Technical Services
  - -Manager, Circulation Services

*Note: Each team includes consultants for computing, libraries, instructional support, and administrative systems.

FURTHER READING

Huber, G., Glick, W., *Organizational Change and Redesign*, Oxford University Press, 1993

Knights, D., Murray, F., *Managers Divided: Organisation Politics and Information Technology Management*, Wiley, 1994.

Stoddard, D., Jarvenpaa, S., Business Process Redesign: Tactics for Managing Radical Change, *Journal of Management Information Systems*, Summer 1995, Vol. 12, No. 1 81-107.

Zuboff, S. *In the Age of the Smart Machine*, Basic Books, 1988.

REFERENCES

Agarwal, R., Krudys, G., Tanniru, "Infusing Learning into the Information Systems Organization," *European Journal of Information Systems*, 1997, 6, 25-40.

Caldwell, B., "We are the Business," *Information Week*, Oct. 28, 1996, p. 36.

Davenport, T. *Process Innovation: Reengineering Work through Information Technology*, MA: Harvard Business School Press, 1993.

Hirshon, A., "Integrating Computing and Library Services: An Administrative Planning and Implementation Guide for Information Resources," CAUSE Professional Paper Series, #18, Boulder, Colo: CAUSE, 1998.

Keen, P., *Shaping the Future: Business Design through Information Technology*, MA: Harvard Business Review Press, 1991.

Schein, E., *Organizational Culture and Leadership*, San Francisco: Jossey-Bass Publishers, 1992.

BIOGRAPHICAL SKETCH

Susan A. Sherer is associate professor of business at Lehigh University where she serves as director of the Information Systems Curriculum and vice chair of the Business Department. Research interests include software failure risk management, information systems risk management, and risk in manufacturing networks. She is the author of Software Failure Risk: Measurement and Management, as well as articles in Journal of Systems and Software, Software Engineering Journal, IEEE Software, Information and Management, and Journal of Information Systems. Sherer received a Ph.D. in decision sciences from the Wharton School of the University of Pennsylvania.

The Telecommuting Life: Managing Issues of Work, Home and Technology

Gigi G. Kelly
College of William and Mary, USA

Karen Locke
College of William and Mary, USA

EXECUTIVE SUMMARY

As the 21st century approaches, the work environment is transforming, driven in large part by technology. For example, technology is challenging ideas about where and when work needs to take place. Technology allows employees to work from home. However, this new distributed work arrangement brings with it many new challenges for both the employee and employer. In this case, we introduce one company that has decided to experiment with the telecommuting arrangement. Through the eyes of one teleworker, many of the benefits and challenges of telecommuting are explored.

INTRODUCTION

One doesn't have to look vary hard at the business world to figure out that something significant seems to be happening with the current workplace. Look at the language we are using to describe our situation (emphases added): "A tidal wave of change is sweeping across the American workplace (Ehrlich, 1994:491); "The last decade, perhaps more than any other time since the advent of mass production, has witnessed a profound redefinition of the way we work (Business Week, 1994:76); and, the very notion of a job itself, is being questioned (Ancona, Kochan, Scully, Van Maanen, & Westney, 1996a). Whether one views the terms in which change is described as dramatic hyperbole or a reasonable representation of what is happening, it does seem that change, dramatic or evolutionary, is indeed taking place. Furthermore, the impetus to change is being felt across a wide variety of organizations, from Fortune 500 companies to government bureaucracies and even to the military.

One innovation in how work is accomplished is telecommuting, and it is increasingly being adopted by organizations. Telecommuting constitutes a fundamental change in where, how, and even when work is accomplished. After this brief introduction to telecommuting, we will present the case of Glenn Smith, an account of one individual's and organization's experience with creating this new work arrangement. We hope that the issues raised will offer insight into the implications of this new work arrangement both to individuals and organizations that might be thinking about the telecommuting option.

Telecommuting or teleworking, as defined by U.S. General Service Administration (1995), "refers to a means of performing work away from the principal office—typically at home or at a nearby telecenter." Telecommuting increases separation from the principal office while simultaneously increasing connection to the home. Telecommuting is not new; it has existed for several decades. In the late 1970s and early 1980s, businesses projected that the physical location of the worker would shift from the central business building to the home. Although there were organizations that pioneered this movement, it was not until the 1990s that telecommuting became a viable, acceptable, and, occasionally, a preferred work option.

A survey conducted by Telecommute America reported 11 million US teleworkers in early 1997, an increase from 4 million in 1990 (Murphy, 1998). Furthermore, conservative estimates for the number of teleworkers in the year 2000 range from 11 to 15 million (Piskurich, 1997) with other estimates indicating growth rates of 18% a year resulting in 20 million U.S. telecommuters by the year 2000 (Nilles, 1997). The GartnerGroup predicts even more involvement of U.S. workers in teleworking situation — by the year 2000, 30 million with more that 137 million involved in some form of teleworking by the year 2003 (Langhoff, 1998). These estimates suggest a growing recognition that telecommuting is a viable work option that organizations must acknowledge. Indeed, as the case will present, the forces driving telecommuting clearly express the interests not only of individuals seeking increased flexibility to reconfigure home and work but also of organizations focusing on economics, environmental regulatory pressures, and technological advancements. With this alignment of organizational and individual interests, the increasing growth in the population of teleworkers is not surprising. Nevertheless, it remains a "new work arrangement," very much in its initial adoption phase.

Let's now turn to Glenn's experience to help us better understand what might happen when an individual completes his work at home, electronically connected to but geographically distant from his principal office. This case portrays the real-life experiences and reflections of one teleworker. It is worth noting that this case tells a telecommuting story from one demographic perspective — a manager-level teleworker who belongs to a dual career family with children. Although some issues raised in this case are unique to this particular situation, many of the issues are faced by the mass of teleworkers.

BACKGROUND: AMERICAN BANK CORPORATION

Kelly Watson is the vice-president for information systems (IS) at American Bank Corporation (ABC) in Chicago. ABC is a large financial institution that has typically been characterized by its conservative culture. However, over the past decade mergers and acquisitions in the financial industry have required ABC to quickly change from an ultra-conservative, "change very slowly" mentality towards new and more progressive ways of doing business. Fortunately, the president and CEO, who are both in their mid-60s realized the need for change and have been instrumental in setting the precedence for making change happen. One recent change in the IS division was the hiring of Kelly Watson 3 years ago as Chief Information Officer (CIO).

In this capacity, Kelly is in charge of the migration of systems from the mainframe legacy environment to a new client server environment, and she faces a number of challenges. First, the IS division is under considerable pressure from its customers to get new systems up and running. Second, even though the IS professional resources in her division have grown tremendously (see Figure 1 for Kelly's staffing structure), she has employee retention problems. The market for IS professionals is so competitive that Kelly faces the constant problem of having to replace staff who have been lured away by competitors, consulting companies, and vendors.

SETTING THE STAGE: FINDING A SOLUTION TO A PROBLEM

Kelly's most recent expression of her retention problems is the imminent loss of one of her most senior development supervisors, Glenn. Glenn's wife, Vicki, had received a very attractive job offer in Cincinnati, where Vicki had grown up. Though Glenn really enjoyed his work at ABC and has developed satisfying work relationships with his manager as well as with Kelly, he feels that it wouldn't be fair to Vicki to deny her this opportunity. Given the number of people who have recently

Figure 1: Organization Chart

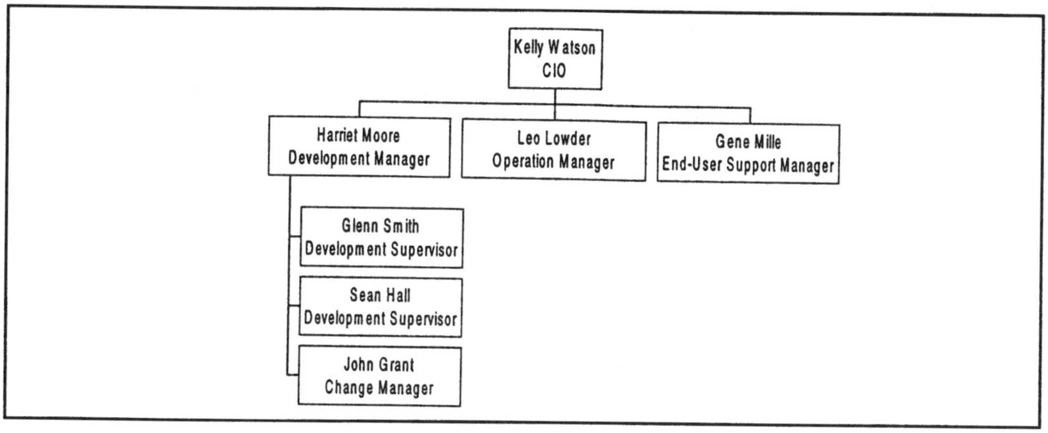

been hired away from the IS division and the fact that he receives calls from headhunters all the time, Glenn knows he'd have little problem finding a comparable position.

Kelly, anxious not to lose yet another experienced development professional (especially someone with Glenn's track record with clients) was prompted to try something new. Kelly had recently come across a number of interesting articles that highlighted the benefits of creating a virtual office. Apparently, companies were increasingly allowing knowledge workers to telecommute. In the financial industry, specifically, companies such as Bank of America, Citibank, and HomeFed bank were all offering some form of telework to their IS personnel, and they were reporting benefits in employee productivity and in reduced operating costs. She called Glenn into her office and broached the subject with him, "Would you be interested in working out a telecommuting arrangement with ABC?" Glenn was at first taken aback, but the idea did have some appeal. Organized and hardworking, he felt that he would have little problem working in a less structured format. Additionally, in his role as a development supervisor, most of his work revolved around projects. Glenn believed that during some phases of the project life cycle he would require close contact with the users; however, his current project was a long way from that point. And, right now, at least, his physical presence was not really a necessity. Certainly, he wouldn't miss the two hours he spent commuting through city traffic every day. On the other hand, Glenn was somewhat concerned with the implications of his telecommuting with his staff, and he knew this would have to be addressed. Kelly asked Glenn to think it over, to speak with Vicki about it, and they would talk again soon.

In the meantime, Kelly met with Harriet Moore, her development manager. It was Harriet who had first informed Kelly that they might be losing Glenn and had urged her to "do everything possible" to keep him. So, while Harriet was uncertain as to how telecommuting would work out — she had never worked under an arrangement like this before — she was anxious to give it a try. She did suggest to Kelly that it would be important to discuss this with the other IS managers. Kelly agreed and convened a meeting with her three managers: Harriet from development, Leo Lowder from operations and Gene Mille from end-user support. She also asked a representative from human resources (HR), Bob Loskins, to attend.

CASE DESCRIPTION: THE DECISION TO TRY TELECOMMUTING

Kelly began the meeting by emphasizing the retention challenges the IS division faced. She informed everyone that they were confronting the imminent loss of yet another experienced professional, this time a development supervisor. She told them that she was giving very serious consideration to allowing the supervisor to telecommute in order to keep him at ABC.

Leo immediately asked Harriet if she knew about this and if she supported the idea. Harriet responded, "Yes, Glenn is really respected by his clients, and I can't afford to lose him right now ... and, who knows, if this works out, it may give us an advantage in attracting and keeping the best people."

Table 1: Benefits of Teleworking

BENEFITS TO ORGANIZATION	BENEFITS TO INDIVIDUAL
* Improve office productivity	* More retained income (reduce personal expense)
* Increase use of new technology	* Increase job satisfaction
* Enhance recruitment and retention of personnel	* Productive and balanced lifestyle
* Increase competitive advantage	* Reduce stress and illness from commuting
* Reduce central office costs	* Integrate work and personal life
* Optimize personnel performance	* Improved family functioning
* Minimize cost and disruption of relocation	* Effective time management
* Improve environmental reputation	* Increase personal safety
* Strengthen Disaster Planning	* Control and design of workplace

Sources: Hodson, 1997; U.S. General Service Administration, 1995; ITD Telecommuting Task Force, 1997

"That's right," interrupted Kelly. "I wouldn't be proposing this if I didn't see it as a strong initiative for not only the division but also whole company. The opportunities with distributed work are really considerable," insisted Kelly, passing around a sheet that outlined the benefits of teleworking (see Table 1).

"But, you're only offering it to Glenn, right now?" Gene retorted, looking at Kelly. "Yes," Kelly replied, "We've never tried anything like this before so I want to move slowly on it...I'm sure there will be some bugs we'll have to work out as we go along."

"Right!" Leo jumped in, "How are you going to maintain control over his work quality? What happens when he has to meet a deadline and he's got a sick child at home with him who's lying around on the sofa? For that matter, what is he going to do at home, set up a separate office or is he going to be working from his kitchen table?"

"And," Gene chimed in, "I know that Glenn's really reliable, but what are you going to do when everyone starts wanting to try telecommuting?"

"Hold on!" Kelly interrupted, "These are all really important points and they illustrate what I was just saying, ... we'll have a lot of issues to work out. But, for right now, can we agree in principle that this is something we want to try and that we'll begin offering it to Glenn?" Gene nodded.

"It's definitely something that is in our interest as a division to explore," agreed Leo looking at the list of benefits Kelly had passed around, "Heck, I might try it myself!"

"What do you think, Bob?" asked Kelly, turning to the HR manager, "You've been pretty quiet during all this?"

"Well, I'm all for trying something new, especially if it's going to put a dent in our hiring and retention problems," replied Bob. "But, our discussion here has made it clear that there are going to be a lot of issues we'll have to deal with. We'll need a policy on telecommuting — I'll start looking into what other companies are doing with it."

"Good, then we have agreement," said Kelly, "We'll move on offering it to Glenn as an experiment. In what kind of time frame do we want to revisit this?"

"How about three months?" Bob offered.

Kelly looked around the room; the others all nodded. "O.K." said Kelly, "Now I only hope that Glenn wants to move ahead with this."

Glenn Tries Life as a Teleworker

Glenn let Kelly know that he would like to give telecommuting a try. The juggling of their respective professional careers was a recurring topic of conversation in the household. Glenn and Vicki agreed that they should do this if for no other reason than this was a working arrangement that they would definitely have to explore at some point in their careers. So why not now?

Initial Set-Up: The First Couple of Weeks

Setting up their new home in Cincinnati, Vicki was thrilled with the prospect of Glenn working from home. Now he would be home when the kids got off the bus and to help out with all the "taxi" and other family duties that had always been her responsibility to manage. Their lives would change in other ways too. Glenn needed an office. Since the living room was rarely used, this space was designated his office. With his own money, Glenn purchased used office furniture. Its cost was considerable, $1,500.00, but since it was to go in the living room, Glenn and Vicki had decided that it was worth buying quality pieces. Also, Glenn wanted a good chair, not a spare from the kitchen.

The phone lines were a little more of a challenge. After spending $120.00 for the initial set-up and an additional $30.00 a month (for voice mail, call waiting, caller-id, and call forwarding), Glenn was ready to connect to ABC. They already maintained two telephone lines for their house (one for the computer and one for voice), but a month into the project, Glenn's work conversations were so monopolizing their voice line that Vicki insisted that they get a third line designated for the family. They carried the cost of this additional phone line. After complaining about the costs that he was carrying, ABC did come through and provided Glenn a laptop computer (56k modem), and they provided Glenn a calling card for the voice long distance. ABC already had an "800" number to support the long distance connection for data. Glenn had asked for an ISDN line, but ABC was not set-up to receive ISDN communications. Actually, ABC only currently supports data connections at the speed of 28.8k. And, Leo Lowder in Operations was not ready to make any additional purchases to support teleworkering at this time.

The hassles of getting set up continued. Connecting to ABC's system from his home was more difficult than he originally thought it would be. His first week of telecommuting was pretty unproductive as the kinks were worked out of connecting his computer to ABC's network. Glenn spent a lot of time trying to get the technical support he needed to get set up. But, the technicians at ABC were swamped with service requests at the home office; and when they did find the time to call Glenn back, they had difficulty understanding and responding to his particular needs. They just hadn't had any experience with helping someone set up a distributed office. Then, Leo Lowder raised some concerns about security because Glenn was dialing directly into ABC's network. While Leo was uncomfortable with the current arrangement, he allowed it to continue for the duration of the experiment. Further, the connection problems persisted. Sometimes the local telephone company was to blame; at other times it was ABC's corporate telecommunication network.

Managing his time productively would be an issue, and this required some structure. Thinking that it would have been helpful to have had some specific guidelines or training courses for structuring his day and for maintaining effective communications links with the corporate office, Glenn, nevertheless, proceeded on his own to make changes he thought would be helpful to his situation (certainly, ABC offered no such programs). Glenn decided to set up a daily agenda of tasks to complete. Because he was a "night owl," he found himself completing tasks in the late evening hours when the house was quiet and he was not disturbed by the phone. Glenn had always been a hard worker and this continued dedication to work did not change with his new work venue. Glenn also set up daily lists of the people he needed to contact. He would spend a considerable amount of time on the telephone as well as maintaining contact through e-mail.

Back at ABC, the initial reception of Glenn's telecommuting highlighted just how much of a departure from traditional work practices this new arrangement was. On the phone, Glenn received a lot of friendly ribbing about how it must be great being able to get up when he wants and do what he wants. Ads running on the TV of teleworkers at home in their pajamas only added to the barrage of stereotypical comments (e.g., "Your golf game must be improving!" or "What's on ESPN now!"). Some of these comments about the informality of working at home were correct. Glenn did enjoy the relaxed dress code and was often on the golf driving range during lunch; however, these offhanded comments concerned Glenn. The monetary costs associated with establishing himself as a teleworker were clear, but the friendly ribbing made him wonder what other price he might be paying?

The Next Two Months: A New Set of Issues

The project Glenn was working on had up to this point been conducive to letting him work at home. During this time frame, he had returned to ABC twice for 4 days. ABC picked up the costs for travel and lodging. These interim visits to corporate office were going to be necessary at least for the short-term and maybe long-term. In addition, the project would soon be starting implementation and Glenn realized that he would have to be on-site more; although, he wasn't sure how much more. He could have done with some help to develop a corporate on-site schedule. Needing to be on-site was not all bad. Sometimes, Glenn felt the isolation and wanted to simply go out to lunch with his colleagues. He was also concerned that he was out of the day to day loop. Though, Glenn was fortunate that Harriett, his boss, acted as an advocate for him in his absence.

The clients Glenn supported also raised some issues. Over the last month or so, there had been times when end-users wanted to have Glenn on-site. To maintain his communications with end-users, Glenn audio-conferenced in on meetings, however, no video was being used. And, while audio conferencing provided a connection, Glenn did not always feel as an equal participant. One observation Glenn made was that if all participants were meeting face-to-face in a room with Glenn present on the speakerphone, he did not feel a real part of the exchanges taking place, regardless of how frequently he "spoke up." By contrast, if everyone "met" on a conference call from their separate offices, the meeting dynamics were more equal and satisfactory.

A more daunting issue for Glenn was learning how to manage from a distance. Glenn's prior management style had been "by walking around." He would stop by his staffs' cubicles to chat and keep abreast of their work. Now, this was impossible to do from his home. Some of Glenn's employees had no problem adjusting to Glenn's physical absence; however, one employee was having difficulty completing his work. Glenn knew that a more formal management routine was going to be necessary; so he set-up weekly audio staff meetings. What else could he do to better monitor his employees' performance? What technology might further facilitate his managerial responsibilities? Glenn was very interested in exploring the available groupware technologies to support virtual teams, but ABC had initiated little use of this type of software (except for e-mail). Furthermore, his ongoing management and project responsibilities kept him so busy (he was hiring additional staff) that he simply didn't have the time to investigate the technology. Glenn also felt he needed a fax machine, but at this time no money had been budgeted for such purchases. Frustrated at not having all the means of keeping in contact that he needed, Glenn reached into his own pocket to buy a fax machine.

On the home front, many of the benefits Glenn believed he would have by working from home did come to fruition. He did have more flexibility and was able to do things with the family that he simply did not have the opportunity to do under his previous working arrangement. Occasionally, he had been able to meet with the children at lunch; he had cared for the children when they were ill; and he was able to be more active in the neighborhood.

The adjustment to working from home, however, was not without problems. A number of issues arose with his family. Glenn started receiving an increasing number of work related calls in the evening when the family was eating dinner or when they were relaxing together. Time and time again, Glenn would have to leave family activities in order to deal with a work-related problem. Glenn and Vicki decided that he would not answer his work-designated phone after 6:00 p.m. That worked for a while until his manager and clients figured it out. If Glenn didn't pick up on his work line, they simply dialed the home number. Sometimes, it felt like he was "on call" 24 hours, 7 days a week.

Additionally, late afternoons when the kids came home from school and Vicki returned from work proved difficult. When the kids got off the bus in the afternoon, they were thrilled that Dad was home and were anxious to talk about their day. Glenn loved being available to hear all the school stories, but, occasionally, the children would rush into his office excitedly shouting, "Dad! Dad! Guess what ..." only to interrupt a telephone call or an urgent task. This disruption was annoying and would lead to conflict as the kids were told to, "Leave the room; Daddy's working!" On rare occasions when Glenn was dealing with a particularly frustrating problem, family members would bear the brunt of his frustration.

Tensions also arose between Glenn and Vicki in the allocation of responsibilities for managing the household. At those times when Glenn had to return to Chicago to interface directly with the home office, Vicki had to single parent, and this put considerable pressure on her. On the other hand,

Glenn's apparent availability at home had also lead to some conflict. On one occasion, Vicki arranged to have some landscaping done. That morning was one of those difficult times when Glenn was having problems getting connected just when he was under a time crunch. The door bell rang five times in the space of two hours as truckers wanted to know where they should dump the mulch, the gardener wanted to know if Glenn had any oil for the tractor because it was low, and everybody wanted to get paid. Glenn called Vicki at work furious that she had presumed on his time in this way. At moments such as this, Glenn wondered whether the new work arrangement really was working out satisfactorily.

Back at American Bank Corporation: The Experiment Receives Mixed Reviews

Results Inside the IS Division

As the three months experimental period drew to a close, the reviews on the telecommuting experiment were mixed. It certainly had taken Glenn a little longer to get set up then either he or Harriet anticipated, and the downtime associated with that put pressure on everyone to catch up the project. Still, once the technology-related bugs were worked out, Harriet and Glenn had both been happy with the latter's overall productivity and work quality. Also, Glenn's current clients were beginning to adjust to Glenn's physical absence and realized that quality products were still being delivered. Some clients did believe that Glenn was going to have to be at corporate more for the upcoming implementation and they were asking Harriet for her assurances that he was going to be physically available.

Unfortunately, the experiment also created some tensions. Several other development professionals approached Harriet. And, when they were told that it was only being offered to Glenn as an experiment for the time, they wanted to know why Glenn was being singled out for this special privilege. Harriet also received a few complaints from some of the division's other internal clients who wondered why they couldn't find Glenn despite the fact that they had dropped by on several different occasions to ask him questions. Additionally, the concern around "special privilege" also cropped up in the other IS functions and highlighted some stresses in the division. For example, operations staffers who as a group were prone to feel that they always had to cover for mistakes that the developers made, complained to Leo that it was unfair that it was someone from development who got "to stay home and relax on the sofa."

Predictably, perhaps, new technology issues were also raised. If this was truly the way the company was going to move, then a major commitment of additional resources was going to be required: new hardware, software, and telecommunication equipment was going to be necessary. With this, it was clear that the IS division would have to figure out a way to provide technical support for specific teleworker problems that would be difficult to solve because they would not be able to go to the site. Leo, especially, had serious concerns about whether this level of investment in telecommuting was warranted at this time.

Results in the Rest of ABC

In HR, Bob was also receiving a lot of inquiries from ABC employees, not only in IS but also in the rest of the company who had heard about Glenn's arrangement and who were also interested in telecommuting. Bob knew that the question of whether telecommuting was going to be a company wide option would have to be answered fairly soon. Additionally, if the telecommuting option was opened up to other employees, then, this would have to be addressed at the company level. His research into what other companies were doing made it clear to Bob that ABC was getting into something much larger than anyone had imagined when they first discussed telework three months ago. For example, exactly how would people be compensated and reviewed? What technology would be supplied by the company? How would worker's compensation work if someone got hurt at home? What new training was going to be necessary and for which groups in the company? Would ABC develop these training courses or would they look for outside consultants who specialized in establishing a telecommuting environment? Bob had gathered some initial statistics that other companies had published regarding their experiences (see Table 2) to share with Kelly. But, he knew

Table 2: Employer's Costs and Benefits per Teleworker

Employer Costs per Teleworker		Employer Benefits per Teleworker	
Computer costs and associated hardware	$3,200	Cubicle Savings $5,000	
New phone lines (ISDN more expensive)	$100 - $500	Absenteeism Decreased	10%
Home office furniture (desk, chair, lighting, supplies)	$1,000	Productivity Increases	15%-25%
Selection and Training	$175	Office space savings	1000 square feet
Moving costs such as materials	$240	Turnover Decreased	5% of salary (costs for searching/training)
Annual Costs (e.g. phone lines)	$1,000	Park Expenses Decreased	
		Environmental savings due to less vehicle commuting	
		Increased Competitiveness	

Source: JALA International, Inc., 1998 and Fitzer, 1997

he would need to determine not only the tangible but also the intangible costs and benefits associated with telecommuting that ABC would incur if it moved ahead.

The Next Step

It was now three months since the IS division had made the decision to experiment with the telecommuting option. As she prepared for her meeting with her IS managers and Bob Loskins, Kelly Watson wondered what the next step was going to be. Given the reactions to the experiment, should they go ahead and make telecommuting more widely available? If they chose this option, what had to happen to ensure its success? Should they put it on hold and delay the decision? Or, should they abandon it as something that was not feasible in American Bank Corporation?

DISCUSSION: CURRENT CHALLENGES OF INTEGRATING WORK, HOME, AND TECHNOLOGY

The preceding case points to a number of issues that highlight the altered bounding of work and home created by a telecommuting arrangement and the influence of new technology. Obviously, Glenn is no longer physically present at work. Glenn's work arrives at the office electronically but he isn't there, thus creating some dynamics that require active attention. This separation or unbundling of the worker from principal work location raises issues not only with regard to immediate work relationships, namely Glenn's boss, co-workers, and subordinates, but also more peripheral relationships, those organizational employees who are acquainted with Glenn or simply with Glenn's new work arrangement. Additionally, Glenn's physical absence from corporate offices also has more long-term implications for his development opportunities and advancement. At the same time, the creation of a workspace in the home bundles together home and family life with work life. Boundaries that facilitated the creation of a distinct way of living that was non work-obliged are no longer present, and their absence raises concerns for work, family, and leisure. All these issues will require attendance and active maintenance on the part of the teleworker. Finally, the technology that allows telecommuting to be viable surfaces new issues that both the organization and Glenn will have to address.

Getting Work Done From a Distance

All parties immediately feel the physical distance between and the separation of the teleworker from the organization involved. Absence from the office creates a void, and the managers, co-workers, and others left behind who are unfamiliar with telecommuting have no experiences from which they can draw to fill it. Consequently, it is easy for the ambiguity associated with what the teleworker might be doing away from the office to be resolved with images of activities that are non-work associated. Perceptions that Glenn is involved in non-work related activities (e.g., watching TV, or playing golf) are understandable, if inaccurate. Second, Glenn's physical separation from those he supervises requires revisiting how he enacts his role as a manager. The interpersonal, informational, and decisional functions that comprise the traditional managerial role typically include a sizable face-to-face component (Mintzberg, 1973). A new way of enacting these roles will be necessary. Third, the potential for deterioration in all work relationships must be acknowledged and proactively managed. And, there is, indeed, evidence to suggest that telecommuting does have a destabilizing effect on work relationships that is expressed approximately six months into the new arrangement. The good news is that a recovery and mutual adjustment seems to be possible after about a year (Reinsch, Jr. 1997).

These three sets of issues clearly point to the need for a "learning and adjustment" period for those organizational members touched by the new telecommuting work arrangement to discover and establish new ways of working. It is clear that training programs need to be offered for managers and teleworkers that address issues such as communications and time management. Additionally, as the case indicated, Glenn had a strong working relationship with his manager, Harriet, and his ability to continue to provide quality contributions to ABC under the new work arrangement was never in doubt. Nevertheless, the issue of ensuring a teleworker's actual and "visible" productivity is central to this new work arrangement.

As indicated earlier, learning how to supervise others' efforts in a distributed mode can be particularly difficult and would seem to require a careful and explicit consideration of both the supervisors' and their subordinates' preferred work styles. This has been a challenge for Glenn and for his employees. Prior to telecommuting, Glenn ensured the productivity and quality of his department's work through an informal management-by-walking-around style. In order to stay on top of his department's contributions and particularly the efforts of those employees who performed better with closer monitoring, Glenn had to find ways to formalize his supervisory activities.

The above discussions emphasize the importance of developing new ways to ensure the contributions of the teleworker and underscore the importance of paying close attention to the network of relationships in which any of us organizational members are embedded. Teleworkers have to make projecting their virtual presence and developing substitute communication channels to stay connected with their relational network a priority (Bjorner, 1997).

Issues of Development and Advancement

Before concluding our discussion of the impact of separation from the principal work site, it is important to raise the question, what happens to a teleworker's advancement opportunities and career path? Certainly the lack of visibility that we have already discussed may cause some problems if not actively managed. As one AT&T teleworker rather crassly put it, separation from the workplace means a loss of "suck up" opportunities (Hequet, 1996). It is important for the teleworker to take responsibility for charting his development path under this new work arrangement, for example by actively seeking and initiating new challenging work projects. However, because this is uncharted territory for most organizations, we simply do not know the answers to this. Will the teleworker have to return to the home office to maintain upward mobility or must the teleworker trade advancement for location flexibility?

Dealing with Confounded Boundaries between Work and Home

Telecommuting muddies the boundaries between work and home, and the telecommuter can cycle continuously among personal, family and work related tasks. This arrangement provides a potential flexibility not only in where, but also, to a degree, when the work takes place. While some aspects of the work, most notably contact and coordination with office workers and clients, need to take place during corporate business hours, the independent tasks that make telecommuting a viable option for a given employee can be completed at any time. The 24 hour / 7 day week, thus, potentially becomes the frame within which work can be accomplished.

This continuous time frame brings both positive and negative attributes for the teleworker. On the positive side is the flexibility to complete work during times when the teleworker is most productive. Additionally, this potential flexibility of where and when to work opens up opportunities for the teleworker to improve their relationship with family members by permitting a more active role in family life (Hill, 1995; Olsen, 1987). For example, being available to pick a child up from school or to coach a sports team whose practice would typically begin 45 minutes before the parent could get home from work. Thus telecommuting allows the teleworker to work at some tasks early or late in the day, saving times in between for family.

On the negative side, imposing appropriate time boundaries around work and separating and conserving the quality of home life, including relationships with the family are critical. Indeed, there is a real possibility that unless specific steps are taken to define specific work boundaries, for example, by establishing a daily work agenda, that one's work life may overwhelm the home, temporally (there is precious little time that work does not impinge on) and emotionally (family members falling prey to work related frustrations). As is true with regard to the workplace, a learning and adjustment period are also needed with regard to the home.

Technical Issues: Connecting the Teleworker

The technology makes it possible to work anywhere any time providing the flexibility increasingly prized by so many of us on how, when, and where we get our work done. In order for Glenn to telecommute, the technical environment had to be set-up. This requires a negotiation between the teleworker and the organization to fully understand what equipment is necessary. Having the proper technology is essential for successful telecommuting (Artz, 1996). At a minimum, the teleworker must have a telephone, with many teleworkers also possessing a computer and a modem. The Internet has provided a low cost means to connect the teleworker to the office. E-mail has become a primary tool for communication within organizations thus facilitating an additional means to keep the teleworker connected. One teleworker made the following comment, " You learn to live by e-mail. That was my water cooler" (Murphy, 1998). Other equipment that may be required include printer, fax machine, copier, pager, answering machine, mobile phone, and additional phone lines. Exactly who is responsible for these costs needs to be addressed. Also, the slow access to the organization's LAN (local area network) was a constant source of frustration for Glenn. It is important to recognize that the phone line has become an umbilical cord that provides the connection of the teleworker to the organization. When there are technical problems (and there will be), this can render the teleworker lifeless. Solving technical problems remotely are often more challenging and procedures need to be established to address these issues.

REFERENCES

Ancona, D., Kochan, T. Scully, M. Van Maanen, J. & Westney, D. E. (1996). Workforce Management: Employment Relationships in Changing Organizations, In *Managing for the Future,* Cincinnati, Ohio: South-Western College Publishing.

Artz, T. (1996). *Technologies for Enabling Telecommuting,* [Online]. Available: http://www.cba.uga.edu/management/rwatson/tc96/papers/artz/artz/toc.htm1998, April 1].

Bjorner, S. (1997). Changing places: building new workplace infrastructures, *Online,* 21(6):8-9.

Business Week.(1994). The New World of Work, October 17:76.

Ehrlich, C. J. (1994). Creating an employer-employee relationship for the future, Human Resource Management, 33(3):491-501.

Fitzer, M. M. (1997). Managing from afar: Performance and rewards in a telecommuting environment, Compensation & Benefits Review, 29(1):65.

Hequet, M. (1996). Virtually working: Dispatches from the home front, *Training,* August:29-35.

Hill, E. J. (1995). The perceived influence of mobile telework on aspects of work life and family life: An exploratory study. Ph.D. dissertation, Utah State University, Logan UT. Dissertation Abstracts International 56(10), 4161A. University Microfilms No. DA9603489.

Hodson, N. (1997). *"Teleworking ... a Different Way of Doing Things,"* [Online]. Available: http://www.teleworker.com/seminar/index.html [1998, April 1].

ITD *Telecommuting Task Force Report,* (1997). [Online]. Available: http://www.itd.umich.edu/

telecommuting/report/ [1998, April 1].

JALA International, Inc. (1998). "Home-based Telecommuting Cost-Benefit Analysis," [Online]. Available: http://www.jala.com/homecba.htm [1998, January 7].

Langhoff, J. (1998). "FAQs about telecommuting...," [Online]. Available: http://www.hanghoff.com/faqs.html [1998, September 26].

Mintzberg, H. 1973. *The Nature of Managerial Work.* New York: Harper & Row.

Murphy, K. 1998. Web Fosters Telecommuting Boom, and Many in the Industry Take Part, internet world. http://www.internet.com, February 9.

Nilles, J. M. 1997. *Making Telecommuting Happen: A Guide for Telemanagers and Telecommuters.* John Wiley & Sons.

Olsen, M. H. 1987. Telework: Practical experience and future prospects. In R. E. Kraut (Ed.), *Technology and the transformation of white collar work*:135-152. Hillsdale NJ: Lawrence Erlbaum Associates.

Piskurich, G. M. 1996. Making Telecommuting, *Training & Development,* 50:20+.

Reinsch, Jr. N. L. 1997. Relationships between telecommuting workers and their managers: An exploratory study, *The Journal of Business Communication,* 34(4):343-369.

U.S. General Service Administration, *Office of Workplace Initiatives.* 1995. Interim Report: Federal Interagency Telecommuting Centers.

BIOGRAPHICAL SKETCHES

Gigi G. Kelly is an assistant professor of Management Information Systems at the College of William & Mary. She holds a B.B.A. from James Madison University, an M.B.A. from Old Dominion University, and a Ph.D. from the University of Georgia. Dr. Kelly teaches in the areas of systems analysis and design, information resource management, and decision support systems. Her research interests focuses on the various effects of IT on the work environment, including telecommuting, virtual group development, and enterprise-wide systems implementations. She has extensive consulting experience in MIS and has published articles in Journal of MIS, Small Group Research, Computerworld, various conference proceedings and book chapters.

Karen Locke, Ph. D., is Aassociate professor of Business Administration at the College of William and Mary's school of business. She joined the faculty there in 1989 after earning her Ph.D. in organizational behavior from Case Western Reserve University. Karen uses qualitative methods to pursue her research interests. Her investigative work focuses on the written construction of scientific work in the organizational studies community, on the examination of emotionality in the workplace, and more recently on the impact of telecommuting on home and work. Karen's work has appeared in journals such as Academy of Management Journal, Organization Science, Journal of Management Inquiry, and Studies in Organization, Culture and Society. Additionally, she has co-authored a book, titled, Composing Qualitative Research, published by Sage.

Leveraging Objects for Mission-Critical Applications

Mahesh S. Raisinghani
University of Dallas, USA

Bruce Adams
Electronic Data Systems, USA

Conceptual integrity is central to product quality.
—Fred Brooks, The Mythical Man-Month

EXECUTIVE SUMMARY

This case study is based on an enterprise-wide consulting project for a financial services firm in a major metropolitan area of the southwest United States. This case addresses the underlying principles (i.e., techniques and processes) and real-world practical application of object orientation (O-O). The objectives of this case study are to reinforce the student's foundation in fundamental O-O concepts, to provide an in-depth example of the application of O-O analysis and design techniques and formalisms and to enable the student to transfer this knowledge to the student's actual work.

The merit of the development process described here, Compass, is that it presents a repeatable process for delivery of client server architectures, object oriented systems, and distributed objects and components. It helps manage three interdependent variables common to most projects, i.e., deliverables, resources, and time, in a cost effective and efficient manner. The concept behind Compass is to integrate the best-proven solutions currently available, drawing upon several best-of-breed approaches used within the Information Services industry.

Background

EDS is implementing a mission-critical business system, which is the cornerstone for the customer's strategic vision. For EDS/customer confidentiality reasons the company will not be named.

The system is being developed for a company that wants to grow its customer base by a focused targeting of niche markets. In order achieve this goal it was deemed necessary to expand the technological capabilities provided by the current mainframe based system. These capabilities would allow the business to reach potential customers in new ways such as the Internet / Intranet, kiosks, and "portable" business offices using laptop computers. The new system was also required to provide flexibility in areas such as new product creation.

The motivation for using the O-O paradigm for software development is illustrated by a closer look at two primary approaches to software development. These are the function-oriented paradigm, where functions are defined first and data is defined later, and the data-oriented paradigm, where data are defined first and functions are defined later. Both these approaches have their drawbacks. The drawbacks of the function-oriented paradigm are as follows:

- The difficulty in maintaining the consistency between the data flow diagram (DFD) and the data dictionary.
- The difficulty in eliminating or minimizing the "ripple-effect".
- The difficulty in software reuse due to tight coupling between the functions, and treating data and functions separately.
- The difficulty in following data flows in structured analysis and building hierarchies of tasks in structured design.

The drawbacks of the data-oriented paradigm are as follows:

- The difficulty in maintaining the consistency between the process model and the snapshot model.
- The difficulty in comprehending the effect of the interaction between the functions which are not explicitly modeled in the process or transaction specification language.
- The difficulty in software reuse since data and functions are treated separately.
- The difficulty in converting an analysis model into a good software design.

The object-oriented (O-O) approach attempts to give a balanced treatment of functions and data. An object is a state and a set of methods that explicitly embodies an abstraction characterized by the behavior of relevant requests and an entity that has the state and functionality. In the O-O context, an object is "an instance of a class" (www.cyberdyne-object-sys.com/oofaq). Thus the class is the fundamental building block of OO software. A class determines everything about an object (Elmasri and Navathe, 1993; Brown, 1997). A class defines a data type where a type consists of both a set of states and a set of operations which transition between those states. A class provides a set of (usually public) operations, and a set of (usually non-public) data attributes representing the abstract values associated with instances of the type. Object-oriented concepts such as inheritance, polymorphism, encapsulation, and data abstraction that lead to compatibility, flexibility, reuse, extensibility, and easier maintenance are the primary reasons for using the O-O paradigm in this case.

Distributed objects are extended objects that are not restricted to a single program and can exist as independent entities and be remotely accessible by other objects. Rather than producing monolithic applications, distributed object systems tend to consist of a number of small units or components. Components are the smallest, self-managing, and independent parts of a distributed-object model that work in heterogeneous environments (Lewandowski, 1998). The key benefits of distributed objects are that they allow the development of scaleable client/server systems by virtue of modularized software that features interchangeable parts and an option to add components for custom solutions in advanced architectures.

Setting the Stage

EDS and members of the existing information systems (IS) staff defined a process to gather defined project metrics for quantification of the results and a formal testing process in which deliverables must pass through a series of "quality-gates" (Q-Gate).

Compass, which was developed by EDS, is being used to provide a formal OO development process and OO methodology support. Like any of the popular O-O analysis and design approaches, Compass should not be seen as a definitive and comprehensive source of guidance for O-O software development. It should be treated as a set of pragmatic guidelines and instructional tools where the focus is on sensible use of formalism and process in laying the foundation for reliable O-O software development

WHY IS ANOTHER PARADIGM NEEDED?

Current research indicates that practitioners are not necessarily using the methodologies as

Figure 1: The Diamond of Change

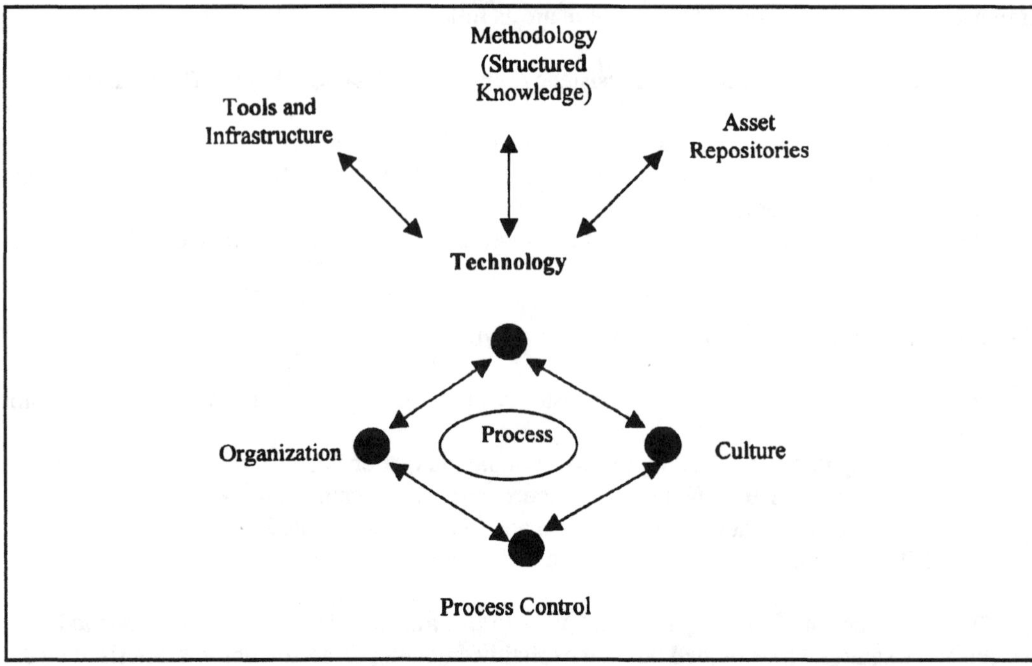

defined by their originators, but rather are customizing these methodologies to suit their application development needs (Hardy et al, 1995, Vlasborn et al, 1995, Vessey and Glass, 1998). The development process described in this case uses the best-practice approach in borrowing from several different methodologies both to overcome the deficiencies of any one methodology and to customize the approach to meet the needs of the project and organization.

More specifically, it addresses the two categories of deficiencies identified by Arnold et al. (1991). These deficiencies have to do with notations used to capture the system under development and/or the process used to drive the development from requirements definition to implementation. Business processes change with almost every project. The diamond of change represented in Figure 1 must be effectively managed in a balanced manner in order to respond to the needs of the business (AlBanna and Osterhaus, 1998). Shorter time to market and software replacement timeframes provide competitive advantage to the organization.

With the value of the U.S. software industry estimated at $92.5 billion and the U.S. currently providing almost 60% of the global software market (Albanna and Osterhaus, 1998), the diamond of change represents the forces shaping the software development environment. Although traditional organization theory assumes that the organization has a static and definable culture, the reality is that it is dynamic and IS managers of responsive, learning organizations need to make increasingly difficult decisions in the related dimensions of process, organization, and culture. The next section reviews the Compass process that addresses these dimensions of the diamond and creates a dynamic balance between them.

The Compass Process

Compass is designed to support iterative OO development activities and to work in conjunction with existing EDS development processes for project management. This allows managers familiar with the existing project management approaches to adapt them to projects using *Compass*.

Compass supports two different perspectives on the project: the manager's perspective (the macro view) and the developer's perspective (the micro view). The manager needs to understand from a scheduling perspective how the project is organized and maps the different phases to a project schedule. The developer is also concerned with the phases but is also concerned with the daily development tasks, such as identifying abstractions, establishing the attendant relationships, identifying the semantics, and creating the various artifacts associated with each phase.

Compass was designed to be a developer's guide focusing on the critical areas in developing OO systems. Examples, templates, and detailed documentation explain the steps in the development process. This documentation includes a detailed view summarizing each step of the process, the inputs, the artifacts produced, and whether or not there is a quality gate inspection required. A "participation matrix" is also provided.

The participation matrix specifies, for each step, a listing of all the necessary project roles and their level of participation in the event. For example, producing an Analysis Behavior model the business analysis may have the primary responsibility, the technical lead the secondary responsibility and the architects might want to be notified of the results.

An important point to bear in mind is that the Compass process is derived from a combination of several popular O-O methodologies and does not represent an entirely new one. It is designed to synchronize and integrate with other existing processes, i.e. for project management and testing, to provide the support necessary for system development.

Project Foundation

The Compass process begins with a project startup phase called Project Foundation. Project Foundation's primary purposes are:

- Establishing the scope of the project
- Creating an architectural vision for the system to be developed
- Planning the organization of the project into Releases
- Assessing and managing risks identified within the project
- Creating the Project Plan
- Producing initial requirements for the project. This will include identification and preliminary specification of the Use Cases for the project. These specifications will be elaborated during the iterations.

Each Release of the overall project has a specific goal designed to stage the eventual delivery of the entire project. Each Release is broken up into an appropriate number of Iterations. Each Iteration functions within a Release in the same way that each Release functions within the overall project. Each Iteration may be considered a miniature project within its Release.

The core of the system will be developed initially and additional functionality will be added in layers. A useful metaphor for describing this "Core/Layer" approach to building systems is the onion. An onion is made up of a central core. The core looks, smells, and tastes like an onion, but a relatively small onion. The core of a system is like the core of the onion. To this core, each successive iteration and release adds layers of functionality onto the growing "onion" system. In this manner, capabilities are added to the core, which grows toward the finished product.

A significant byproduct of this process is that portions of the system are available very early in the lifecycle of the project. This allows the customer to provide early feedback and, if appropriate, portions of the system may be deployed to production creating a "staged" deployment.

Project Release

The Foundation phase is followed by one or more Releases. A Release produces a completed project or a shippable version of the software. Each Release is comprised of one or many iterations. Iterations are driven by Use Cases that are assigned to teams. Each release of the overall project has a specific goal designed to stage the eventual delivery of the entire project.

Iteration Phases

Each Iteration is made up of four phases, which correspond, to the traditional "waterfall" approaches: Conceptualization, Analysis, Design, and Evolution. Rather than being a sequential process, there may and should be considerable overlap among the phases.

For each new iteration there is a point at which the vision for the subsequent work is established.

Booch (1994, 1996) refers to this as "Conceptualization". The purpose of this step is to establish the "core requirements" and architectural vision. Establishing success criteria and performing risk management activities are also crucial artifacts of Conceptualization.

The purpose of analysis is to provide a description of the problem. The modeling during this phase will focus on abstractions that form the vocabulary of the problem domain.

The objective is to build a model of what the system does (its behavior) rather than how it does it (its form).

In practice, it is absolutely imperative that Analysis be kept separate from Design. Many developers tend to prefer the more technical design and coding phases. Unless there is a distinct Analysis phase, the tendency is for developers to go directly to design and coding. If the underlying Analysis models and the supporting requirements are weak, this in turn weakens the underlying semantic model. As stated earlier, this semantic model provides one of the major benefits of an OO approach.

The Design phase is begun as soon as there is a reasonably complete set of Analysis models for an iteration. The Design phase can be considered a process of expanding or extending the work that was begun in the analysis phase. The models developed in analysis are extended to include design heuristics and notations. Rather than switching from one type of modeling to another, during design the analysis models are transformed into a design.

Evolution is the process of taking the design artifacts and producing working components of the system. Included in this process are the necessary testing steps to ensure accuracy and conformance to established success criteria. Also during evolution the deployment plan that will be used to place the system into production is developed

Finally, the tested release package is deployed according to the deployment plan and turned over to production support.

Compass Architecture

Compass revolves around the 4+1 View of Software Architecture illustrated in Figure 1. Each view addresses a specific area of interest to stakeholders. The four views of the architecture are organized around key use cases, which constitute a fifth view. Each view is described by appropriate notation that represents the view's key aspects.

The 4+1 Views can be characterized as follows:

Logical View — describes how the software design supports the requirements of the system. This is captured as various O-O artifacts such as domain class models, dynamic models, and state transition diagrams.

Process View — captures the runtime aspects of the system, such as threads of control, process / node deployment, and the resource requirements of the systems.

Physical View — encompasses network topology, hardware, system software, transport mechanisms, data links, and DBMS.

Figure 2: The 4+1 View of Software Architecture

See Appendix A for further details on the Compass methodology

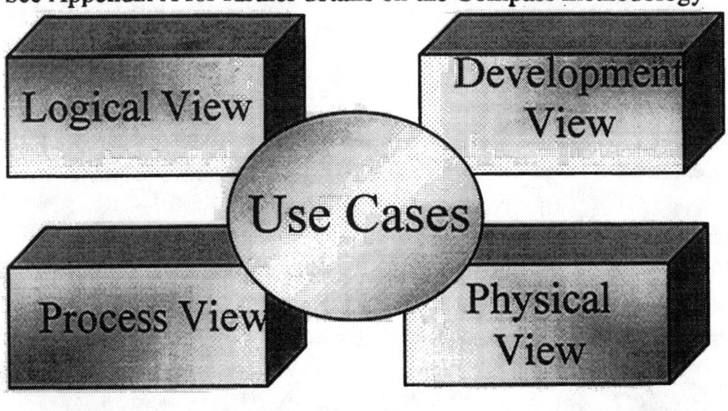

Development View — describes the development environment, tactical policy, and the architecture itself.

Use Cases—captures requirements and both demonstrates and validates the convergence of the four views.

Project Description

The project is being developed using a CORBA based architecture running on UNIX servers. Users have PCs running NT clients running Java applets. The system also provides access to key mainframe legacy systems.

The system provides sales support and customer servicing for financial products that represent the core business of the company. The system will be deployed in over one thousand branches and supporting between five- and nine thousand users.

The system is expected to be relatively long lived, supporting the business over the next decade and possibly longer. An object-oriented approach was selected since much of the benefit from OO occurs in the maintenance phase of the lifecycle. This is due to several factors including the ability to support abstractions that provide encapsulation of data and behavior in objects.

An OO approach also supports a rigorous modeling effort that enables the establishment of a semantically accurate representation of the business. In fact, some early object oriented systems were developed to support real world simulation applications. Today, practitioners use OO approaches to develop systems that simulate the real world business that the system is supporting. This allows key workflows, products, relevant parties, and the accounting model to be supported by software that closely parallels the behavior of the business.

Such a semantically rich system is more adaptive to change than the current legacy system. It will also be more conducive to new and more flexible distributed object (Lewandowski, 1998) and network approaches, such as Intranets and Extranets. This will greatly enhance the range of employees that have access to the system and support a large amount of information needed for the financial transactions, distribution of objects across the network, and high performance of the system with servers of thousands of pages, frequently changing content and references.

Utilizing a web front end and a standard browser gives users greater access to key business systems. These systems are currently accessible only through character mode terminals on a local area network. This will reduce overhead costs for current and new sites requiring system access to support the business.

This approach will also enable new sales capabilities such as allowing direct dialup access to a mobile sales force equipped with laptops and a modem. This allows access to core business systems from a wide range of locations: customers' homes, kiosks at a special event, essentially anywhere with a telephone line.

These and other additional business objectives in developing the system can be summarized as follows:

- To provide a simplified and more intuitive interface that would reduce training costs and allow new employees to become productive more quickly.
- To provide a standardized architecture that would allow integration of the current product lines.
- To increase productivity with a modern system that would allow for greater flexibility in sales support.
- To provide a system that would allow for greater market penetration by utilizing technologies, portable computers for sales people in the field.
- To provide a system that would allow for faster response to market trends providing a competitive advantage in the market place.
- To establish an architectural approach providing a highly distributed network with a web-enabled thin client, fat server processing scheme supporting an enterprise wide Intranet.

Current Status

The first phase of software has been released in a pilot deployment. Based on the continued success of the pilot, the system will be rolled out in a phased release to the existing field sites. The remaining application that make up the overall system will be completed and deployed as pilots and then rolled out in a phased manner.

Successes and Failures

- The organization has developed a common software framework that can be used to support the evolution of the core business support applications.
- The first phase of the business applications has been delivered on schedule and has been deployed in pilot. It is scheduled to be phased-in to production in the near future.
- A solid semantic model of the business has been developed which can be reused in the development of additional applications in the financial domain.
- The customer is gaining a system that supports the technology necessary to the corporate strategic vision.

In addition to the EDS OO developers, the customer wanted IS staff to help with the development of the system. This knowledge transfer was viewed as critical for the long-term success of the project. The customer's IS staff members assigned to the project were a mix of experience levels: interns and new hires along with developers having years of experience on the legacy system.

Also the OO mentors have varying degrees of ability in the knowledge transfer. The skills that make a good OO developer are not the same skills that make a good trainer or mentor. The project has had mixed results with mentoring. It was found that some of the developers are better suited to strictly technical work. This did not present a problem once the resources were properly aligned.

Finally, factors contributing to the lack of organized corporate O-O strategies include poor reuse statistics, lack of shrink-wrap business components, and an overall disenchantment with in-house development. The team members involved in this project placed a strong emphasis on assuring the stakeholders that each of these factors had been addressed in this solution. For instance, new techniques for estimating project size and tracking development (i.e., a Q-gate process) were developed to manage the iterative process of O-O software development.

CONCLUSION

In conclusion, it is crucial to get the "buy-in" from the top management for successful implementation of an enterprise wide project. Since it is difficult to acquire systems engineering thinking and practice without continuous training and continuous learning from the lessons of both successful and failed projects, a systematic learning process that changes systems engineering thinking and improves the process is extremely important.

Although the Compass process is risk adaptive as opposed to risk aversive, there is intrinsically more risk in the evolutionary approach. Although the number of iterations necessary cannot be known in advance, it is advantageous to use a faster modeling technique such as the one described in this case study due to its inherent flexibility to mirror the rapidly changing real world.

APPENDIX A

Objective

Compass

Objectives

- Provide a Repeatable Process for Delivery of
  - Client Server Architectures
  - Object Oriented Systems
  - Distributed Object Systems
- Provide JIT Project Direction
  - Outline for project
    - activities
    - products
    - concepts
    - objectives
  - Detailed information for each phase

EDS

Objective

Compass

The Approach

- The Compass Approach Is:
  - Customizable and Adaptable
  - Streamlined
- The Focus Is On the Critical Path Through the OO Lifecycle
- Compass Utilizes Existing Methodologies

(Booch/Jacobson/Rumbaugh, UML, SLC, PM2)

EDS

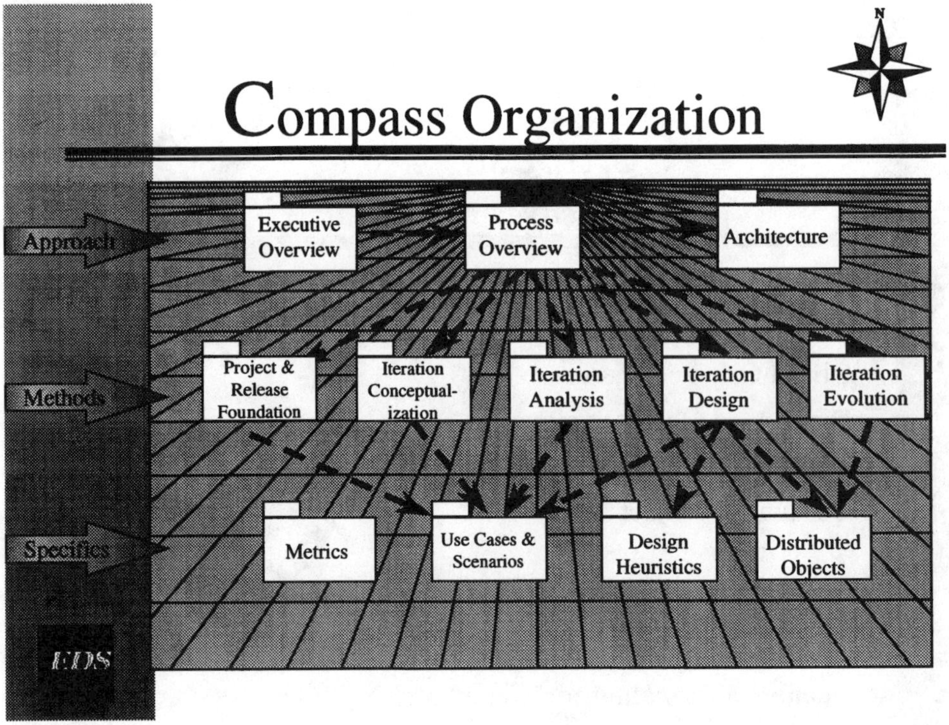
N
Compass Organization
Approach
Methods
Specifics
Executive Overview
Process Overview
Architecture
Project & Release Foundation
Iteration Conceptual-ization
Iteration Analysis
Iteration Design
Iteration Evolution
Metrics
Use Cases & Scenarios
Design Heuristics
Distributed Objects
EDS

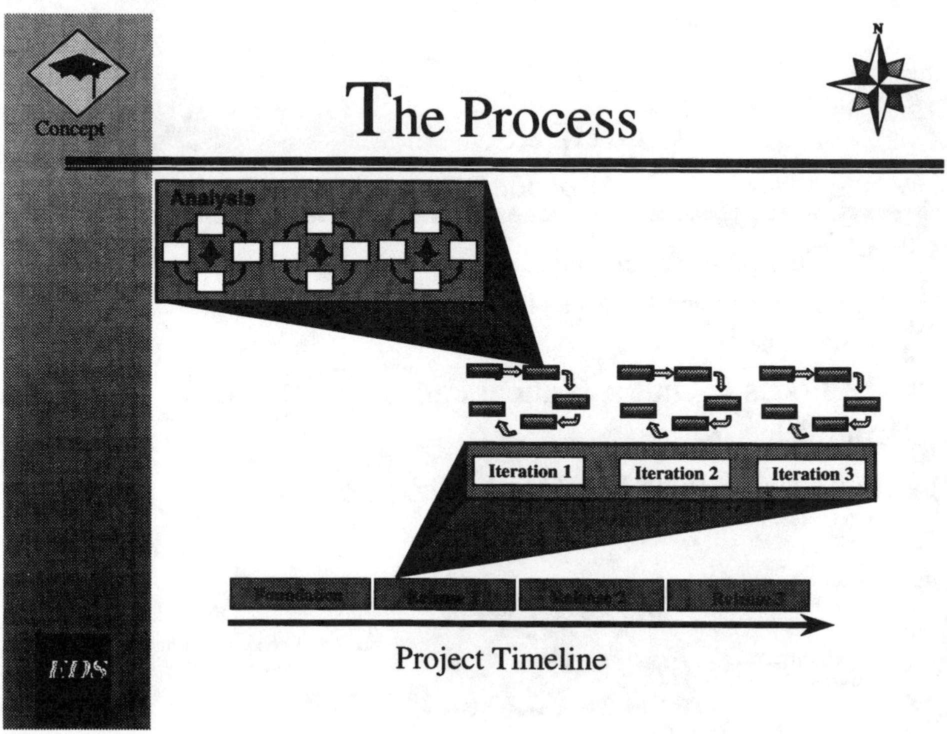
N
Concept
The Process
Analysis
Iteration 1
Iteration 2
Iteration 3
Project Timeline
EDS

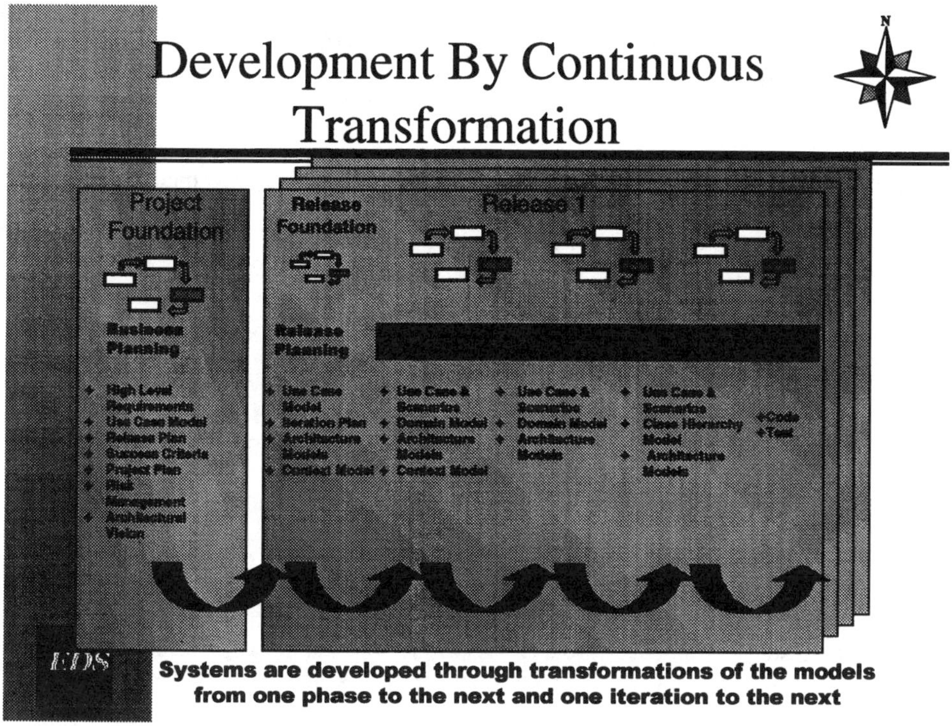
Development By Continuous Transformation
Project Foundation
Business Planning
High Level Requirements
Use Case Model
Release Plan
Success Criteria
Project Plan
Risk Management
Architectural Vision
Release Foundation
Release 1
Release Planning
Use Case Model
Iteration Plan
Architecture Models
Context Model
Use Case & Scenarios
Domain Model
Architecture Models
Context Model
Use Case & Scenarios
Domain Model
Architecture Models
Use Case & Scenarios
Class Hierarchy Model
Architecture Models
Code
Test
EDS
Systems are developed through transformations of the models from one phase to the next and one iteration to the next

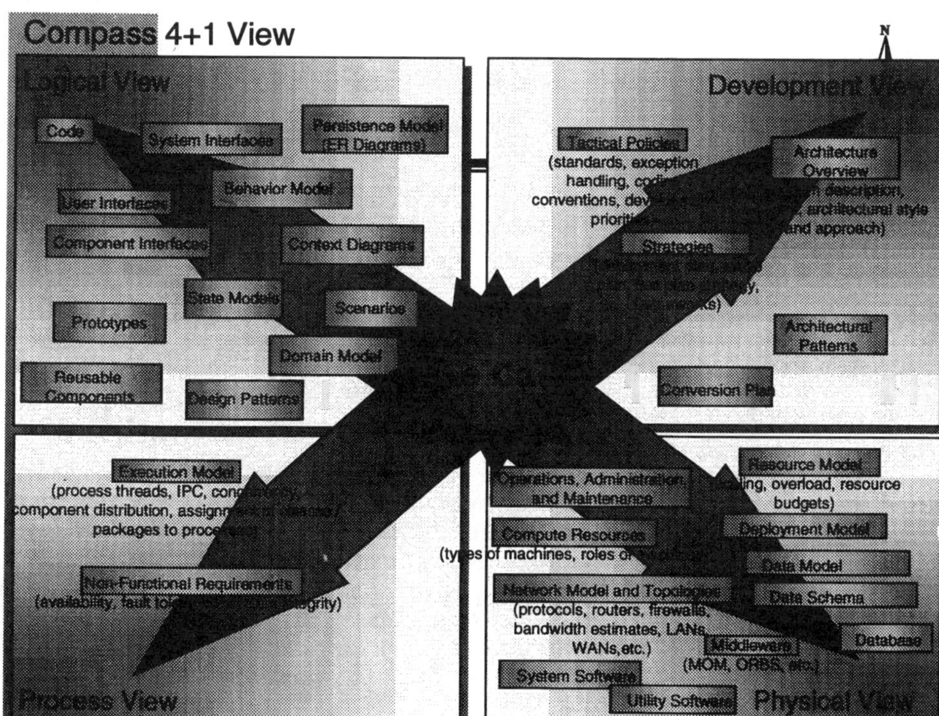
Compass 4+1 View
N
Logical View
Code
System Interfaces
Persistence Model (ER Diagrams)
Behavior Model
User Interfaces
Component Interfaces
Context Diagrams
State Models
Scenarios
Prototypes
Domain Model
Reusable Components
Design Patterns
Development View
Tactical Policies
Architecture Overview
Strategies
Architectural Patterns
Conversion Plan
Execution Model
Non-Functional Requirements
Process View
Operations, Administration and Maintenance
Resource Model
budgets)
Compute Resources
Deployment Model
Data Model
Network Model and Topologies
Data Schema
bandwidth estimates, LANs,
WANs,etc.)
Middleware
Database
(MOM, ORBS, etc.)
System Software
Utility Software
Physical View

ENDNOTE

[1]Details have been disguised to conceal the identity of the company and individuals involved in this case. The case is, however, an accurate depiction of object-oriented modeling using the Compass development process.

REFERENCES

Arnold, P., Bodoff, D., Coleman, H., Gilchrist, H., and Hayes, F. (1991). Evaluation of five object oriented development methods. In *Journal of Object-Oriented Programming: Focus on Analysis and Design,* New York: SIGS Publication, Inc., 101-121.

AlBanna, S. J. and Osterhaus, J. (1998). Meeting the Software Challenge: A Model for IS Transformation. *Information Systems Management,* Winter, 15(1), 7-15.

Brown, D. (1997). *Object-Oriented Analysis-Objects in Plain English.* New York: John Wiley & Sons, Inc.

Booch, G. (1996). *Object Solutions: Managing the Object-Oriented Project.* Addison-Wesley.

Booch, G. (1994). *Object Oriented Analysis and Design,* 2nd Edition, Benjamin Cummings.

Coad, P. and Yourdon, E. (1991). *Object-Oriented Analysis.* Second edition. New Jersey: Prentice-Hall,. Inc.

Coad, P. and Yourdon, E. (1991). *Object-Oriented Design.* New Jersey: Prentice-Hall,. Inc.

Coleman, D., Arnold, P., Bodoff, S., Dollin, C., Gilchrist, H., Hayes, F., Jeremaes, P. (1994). *Object-Oriented Development-The Fusion Method.* New York: Prentice-Hall, Inc.

Elmasri, R. and Navathe, S. (1993). *Fundamentals of Database Systems,* II Edition, Benjamin/Cummings.

Hardy, C. J., Thompson, J. B. and Edwards, H. M. (1995). The use, limitations, and customization of structured systems development methods in the United Kingdom, *Information Software Technologies,* 37(9), September.

Kruchten, Philippe (1995). The 4+1 View Model of Architecture, *IEEE Software,* November

Jacobson, I., Christerson, M., Jonsson, P., and Overgaard, G. (1992). *Object oriented Software Engineering: A Use Case Driven Approach.* Reading, MA: ACM Press/Addison-Wesley.

Jacobson, Griss, and Jonsson (1997). *Software Reuse—Architecture, Process, and Organization for Business Success,* Addison-Wesley.

Lewandowski, S. M. (1998). Frameworks for Component-Based Client/Server Computing, *ACM Computing Surveys,* 30(1), 3-18.

Rumbaugh, J. (1994). Object-Oriented Modeling and Design. In *Proceedings, Object Expo 1994.* New York: SIGS Publications.

Rumbaugh, J., Blaha, M., Premerlani, F., and Lorenson, W. (1991). *Object-Oriented Modeling and Design.* Englewood Cliffs, NJ: Prentice Hall, Inc..

Vessey, I. and Glass, R. 1998. Strong Vs. Weak Approaches to Systems Development, *Communications of the ACM,* 41(4), April, 99-102.

Vlasborn, G., Rijsenbrij, D. and Glastra, M. (1995). Flexibilization of the methodology of system development. *Information Software Technology,* 37(11), November.

BIOGRAPHICAL SKETCHES

Mahesh S. Raisinghani is an Assistant Professor of Information Systems at the Graduate School of Management, University of Dallas. Professor Raisinghani's research interests are in object-oriented database management systems, group dynamics/ knowledge management in virtual teams, strategic impact of electronic commerce on organizations, and global information systems. Professor Raisinghani's previous publications have appeared in the Journal of Information Systems Management, International Journal of Information Management, International Journal of Materials and Product Technology, Journal of Electronic Commerce, Electronic Commerce World, Minority Business News USA and in several international, national and regional conference proceedings.

Bruce Adams has been with Electronic Data Systems since 1986. He is a System Architect specializing in Object-Oriented systems development and is currently working in the Financial Industry Group. In the past he has worked as a consultant at AT&T, AT&T Bell Labs, Western Union, and GMAC. He is the co-author of EDS' Compass development process.

Index

Measuring Information Technology Investment Payoff: Contemporary Approaches

NEW!

Mo Adam Mahmood, University of Texas El Paso
Edward Szewczak, Canisius College

It would seem that business investment in information technology (IT) is at root no different from business investment in anything else. After a careful consideration of the costs of the investment and its anticipated benefits, a decision is made as to whether the benefits of the investment outstrip the costs and by how much. If the benefits are competitive with other investment alternatives (say, a major marketing campaign), then the business will commit financial resources to the IT proposal. Otherwise it won't. This decision making process is at the heart of capital budgeting. Senior executives have been making IT investment decisions for well over three decades.

So why is the measurement of IT investment payoff so difficult and controversial? Why do we need a book dealing with contemporary approaches to measuring IT investment payoff? Why have earlier approaches to measuring IT investment payoff proven unsatisfactory? In what respects have earlier approaches fallen short? Do we need to scrap earlier approaches entirely or can we find important improvements to these approaches such that they can be newly applied to effectively measure IT investment payoff in ways that are convincing to senior management?

Measuring Information Technology Investment Payoff: Contemporary Issues will help you to find improvements in existing methods for measuring IT investment payoff as well as to find new, innovative methods for addressing the value of emerging IT. *ISBN 1-878289-42-X (hard cover) • US$139.95 • 568 pages • Copyright © 1999*

An excellent addition to your library

Idea Group Publishing

1331 E. Chocolate Avenue, Hershey, PA 17033 USA
Tel: 1-800/345-4332 (USA) or (717) 533-8845
Fax: (717) 533-8661

Order this book or learn about other IGP titles at http://www.idea-group.com